Micah Warren
Bitcoin: A Game-Theoretic Analysis

Also of Interest

Probability and Statistics. A Course for Physicists and Engineers
Arak M. Mathai, Hans J. Haubold, 2017
ISBN 978-3-11-056253-8, e-ISBN (PDF) 978-3-11-056254-5,
e-ISBN (EPUB) 978-3-11-056260-6

Geometry and Discrete Mathematics. A Selection of Highlights
Benjamin Fine, Anja Moldenhauer, Gerhard Rosenberger, Annika Schürenberg, Leonard Wienke, 2022
ISBN 978-3-11-074077-6, e-ISBN (PDF) 978-3-11-074078-3,
e-ISBN (EPUB) 978-3-11-074093-6

Game Theory and Partial Differential Equations
Pablo Blanc, Julio Daniel Rossi, 2019
ISBN 978-3-11-061925-6, e-ISBN (PDF) 978-3-11-062179-2,
e-ISBN (EPUB) 978-3-11-061932-4

Characters of Groups and Lattices over Orders. From Ordinary to Integral Representation Theory
Alexander Zimmermann, 2022
ISBN 978-3-11-070243-9, e-ISBN (PDF) 978-3-11-070244-6,
e-ISBN (EPUB) 978-3-11-070255-2

Blockchain. Technology and Applications for Industry 4.0, Smart Energy, and Smart Cities
Matevž Pustišek, Nataša Živić, Andrej Kos, 2021
ISBN 978-3-11-068112-3, e-ISBN (PDF) 978-3-11-068113-0,
e-ISBN (EPUB) 978-3-11-068120-8

Micah Warren

Bitcoin: A Game-Theoretic Analysis

DE GRUYTER

Mathematics Subject Classification 2020
Primary: 68M14, 91A80, 91A60; Secondary: 91A15, 91A10, 91A12, 91D30, 91A28, 91B50, 68W15, 60C05, 60F05, 65L12, 60J20, 60G40, 37N40

Author
Prof. Micah Warren
University of Oregon
Department of Mathematics
318 Fenton Hall
Eugene OR 97403
USA
micahw@uoregon.edu;
https://indieweb.social/@micah

ISBN 978-3-11-077283-8
e-ISBN (PDF) 978-3-11-077284-5
e-ISBN (EPUB) 978-3-11-077305-7

Library of Congress Control Number: 2022950649

Bibliographic information published by the Deutsche Nationalbibliothek
The Deutsche Nationalbibliothek lists this publication in the Deutsche Nationalbibliografie; detailed bibliographic data are available on the Internet at http://dnb.dnb.de.

Cover image: Vertigo3d / iStock / Getty Images Plus
Typesetting: VTeX UAB, Lithuania
Printing and Binding: LSC Communications, United States

www.degruyter.com

Preface

The question of how one party can transfer money to another party without physically exchanging money is an old problem. The most common solution, available for hundreds of years, is to use a trusted third party, like a bank. If a bank is mutually trusted and has custody of one person's money, this person can write a check or by some other means direct the bank to pay their counterparty. A check can be thought of as simply instructions for the money's custodian to do something with the money.

Such relationships require trust. If the two parties do not use the same bank, there must be a relationship between the banks. If the banks are in different jurisdictions, this can be an obstacle.

For many reasons, individuals may want to transmit money without the use of a trusted third party. They may want to do this at a distance. They may also want to do this with some level of anonymity. Suppose a network of individuals forms around this goal: to be able to send money to each other, without doing so in person. Keeping in mind the notion of a bank cashing a check, there are issues that must be overcome:

- There must be a way for everyone involved in the network to know what money is good, at any given time.
- Parties must be able to "sign" payments in a verifiable way. That is, there must be some way to verify that the instructions to transfer money are coming from the owner of the money.
- There must be a way to ensure that a payment is not reused repeatedly.

Cryptography (a subdiscipline of number theory) solves the second issue quite handily. A blockchain attempts to solve the other two problems, with probabilistic assurances.

The goal of this text is to offer some accessible game-theoretic background for analysis of proof-of-work blockchains. Much of the current game-theoretic understanding is found in podcasts, videos, blog posts, and uncurated forum discussions and is poorly formalized. On the other hand, quite a bit of academic research exists, but much is esoteric and not written with immediate applications in mind. This textbook is a brazen attempt to bridge this gap. Deeper game-theoretic issues are discussed at length with the goal that an undergraduate student can understand their applicability.

The background material provided in the first three chapters includes an overview of how proof of work functions, followed by an overview of basic concepts from probability theory and relevant notions from game theory. In the remainder we will discuss the implication for Bitcoin.

This textbook began as a syllabus for the Clark Honors College at the University of Oregon, proposed in February of 2020. As the material was developed it became clear that there was much more than would fit in a single-quarter undergraduate course. The current version contains material far beyond the scope of the original course and seeks to serve several purposes:

- offer a comprehensive textbook for such an upper-division university course,

https://doi.org/10.1515/9783110772845-201

- establish a foundational reference framework for discourse around Bitcoin's game theory,
- facilitate descriptive conversations surrounding Bitcoin, rather than the normative conversations that dominate the public discourse, and
- provide a relatively thorough reference for journalists, policy makers, and others who may have little exposure to blockchain technology but would like to understand the underpinnings of blockchain security.

I have tried to make it self-contained as much as possible. Many chapters require some understanding of college-level mathematics, some do not. Some chapters will go deep into the weeds of the mathematics, while some chapters (the final chapter, for example) offer more of a bird's-eye view.

Acknowledgments

Much of the research for this book was done while prodding social networks for information and feedback. I appreciate all discussions I have had over the last couple of years, in particular those that involved disagreement. I want to thank: Andrew Bailey, Troy Cross, Joe Kelly, Mike Kelly, Moonsettler, Matthew Pines, Murch, Mario Gibney, Ruben Somsen, Ricardo Perez-Marco, blk014, Alex Gladstein, Bennett Tomlin, Paul Butler, Shinobi, Henry Magram, Ari Paul, GoatFeng, Joey Levine, Cyber Hokie, Bradley Rettler, Craig Warmke, Tony Arcieri, Russell Miller, Andrew Wertheim, Ravi Culleton, Jack Ronaldi, Alex B., Bob McElrath, Enrico B., William Luther, Mark Goodwin, alpacasw, Gael Sánchez Smith, Mr. Hodl, Bryon Bothun, Rocket, Andy Smith, Margot Paez, Javier Gonzalez, Fool, Jal, smallpatatas, Russell Rukin, Brandon Quittem, David Strayhorn, David Zell, and Vlad Zamfir.

Despite the pandemic I was able to have useful in person discussions with Troy Cross, Matthew Pines, Peter McCormack, Danny Knowles, David Levin, and Jiabin Wu.

I also appreciate feedback from and email conversations with István András Seres, Yuval Kogman, Dhruv Bansal, Henry Magram, and Dong Le.

https://doi.org/10.1515/9783110772845-202

Contents

1 Basic of cryptographic payments

We begin with an extremely short overview of cryptographic signatures. This is not meant to be a thorough overview of cryptographic signatures; we only aim to describe the nature of the cryptography used in Bitcoin.

1.1 Preliminaries on cryptographic signatures

1.1.1 Binary and hexadecimal

The standard number system most of us are familiar with has base 10. This system requires 10 digits and allows us to represent numbers using these digits, for example,

$$328 = 3 \times 100 + 2 \times 10 + 8 \times 1.$$

Each digit corresponds to a higher power of 10. For example, written as powers of 10, we have

$$43{,}729 = 4 \times 10^4 + 3 \times 10^3 + 7 \times 10^2 + 2 \times 10^1 + 9 \times 10^0$$

(recall that $10^0 = 1$).

There is nothing sacred about the number 10, other than perhaps it corresponds to the number of fingers (which are also known as digits) that most people have. A computer does not have 10 fingers; instead computers build data with two digits, 0 and 1. So computers prefer to store numbers in binary representation. Binary numbers are strings that consist of only 0 and 1.

An example of a binary number is

$$1001 = 1 \times 2^3 + 0 \times 2^2 + 0 \times 2^1 + 1 \times 2^0,$$

which is more familiar to base 10 users as the number 9. To count in binary, starting with 0, we have

$$0, 1, 10, 11, 100, 101, 110, 111, 1000, 1001, 1010, \ldots .$$

A 4-digit binary number represents any number between 0 and 15 (so $2^4 = 16$ total numbers). An 8-digit binary number corresponds to a number between 0 and 255, giving 2^8 possibilities.

Since computers use the binary system, when dealing with integers, it makes architectural sense to deal with sets of integers that are powers of 2. We can compress this further. Noting that $2^4 = 16$, we can convert each 4-digit binary number into a number between 0 and 15. In order to do this, we take the standard numbers, $0, 1, \ldots, 9$, and then

https://doi.org/10.1515/9783110772845-001

take a, b, c, d, e, f to represent the remaining six. This gives us 16 digits, so we can write numbers in base 16, also known as a hexadecimal system.

Binary	Hexadecimal	Base 10
0000	0	0
0001	1	1
0010	2	2
0011	3	3
0100	4	4
0101	5	5
0110	6	6
0111	7	7
1000	8	8
1001	9	9
1010	a	10
1011	b	11
1100	c	12
1101	d	13
1110	e	14
1111	f	15

The hexadecimal system is especially convenient when writing huge numbers.

Hex	Base 10
$ffff$	65,636
1000000	16,777,216
$ffffffff$	4,294,967,295
100000000	4,294,967,296
$ffffffffffffffff$	18,446,744,073,709,551,615

Note that a 256-digit binary number, which will correspond to a number between 0 and $2^{256} - 1$, can also be represented as a 64-digit hexadecimal (because $2^{256} = 16^{64} \approx 10^{77}$). Many of the frequently used algebraic functions involved in cryptography involve 64-digit hexadecimal numbers.

1.1.2 Hashing function

A hashing function takes a message, string, or number and converts this into a number in a way that is practically impossible to invert. For example, the RSA hashing algorithm works as follows. Take a pair of large numbers (n, e). Then given a message m, which will be encoded as a number between 0 and n, compute

$$E(m) := m^e \bmod n, \tag{1.1}$$

where the notation

$$a \bmod b$$

(read "a modulo b") means "take the remainder when dividing a by b." For example,

$$19 \bmod 4 = 3,$$
$$27 \bmod 10 = 7.$$

We also use notation

$$p \equiv q \bmod n$$

to denote that "the remainders p and q after dividing by n are equal." In Example 1.1, it is easy to see that for large numbers e and n, the hashed value will be nearly impossible to guess. In other words, while it is easy to directly compute (for a computer built after 1975) in the forward direction, say

$$4{,}219^{1{,}687} \bmod 67{,}108 = 43{,}511,$$

it would be more difficult to go backwards and determine the value of m by solving the equation

$$m^{1{,}687} \bmod 67{,}108 = 43{,}511.$$

It is easy to verify that 4,219 is a solution, but it is not easy to find 4,219 without going through and checking all of the possible values for m until 4,219 is identified. Because the numbers in this example are relatively small, it would take most computers less than a few seconds to find 4,219 by brute force. However, if we were to take n and e to be numbers on the order of 10^{77} (note that 10^{82} is an estimate for the number of atoms in our observable universe) it would be difficult to find m by brute force.

Modular arithmetic works very much as classical arithmetic, that is,

$$pq \bmod n \equiv (p \bmod n)(q \bmod n) \bmod n$$

and

$$(p + q) \bmod n \equiv ((p \bmod n) + (q \bmod n)) \bmod n,$$

and also

$$p^x \bmod n \equiv (p \bmod n)^x \bmod n.$$

1.1.3 Cryptographic signatures

Bitcoin uses the Elliptic Curve Digital Signature Algorithm (ECDSA) to verify that messages sent over the network are from whom the messages claim to be from. The process is based on the idea of public-key (or asymmetric) encryption. In general, public-key encryption schemes use a key pair (A, B). The private key A is a secret held only by the owner of the key, while the key B is public, published by the owner of the private key. For encryption, messages that are encrypted using B can only be decoded using A. For signatures, the fact that a message was signed by the holder of the secret key A can by verified using the key B.

The details of ECDSA involve some heavy concepts from number theory and algebraic geometry that are irrelevant to this discussion, but the principle can be illustrated using the ElGamal signature scheme, which is quite similar and illustrates the nature of the signature method. The reader may safely skip this section and proceed to §1.1.4.

To come up with an ElGamal scheme, we begin with a very large prime number p and find a primitive root b modulo p, which means that every number y between 1 and p can be written as

$$y = b^l \bmod p.$$

Methods exist to find such primitive roots. The value l is called the *discrete logarithm of y with respect to the base b*. In practice, it is very hard to determine l given y, but easy to compute y once l is chosen.

Suppose Alice would like to send messages to Bob over a network which might include interlopers who could tamper with the message. Bob would like proof that the message was from Alice and was exactly the message that Alice intended to send. To do this, Alice generates a key pair as follows; choose l an integer in $(1, p - 2)$, then compute $y := b^l \bmod p$, and then publish the number y. We can assume that Alice was able to get the key y to Bob in some sort of secure way ąt some point in time.

Now Alice has a message m. She computes a hash of the message $H(m)$, through some hashing function H. The number of digits in $H(m)$ could be quite long, so Alice truncates this by taking the first digits of $H(m)$ so that the resulting number has no more digits than p. Now Alice picks a random number k in $(1, p - 2)$ and computes

$$r := b^k \bmod p. \tag{1.2}$$

It is important that k is kept secret and that it is generated randomly and then discarded; if Alice is not careful and publishes k or uses the same k to sign different messages, others may use this information to triangulate the private key l.

Now Alice solves the following equation for s:

$$sk + lr \equiv H(m) \bmod (p - 1),$$

that is,

$$s = (H(m) - lr)k^{-1} \bmod (p-1).$$

Here, k^{-1} is the unique number between 1 and $p-2$ such that

$$k \times (k^{-1}) \equiv 1 \bmod (p-1).$$

Now Alice publishes the message m and the signature (r, s).

When Bob receives the message, he checks the following condition:

$$b^{H(m)} \equiv y^r r^s \bmod p.$$

If this is true, the message was signed by Alice (with very close to absolute certainty, barring the possibility that someone stole her key). If not, something went wrong along the way.

Why does this work? For any value v in between 1 and p a famous theorem from number theory known as Fermat's little theorem asserts that

$$v^{p-1} \equiv 1 \bmod p.$$

So because

$$sk + lr = H(m) + N(p-1)$$

for some integer N we have

$$b^{sk+lr} = b^{H(m)} b^{N(p-1)}.$$

Now modular arithmetic works like classical arithmetic, so

$$b^{sk+lr} \equiv b^{H(m)} b^{N(p-1)} \bmod p,$$

but because $b^{N(p-1)}$ is a $(p-1)$ power, we have

$$b^{N(p-1)} \equiv 1 \bmod p.$$

Hence

$$b^{sk+yr} \equiv b^{H(m)} \bmod p.$$

But

$$\begin{aligned} b^{sk+lr} &= (b^k)^s (b^l)^r \\ &= r^s y^r. \end{aligned}$$

The security of this methods relies on the fact that provided p is quite large, it will be difficult to come up with a message m and a signature (r, s) that satisfies $b^{H(m)} \equiv y^r r^s \bmod p$, given that y has been published, without knowing l. Also, it will be very difficult to determine l knowing y.

ECDSA is similar in nature, but it is a bit more complicated and is believed to be quite secure. This belief is backed by the fact that brilliant mathematicians and computer scientists have spent hundreds of thousands or millions of hours attempting to find weaknesses over recent decades. We do not see the cryptography as a weakness in Bitcoin that will hold Bitcoin back any time in the coming decades. This being said, bad implementation of good theory is and has always been a problem: ECDSA has a history of events in which keys have been exposed when wallet software failed to properly generate a random value analogous to the value k in (1.2). Neither game theory nor number theory accounts for poor implementation.

More generally, beyond allowing for signatures, a public-key (or asymmetric) encryption scheme allows users of an insecure network to communicate securely. There are a number of functioning schemes for this. Typically, they work by generating two passwords p and q that have a special relationship: If you encrypt a message with p it can be decrypted with q. While the pair (p, q) is easy to generate, it would be very difficult to determine q simply by knowing p. What this allows one to do is publish the password p. This will be called the "public key." If you keep q a secret, anybody will be able to send you a message encrypted with p and only you will be able to decrypt it, because only you know the decryption password q. Typically, these keys p and q are very long. A Bitcoin address is a 160-bit hash of a public key.

1.1.4 Takeaway

It is possible for anyone (using basic software) to generate a public and private key pair. Using the public key, anybody can verify that a message was encrypted with the private key, without knowing the private key. This allows the opportunity for Bob, whose public key is known, to sign a message, saying "I would like send Alice 1 bitcoin." Everybody can read the message and verify it is from Bob.

In the most basic sense, this is how cryptographic payment systems work. If Bob has a bitcoin, he can sign a message declaring that he is transferring that bitcoin to Alice. Now everybody seeing this message can attest that the bitcoin that once belonged to Bob now belongs to Alice. Subsequently, if Alice would like to send that bitcoin elsewhere, she simply signs a message, saying, "The bitcoin I received from Bob, I now send to Charlie."

Technically, Bitcoin uses what is called a UTXO protocol, meaning it is not account-based (like, say, a bank) but rather transaction-based. This means that instead of having an account number, one has one or more private keys that control the funds in certain transactions. Each transaction must come from a previous transaction, not from an account. This means that each fraction of a bitcoin can be traced backwards all the way to where it was created.

In practice, multiple public keys can be generated from a private key. This allows a user to create multiple receiving addresses, which allows for more anonymity as external viewers do not know that the different public keys are controlled by the same private key.

1.2 Blockchain protocols

The ability to use cryptography to sign messages has been available for decades. Naturally, the question follows of how to use this method for transferring ownership. By themselves, digital signatures are not enough to create a system of money. The most obvious problem is what is known as the *double-spend* problem: Suppose Bob sends a digital coin to Alice, and then later that day Bob sends the same digital coin to Charlie. Both transactions appear to be valid on their own. Wherever Bob received his digital coin, he is assigning it now to Alice, and he is also assigning that same coin to Charlie, hence the term "double-spend."

In order to solve this problem, there has to be a way to determine which transaction is valid, in such a way that everyone in the network agrees.

1.2.1 Consensus

Suppose a group of friends would like to meet up at a restaurant. In order to do this, they must communicate with each other to ensure that they all know at which restaurant they are going to meet. Consensus is not simply the act of negotiating. While some of the friends may have different food preferences, consensus is achieved when, and only when, all of the friends know which restaurant they are headed to. Consensus does not make any normative judgments.

To illustrate the difference between consensus and negotiation, suppose friends (without cell phones) plan on meeting up tomorrow for a drink. They make the following decision. If it is sunny and nice, we will meet at the patio at Magoo's. If it is cold and rainy, we will meet at the lounge of The Moon Temple. As the time approaches, a few clouds form, and a few small raindrops fall from the sky. While the patio at Magoo's may still be a suitable place, what the friends face now is a consensus problem. They need to determine a way in which all parties can decisively determine if it is raining. They must decide "rain" or "sun." This can be difficult for a number of reasons. For example, if a single raindrop is to be interpreted as "rain," then it is always possible that one person will detect rain while another will not. It is not always clear that everybody will feel rain if the rain is sparse.

Consensus protocols are everywhere. When the quarterback for a football team steps up to the line of scrimmage, the hand signals and code calling is for the immediate concern that the football team all agrees on the same play, not to worry about the optimal play. If the whole team agrees on what play is going to be run, they have a chance. If not, the next play will likely be a disaster. Similarly, sports teams use referees or umpires as a consensus mechanism. Clearly a baseball game could not go forward if the two teams did not agree as to whether a player was safe or out.

Similarly, the modern social contract of law and order uses a consensus protocol. While we may not all agree on whether the law is just, moral, equitable, or optimal, we can at least hope to agree on (1) what the law is and (2) what procedure is required to change the law. The social contract is based on the idea that while we may not agree on what the law should be, at least we agree on what the law is.

A classic question of consensus is the Byzantine generals problem: Suppose that several divisions of an army are camped around a city. They must come to an agreement on how to launch an attack. They send messengers from camp to camp, but must face the possibility that the messengers are waylaid or compromised. As for the football team, their success depends on if they can agree on an attack. A poorly coordinated attack is bound to fail.

For applications on a network, nodes often must agree on a binary proposition in order to proceed. In the double-spend example, the network does not really care whether Bob has sent a specific coin to Alice or that same coin to Charlie. What the network wants is clarity. Either the transaction to Alice is valid or the transaction to Alice is invalid. As long as the decision has been made, the network can continue its business. Alice can then spend the coin, and Charlie and his potential counterparties know that he cannot spend the coin.

Generally, consensus protocols are not concerned with the agreement being just, factually correct, optimal, moral, or anything else. The fundamental goal of consensus is to come to an agreement, not to make a judgment on which transaction "should" be the right one. The object is to determine, in a relatively short time, whether a given transaction is valid or not. Once a transaction is determined to be valid, it is desirable for everyone to expect that it will remain valid forever.

The simplest means to consensus is often to choose a leader. For example, on a football team, the quarterback calls the plays, and nobody argues. This is efficient, but is obviously antithetical to the nature of a decentralized payment network like Bitcoin. If there was a leader, the network could be shut down by eliminating the leader, introducing a single point of failure. A leader could also choose to abuse their power, for example by accepting bribes from Bob or Charlie. A decentralized network does not want to rely on the continued benevolence of one entity.

1.2.1.1 Fun problems

Exercise 1.1. Consider a network consisting of an unknown number of nodes, say $3 < N < 1{,}000$. At some point in the future, each node will be handed a random bit 0 or 1.

The network then has to all decide collectively on 0 or 1. Devise some protocols the nodes could agree upon ahead of time for choosing 0 or 1, subject to the following criteria:

- All nodes are treated the same way at the start. That is, we cannot pre-assign a leader.
- Nodes can broadcast information to the group, however, they cannot assume the information is received. Sometimes the message is not received by everybody.
- It must be possible (depending on the random bits) for the agreement to be either 0 or 1 (that is, we cannot simply decide ahead of time the answer will be 0).

Exercise 1.2. The prisoners' light bulb problem: There are 100 prisoners in solitary cells. There is a central room with one light bulb; this bulb is initially off. No prisoner can see the light bulb from his or her own cell. Every day, the warden picks a prisoner at random, and that prisoner visits the living room. While there, the prisoner can toggle the bulb switch if he or she wishes, or not. Also, the prisoner has the option of asserting that all 100 prisoners have visited the room. If this assertion is false, all 100 prisoners are shot. However, if it is indeed true, all prisoners are set free. Thus, the assertion should only be made if the prisoner is 100 % certain of its validity. The prisoners are allowed to get together one night in the courtyard, to discuss a plan. What plan should they agree on, so that eventually, someone will make a correct assertion?

1.2.2 Proof of work

In 2008, in the paper "Bitcoin: A Peer-to-Peer Electronic Cash System," Satoshi Nakamoto (presumably under a pseudonym) proposed a solution to the double-spend problem using a blockchain and proof of work.

A blockchain, as the name suggests, is simply a sequence of blocks. The blocks necessarily go in an order. The genesis block, which contains an initial accounting of the coins held by owners of public keys, is usually labeled as block 0. The next block will be block 1, and so forth.

Each block is made up of a list of transactions, plus some header information including a cryptographic checksum and the hash of the previous block. Necessarily, the transactions must be valid to be included in the block. In order to be valid, the transactions in block $N+1$ cannot contradict any transactions contained in blocks 0 through N. This ordering solves the double-spend problem. If Bob sends a bitcoin to Alice in block 2,178, he cannot then send the same bitcoin to Charlie in block 2,192 – such a transaction will be considered invalid because it contradicts a previous block.

So essentially, the Bitcoin blockchain is simply a very long database of every Bitcoin transaction that is considered valid. Necessarily, these transactions are not contradictory when they fit on the same blockchain.

At first, this may sound like a straightforward mechanism. Simply order the transactions and then all will be clear. However, in practice this is not simple. There needs to be a very decisive way to determine the next block so that all parties in the network can agree on what the next block is. If each node could submit their own block, this could spell potential trouble, especially as nodes throughout the world attempt to communicate. There would be no consensus, and the network could not move forward. So the Bitcoin whitepaper [38] uses what is called proof of work, based on the following principles:

- Whoever creates the block earns a reward (in bitcoin) for creating the block. This incentivizes parties to create the block.
- It is somehow expensive or difficult to create each block.
- Blocks are so difficult to produce that they only appear about every 10 minutes.

The block producers are called miners. To produce a block, a miner does the following:

1. Compile a list of valid transactions compatible with the current blockchain.
2. Generate a large random number, called a nonce ("number used only once").
3. The nonce is combined with the list of valid transactions, other header information, and hashed. The hashing function takes the input and generates an integer, often written as a hexadecimal number in a very large range, which looks something like

$$\text{b0d01118e5480f5df5c97eea94405c80bac5cd3c8fa638541aa6dcfa773cc76b.} \qquad (1.3)$$

4. The miner checks to see if the number is below a certain known threshold called the *target*. Usually this means it has quite a few leading zeros, and looks something like

$$\text{0000000000000000000021216ad71fc9bfb4f9f78cf0704cbc52eabe5c84e4a97.} \qquad (1.4)$$

   If the number is small enough, below the target, then the miner rejoices, having created a new block. They then broadcast the block to the network. Nodes verify that the hash of the block beats the target, and this becomes the next block on the blockchain. In creating the block, the miner gives themselves a subsidy of a specified number of bitcoins (called the *block subsidy*), which as of 2022, is 6.25 bitcoins, plus any fees provided in transactions.
5. If the number created by the hash function was not below the threshold, return to step 1 or 2 and repeat.
6. Repeat trillions of times until finding a hash that meets the threshold or until another miner broadcasts that they have found a block. In this case, add the latest block to the blockchain and go back to step 1.

The number (1.3) is more or less a random string of digits, while the number (1.4) leads with quite a few zeroes (this was an actual hash of a block in August 2022). Because the hash function is generating numbers that are more or less random between 0 and $2^{256} = 16^{64}$, the chance of getting 19 leading zeros is about 1 out of $16^{19} \approx 75 \times 10^{21}$. This means that if you perform 75×10^{21} hashes, the expected number of hashes with 19 leading zeros is 1. If a miner computes a hash that is below the target difficulty, this is called a *proof of work*.

Note on terminology: The noun "proof of work" is a hash which beats the target difficulty. "Proof-of-work" is an adjective, describing a specific consensus protocol which uses proofs of work, while "proof of work" is also a noun representing the general innovation. Sometimes proof of work is used synonymous with *Nakamoto consensus*, but technically these are not the same thing; the latter is a blockchain consensus protocol which uses proof of work as a fundamental ingredient. We use the term Nakamoto consensus to denote the consensus protocol as defined in the Bitcoin whitepaper [38]. There is currently no consensus as to what the term means [52].

As of September 2022, the *hashrate*, that is, the total number of hashes that all miners are producing is about 210 million TH/s (terahashes per second). Recalling that a terahash is a trillion hashes, this means that about 210,000,000,000,000,000,000 hashes are being performed every second. There's 600 seconds in 10 minutes, so we conclude that $210{,}000 \times 10^{18} = 210 \times 10^{21}$ hashes are performed every 10 minutes. Note that 210×10^{21} is roughly on the order of the number 75×10^{21}, which is how many hashes it takes so that the expected number of hashes whose first 19 digits are zero is at least one. As the hashrate increases, so does the target; see §1.2.3.

Miners often organize into *mining pools*. Because a block is only rewarded about once every 10 minutes, if there were 100,000 individual miners mining, the average miner would win a block about once every two years. It makes more sense then to pool resources and split the revenue, so in practice there will be dozens of mining pools that control a large enough fraction of the computing power so that they can expect rewards on a regular basis.

Each pool has its own hashrate, and the fraction of the total hashrate commanded by that pool determines their probability of winning the next block. For example, if Pool A is producing 14 million TH/s and the total hashrate is 210 million TH/s, this pool has a probability of

$$p = \frac{14 \times 10^6 \text{ TH/s}}{210 \times 10^6 \text{ TH/s}} = \frac{1}{15}$$

of winning any given block, or about a block every 2.5 hours.

The above quotient is as simple as it looks: The probability of winning the next block is always going to be given by the fraction of hashrate the pool controls. This simple expression does not unveil the mathematics of independent Poisson processes that lies underneath, but is somewhat intuitive and underpins the "fairness" assumption in proof of work: Your probability of getting the next block is equal to the fraction of the hashrate that you control.

A *Poisson process* is a randomly occurring event such that the average frequency of the event over time can be predicted; however, the event is not expected to happen periodically with any regularity. Each occurrence of the event is independent of previous occurrences. A natural example is earthquakes: If there is a 2 % chance of a 7.5 magnitude earthquake to occur in California in the coming year, this means that in a 50-year period, we expect to see one earthquake, on average. This does not mean, however, that if we have not had an earthquake in the last 40 years the probability has gone up. The probability of an event governed by a Poisson process happening in the next year is the same as last year, whether or not the event happened last year. Similarly, someone who plays lottery every day has the same chance of winning every day and does not build up an expectation of winning. For more details see §2.7.

Mining works the same way. Each second a miner is producing trillions of hashes, each of which has a low probability of being below the difficulty threshold. Each hash output is independent of the hash that came before it or after it, or any of the hashes produced by competing miners. For a mining pool that controls 1 % of the hashrate, they would expect on average to produce a winning hash every 1,000 minutes.

1.2.3 Block difficulty adjustment

Computers become faster every year, and bitcoins have on average become more valuable over time. So the Bitcoin protocol readjusts the *difficulty* every 2,016 blocks. The difficulty measures your odds of computing a hash below the threshold. The target is a number inversely proportional to the difficulty. If the target decreases by a factor of 2, this corresponds to the difficulty increasing by a factor of 2. Every 2,016 blocks, the average time to compute these blocks is computed. If this number is less than 10 minutes, then the difficulty is multiplied by a factor larger than 1, with the idea that in the upcoming blocks, the average time will stay close to 10 minutes. Similarly, if the average time is more than 10 minutes, the difficulty is lowered (or equivalently, the target threshold is raised).

While the number of blocks produced in a short time period may vary, the expected allocation to each miner is fixed during the difficulty period. So even if one mining pool is quite lucky and produces multiple blocks within a few minutes, the other miners' chances of getting a block are not diminished.

So in a fixed time period, the miner's competition is not so much against each other, but against randomness. However, thanks to the difficulty adjustment, over longer periods of time, the competition is with the average miner in recent history.

1.2.4 Longest-chain/heaviest-chain rule

Occasionally, two blocks may be produced independently, both appending different blocks at the same height of the blockchain. In this case, there is what is called a *chain split*. Miners then have two options: build on one chaintip or build on the other. The consensus in Bitcoin is that "the longest chain is the valid chain," as it was described in the original whitepaper by Satoshi. This means that as soon as any miner produces a block that extends one of the competing blocks, that chain becomes “the” blockchain. At this point, most miners will ignore the other chains and all mine on “the” chain.

There is a technical issue here: “Longest chain” cannot literally refer to the chain with the largest number of blocks. If this were the case, it is feasible for a split to nearly catch up while using much less work by abusing the difficulty adjustment. If there was to be a chain split and one chain was mined by only 1 % of miners, while the other chain was mined by 99 % of miners, even after many years the number of blocks would be roughly equivalent due to difficulty adjustment. So what is necessary is a more precise quantity called *chainwork*. Chainwork is the total sum of all difficulties overcome in order to create the particular chain.

The chainwork accumulated with each block does not depend on the particular hash of the block, only the difficulty that was surmounted. This means that two blocks that are produced at the same time will have the same accumulated chainwork, regardless of which block enjoyed a lower hash. If this were not the case, a block produced with hash falling just below the target would be vulnerable to being scooped by a block which equalized the chain length but had a lower hashrate. This works against the consensus.

The protocol also has a convention of using the *first-seen rule*. Miners, by convention, will take the first valid block at a certain height they see and begin extending that. There is no “rule” requiring this or a way of checking that other miners are following the first-seen rule, but this is the default behavior that is coded into the reference implementation. Using the first-seen convention, chain splits that naturally occur when two blocks are produced at about the same time will usually be sorted quickly.

1.2.5 Halving

When writing a block, a miner credits themselves with an amount of bitcoin, called the *block subsidy*. This amount changes: Approximately every 4 years, the automatic reward for producing a block is cut in half. In 2009, the block subsidy was 50 bitcoins. This was cut in half in November 2012 to 25, then to 12.5 bitcoins in July 2016, and most recently to 6.25 in May 2020. Note that 4 years is measured in blocks, namely 210,000 blocks. Because the blocks often are produced at a slightly higher rate than one every 10 minutes, the halvings can run slightly ahead of schedule while the difficulty is increasing. The block subsidy is the only way that new bitcoins are produced. The number of bitcoins produced in between July 2016 and May 2020 was $12.5 \times 210{,}000 \approx 2.6$ million. The number produced between May 2020 and the next halving will be $6.25 \times 210{,}000 \approx 1.3$ million. This number will continue to be cut in half every 4 years. This means that the number of bitcoins mined forms a geometric series:

$$50 \times 21{,}000 + \frac{1}{2} \times 50 \times 21{,}000 + \frac{1}{2} \times \frac{1}{2} \times 50 \times 21{,}000 \ldots$$

or in sigma notation

$$\sum_{k=0}^{\infty}\left(\frac{1}{2}\right)^k 50 \times 21{,}000$$
$$= 50 \times 21{,}000 \sum_{k=0}^{\infty}\left(\frac{1}{2}\right)^k.$$

You may recall the formula for geometric series:

$$\sum_{k=0}^{m} x^k = 1 + x + x^2 + \cdots + x^m = \frac{1 - x^{m+1}}{1 - x},$$

which can be verified with polynomial arithmetic. Now because

$$\left(\frac{1}{2}\right)^{m+1} \to 0 \quad \text{as } m \to \infty,$$

we can take a limit:

$$50 \times 21{,}000 \sum_{k=0}^{\infty}\left(\frac{1}{2}\right)^k = 50 \times 21{,}000 \times \frac{1}{1 - \frac{1}{2}} = 21 \times 10^6.$$

Thus the total supply of bitcoins ever produced is 21 million. This supply cap is celebrated by many Bitcoin users as one of its most distinguishing features. After 30 more halvings, the block subsidy will be

$$6.25 \times \left(\frac{1}{2}\right)^{30} \approx 5.8208 \times 10^{-9} \text{ bitcoins.}$$

This should happen a little before the year 2140. Because the smallest denomination of bitcoin (called a *satoshi* or *sat* for short) is 10^{-8} bitcoins, it will be impossible, by current accounting, to continue to pay miners a block subsidy, and the block subsidy will cease. It is expected that transaction fees will fund miners, as will be discussed in Chapter 11.

Exercise 1.3. What would the total supply of Bitcoin be if the same halving schedule was used but the initial block subsidy was 1,000 bitcoins?

1.2.6 Blocksize

The blocksize of a Bitcoin block is capped at 1 MB. This limit was put in place by Satoshi in 2010 shortly before they disappeared. This limit was the subject of acrimonious debate until 2017. Forks of Bitcoin were created (such as Bitcoin Cash) that allow for larger blocks. These forks were not favored by most Bitcoiners so they are relegated: Bitcoin in this text (and most other places) refers to the chain with 1 MB block limits that has the largest *market cap*, which is simply the sum of the market prices of all coins that have been created. While much of the game theory remains the same for Bitcoin forks such as Bitcoin Cash, the larger blocksize does create a different regime.

1.2.7 Why is it called "proof of work"?

The goal was to create a significant barrier to writing the block, so that only miners who put in enough effort or spent enough electricity would receive the reward. Without proof of work, it is not easy to quantify how much work participants are contributing, especially for a network intended to be open to potentially anonymous parties throughout the world. It does not make sense for the miners to "prove" they have computed trillions of mathematical hashes by spitting out all of the hashes, as this would take just as much effort to verify. So instead, proof of work asks miners to repeat a computation trillions of times, and only report to the network when they find the needle in the haystack. If the odds are astronomically small each time you repeat the calculation, then probability theory says that you must repeat the calculation an astronomical number of times to have any decent chance of winning the block. So, it is not a mathematical "proof." It is a mechanism such that your probability of getting the next block is exactly the fraction of the hashrate that your computer provides.

1.2.8 Rise of ASICs

In Bitcoin's whitepaper [38], the stated goal was "one CPU one vote." Effectively, each of the CPUs of Bitcoin users would contribute roughly equal weight in the mining process. However, this is no longer the case. Since proof of work measures hashes produced, it can be much more efficient to design a computer chip that is optimized to do nothing but produce hashes. Application-specific integrated circuits (ASICs) were introduced into the Bitcoin mining game in 2012. These machines hold dozens or hundreds of chips which perform SHA-256 hashes, along with a brigade of fans or some other means of dissipating heat. The machines are in finite supply and are often expensive to obtain, especially when Bitcoin is surging in price. Before 2012, many bitcoins were mined on laptops. Since then, the raw power of ASICs makes mining Bitcoin on a home computer nearly impossible. This has significant impact on the game theory of Bitcoin. The ability to produce blocks is restricted to the subset of individuals and businesses who have obtained these machines.

Exercise 1.4. Suppose your home computer produces 10 megahashes per second. If the total hashrate of the network is 200 million terahashes per second, compute how many of your computer you would need to have a 0.1 % chance of finding the next block.

1.2.9 Layer 2 networks

The Nakamoto consensus described in this chapter determines consensus over the base layer, "layer 1" of Bitcoin. By nature of the protocol, all layer 1 transactions are broadcast to the entire network and recorded in a blockchain. Everyone must agree this blockchain is the ledger defining the state of all transactions. This somewhat clunky mechanism is necessary for the decentralized protocol to function. Bitcoin payments can also be sent via *layer 2* protocols. Layer 2 protocols such as the Lightning Network refer to the blockchain, but use cryptography in a clever way to allow users to transact with one another in a channel-based way; this does not require confirmation by the entire network and allows payments to happen nearly instantly. The Lightning Network is fascinating and promising, but requires more research and more use. As of August 2022, under 5,000 bitcoins of liquidity can be seen on the public Lightning Network [1]. Due to the peer-to-peer nature of the Lightning protocol, it is impossible to determine precisely the volume of payments.

In common parlance, the term "transaction" is sometimes used interchangeably with "payment." This is not the best practice; we will use the term *transaction* to refer to layer 1 transactions that are recorded in the public blockchain. The term *payment* is reserved for a more general class of Bitcoin use that may occur on or off the layer 1 blockchain (cf. [4]).

1.2.10 Cursory analysis

The objective with proof of work is to make it both costly and rewarding to create the block, so that block producers will have much to lose if they are not careful following the protocol. A proof of work demonstrates to the network that a miner has incurred costs. The network rewards the miner by the *block reward* (the sum of transaction fees and the block subsidy). Because it is costly to get involved in mining, miners will not do this casually. The reward mechanism creates an economic market which will trend towards equilibrium. If the block reward is worth more than the expected cost of finding a block, more miners will join in the hunt and pursue the reward. If too many miners are pursuing the reward, the expected profit will drop and miners will devote less effort.

The security is based on the following: If a transaction where Bob sends Alice some bitcoin is confirmed in a block, say block N, no subsequent block will allow Bob to send that same bitcoin to Charlie. So if Bob signs a transaction sending Charlie that bitcoin, the only way the network will honor that transaction is if a miner decides to build a new block N, creating competing chains. If the latter chain becomes longer than the former, because more miners build on it, then the bitcoin will belong to Charlie.

Because it requires much work and some luck to create a block, miners, in absence of some other motivation, will almost always build their blocks on the longest chain, where they know their block is considered "the" latest block the moment that it is created. Creating a chain split comes with a chance of ending up on the losing branch, in which case the efforts are all for naught. Miners who produce the chaintip are assured that each other miner is actively competing to build on the chaintip, so this chaintip is likely to be part of the chain that exists into the future. Because the proof of work is cumulative (heaviest-chain rule), branches of chain splits that are not extended (called *uncle* or *stale* blocks) fall hopelessly behind in chainwork very quickly.

1.2.10.1 A 51 % attack

If a single miner, or mining pool, were to be consistently generating more than 50 % of the hashrate, they would have the ability to go back to a previous block, create a chain split, and then, by producing blocks at a faster rate than the rest of the network, create a longer chain which would eventually be considered the heaviest chain. The consensus mechanism of Bitcoin is based on the belief that this will not happen, or will happen infrequently and change only very few recent blocks.

If a party is successful in splitting a chain and creating a branch with more chainwork, so that one or more blocks that were previously on the chaintip are no longer considered valid, this is called a *reorg*. Deep reorgs will supplant many blocks and are considered harmful, while shallow reorgs only supplant very recent blocks and are considered more of an annoyance to the miner who recently mined the blocks. Even so, shallow reorgs are quite rare and recently these have not exceeded one block.

We use the term *supplant* to describe the act of replacing a block that was formerly valid. The verb *orphan* is often used; however, this technically is not the best choice of word: A block becomes orphaned not when its parent dies, but rather when the family name is carried on through a sibling; hence the block may become an uncle to the valid chain. For this reason the verb *uncle* is often used as well.

1.2.11 Proof of stake

Proof of work provides a way for distributed and possibly anonymous parties to prove they have spent some money or effort. Those that prove this, while following the protocol, are rewarded. This keeps random or malicious players from messing with the network, as doing so comes with non-trivial cost. But what if that thing that is expensive is not something consumed, like electricity, but a digital cryptocurrency or asset itself, whose ownership can be guaranteed cryptographically? This is the idea behind proof of stake. By only allowing users of the network who own private keys to write blocks, the network can hope to keep out malicious actors. Those with a vested interest in the network are less likely to try to cause the network harm. By paying rewards for the block creators, there is then incentive to hold the coin.

Requiring proof of an internal resource has advantages and disadvantages. One advantage is that the availability of the resource is not influenced by external factors like the availability of electricity or mining equipment. Another is that it does not produce huge externalities that may sour the brand, for example, Bitcoin consumes on the order of 100 terawatt-hours of electricity per year, which gives some who are concerned about the environment pause. On the other hand, the fact that the security of the network is not separated from the functioning of the network can open more complicated security issues. The surface area for attack may allow for more vectors, in particular more financial attack vectors. Because staking units can be bought, sold, and transferred instantaneously, proof of stake lacks a certain physical security found in proof-of-work protocols. Proof-of-stake protocols, without incentives against it, appear to be likely to trend towards control by a handful of parties, increasing the aperture for undue influence. A full discussion of vulnerabilities and strengths of proof of stake is well beyond the scope of this volume.

1.2.12 Nodes and miners

Bitcoin requires several layers to function. Miners are in a certain sense extrinsic to the network, providing a service for a fee. The most basic skeleton of the network itself is formed by the nodes. The nodes are the set of computers who are all running the Bitcoin protocol, communicating with other computers on the network, rebroadcasting transactions, rebroadcasting valid blocks, and validating incoming transactions. Ultimately, everyone who uses the network does so by accessing the network of nodes. Nodes reject transactions or blocks that do not follow the protocol, while valid transactions and blocks are relayed on. Miners typically will need to function at least partially as nodes, but most nodes do not function as miners. The network cannot function without both. Nodes provide connectivity, while miners provide difficult to forge proofs of work needed to maintain consensus. While necessary, nodes can only verify and transmit information, and they can cheaply be replaced.

Intentionally, nodes are inexpensive to run. This is by design and is a feature of Bitcoin. The 1 MB blocksize, the 10 minute expected block times, and the choice of UTXO accounting are all design parameters that allow an individual without substantial computing resources to run Bitcoin software and participate in the network directly.

1.2.13 Soft forks and hard forks

Typically, the terms *soft fork* and *hard fork* will refer to changing of Bitcoin's rules in some way at a given time. The new rules will be applied to all transactions or blocks moving forward. If the new rules are more restrictive, we call this is a *soft fork*. If the new rules are less restrictive, we call this a *hard fork*. It is important to point out that the meaning of "restrictive" here can be subtle: It means that the new rules will reject a strictly larger set of transactions or blocks. This does not mean that the new rules are less expressive: Typically they will be more expressive. So even though soft forks are "backwards compatible" in the sense that new blocks will not be rejected by old nodes as invalid, the old nodes might not interpret the functionality of the new transactions or blocks.

Example 1.1. There is a blocksize limit of 1 MB per block. When Bitcoin Cash created a new rule that allowed larger blocks, this was a hard fork. New blocks created under these rules are rejected by legacy nodes.

Example 1.2. Soft forks have been proposed that allow transactions to require more data to become valid. While these offer more functionality, they accomplish this by creating a non-trivial set of conditions under which transactions can be rejected. Any transactions allowed under the stronger rules would be allowed under the weaker rules, so this is backwards compatible.

Hard forks are considered more dangerous; in the situation of a hard fork, any nodes who are running older software would reject the new blocks, leading to confusion on the network. Soft forks do not force old nodes to update immediately; they will continue to evaluate valid blocks as valid, even though they may not be rejecting blocks that should be rejected. This is OK; other nodes and, more importantly, nearly all miners (who have a vested interest) will be using the new rules and only build blocks according to the newest rule set.

Occasionally there are disputes about proposed rule changes. Bitcoin, like everything else, is vulnerable to political acrimony when it comes to rule changes. Hard forks and soft forks, threatened or implemented, are often used as negotiation maneuvers; see [5] for one account of the 2016–2017 debate over increasing the size of blocks, which led to Bitcoin forking into multiple chains.

1.3 What is Bitcoin and why does it work?

Cryptography allows anyone to cryptographically send and verify messages regarding ownership. A transaction can have a valid cryptographic signature, but this transaction is worthless unless the network has some agreement on its uniqueness. The transaction itself need not be spendable elsewhere and the network must agree on this. To provide the uniqueness, we need a blockchain, which functions as an ordering of transactions, preventing incompatible transactions from occurring. There can be multiple blockchains at a given time, each specifying an ordering of transactions, thus these blockchains also face the problem of uniqueness. The choice of a decentralized leader via the heaviest-chain rule (Nakamoto consensus) provides a default Schelling point used for choosing which out of all possible blockchains is "the" correct one. Nakamoto consensus is designed so that this role of selecting "the" correct blockchain is unlikely to be usurped by any single party. Without 51 % of the hashing power, miners will continue to build on the longest chaintip, giving this chain the most chainwork. The amount of work on this chain distinguishes it and provides a clear beacon; all will regard this chain as correct. While temporary disputes may exist at the very tip, the history becomes well established within a short time period.

Stepping back, Bitcoin is a collection of proposed loose and non-binding agreements between sets of parties in an ongoing negotiation over distribution of buying power covering time and space. This is only a non-binding agreement and is only as valid as the consensus of those that agree to it. Nothing enforces it but beliefs about the likelihood that other parties will also assign value to a particular chain. There are many chains that call themselves "Bitcoin," and there are many cryptocurrencies which lobby for value. Each of these is a separate non-binding agreement by members of a loose coalition who choose to ascribe buying power to the coins apportioned in their chain.

The obstacle overcome by Nakamoto consensus is the ability of a decentralized coalition to form without ceding power to a trusted third party. By agreeing to the heaviest-chain rule, participants can opt in to a novel type of arrangement.

The larger game, of course, is the distribution of power. Participants in a Bitcoin network believe that the move of a committing together with other participants is a strong move that will accrue them more power as the game unfolds. Bitcoin is an attempt to claw back some of the power that is yielded when one uses sovereign currencies. If a government or other entity has the option to alter the supply of their fiat currency, this is something they can use to their advantage. By forming a loose coalition which retains value according to a ledger not controlled by a state entity, players in this game believe they are advantaging themselves in the long run. Similarly, by forming a peer-to-peer network directly that affords for transfer of value between participants, users can disintermediate parties such as banks that would have power to censor payments or to extract rent. As we will see, this can be a high-stakes game that is fraught with unpredictability and different types of counterparty risk. Instead of relying on a single party (say a bank or nation-state) to transfer value or maintain a currency's value, one is relying on a vast network with heterogeneous interests, preferences, and vulnerabilities. The agreement to adhere to a ledger that can be audited by anyone with a computer also offers users some protection. Bitcoins cannot appear out of thin air. On the other hand, dollars spent in the everyday economy are often proxies or promises which can be multiplied by issuing debt via private arrangements.

1.4 More on mining

Before setting up mining games, we offer more details about how transactions are structured.

1.4.1 UTXO

Bitcoin uses UTXO-based ("unspent transaction output") accounting. Bitcoin does not operate based on accounts, like a bank. Rather, Bitcoin operates based on transactions. Each transaction references an input transaction and creates outputs. The transaction directs the bitcoin in the input to an output address, which is a public key or a hash of a public key. When the owner of a corresponding private key would like to spend this output, they create a new transaction, by cryptographically demonstrating they have the private key, and hence are entitled to spend the output, taking this output as the input for a new transaction. This transaction directs the inputs to further outputs.

If part of the output is not specified, it can be claimed by the miner who mines the transaction. This transfer of bitcoin to a miner is called a *transaction fee*. If the owner of the private key does not intend to spend the entire output, they may send the remainder back to themselves, often to another address which they generate. This is one way in which UTXO is different from account-based accounting – the output of a transaction must be spent when the new transaction is created, even if that means sending the transaction back to yourself. The term "unspent transaction output" refers to the transactions that have not been spent; these have outputs that are spendable.

As can be verified by using any online blockchain explorer (the reader is encouraged to do so), most transactions will have at least two outputs. The difference between the value of bitcoin in the input and the value in the outputs will be the fee – this can also be verified by randomly perusing transactions on any blockchain explorer.

One thing to note is that the transaction structure exists independently of the blockchain. That is, each transaction has an ID which does not depend on the block in which it was confirmed. Apart from a given blockchain, there can be many "valid" transactions spending a given output. Here "valid" means only that the transactions are properly signed, are sourced from a valid input, and follow any other basic rules (such as, the total output value must not exceed the total input value). The function of the blockchain is to ensure that there is consensus on which transactions have been accepted, in the case that multiple transactions spend the same output.

Apart from the blockchain, there is a simplicity in validating transactions with UTXO; it requires only looking at the preceding transaction. Each transaction can be traced to previous transactions, and so forth, and this requires no information whatsoever about transactions that occur on other parts of the blockchain. If an account-based system were used, the system would have to keep track of all possible inputs and outputs. Instead, the role of nodes and miners is to first (1) check the transaction is valid and next (2) check that the source transaction output is still unspent. In practice, a node that has been following the chain can discard all the spent transactions and keep only the unspent transactions in memory; any transaction that references an already spent transaction will be rejected as invalid and can be ignored.

In creating each block, the miner creates a transaction giving themselves coins for the reward. The source (which is essentially thin air) is called "the coinbase" (note that while there is a publicly traded corporation that unfortunately shares this name, "coinbase" refers to the transactions that mint new bitcoins, not a corporation paying miners to mine). Fees also are included in the coinbase transaction. Coinbase transactions are locked for 100 blocks, so that miners cannot immediately spend bitcoins which they earn. The 100-block period is a somewhat arbitrary finality parameter. (If there is a deep reorg, say 50 blocks, most transactions could be easily reconstructed, because the UTXO exists independently of the recent choices of the blockchain and points to transactions that happened well in the past, so a 50-block reorg would not completely destroy the ledger. However, any coinbase transactions are the exception: A reorg would destroy the original UTXO creating the transactions and hence all the subsequent transactions would be orphaned and no longer part of the canonical chain. Thus as an added safety precaution to minimize the damage during a deeper reorg, the 100-block finality period is enforced.)

Nodes typically pool and pass along transactions that have been validated. These transactions form the memory pool (mempool) of valid transactions that could be placed on the next block. Usually these are prioritized by fees.

It is possible for the creator of a transaction to create another transaction with the same output, but with a larger fee. This is not necessarily considered a double-spend, because the goal might not be to maliciously spend the money twice – rather the goal is to increase the priority of the transaction so that it will be included by the miners. This move is called "replace-by-fee," often abbreviated RBF. This does add an incompatible transaction to the tree, but this is OK; only one transaction will be on the accepted blockchain.

Remark 1.1. We use the term "valid" as an absolute fact describing whether a transaction is signed correctly, follows protocol, and points to a previous valid transaction. Thus validity is a mathematical proposition that is either true or false and cannot change with time. This is completely agnostic of the underlying blockchain which has to accept and confirm the transactions. Valid transactions are valid apart from "the" blockchain or any blockchain. The question of which blockchain is accepted as "the blockchain" is relative; accepted by whom and at what time? In most cases the economic majority has clear interest in agreeing on which blockchain is "the" blockchain, but this is not an absolute ground truth that can be evaluated universally.

Exercise 1.5. Which of the following represents a hard fork and which a soft fork?

1. The protocol begins rejecting all blocks containing transactions that spend from a blacklisted (but otherwise valid) transaction.
2. The protocol requires that all mining fees exceed 0.1 % of the transaction.
3. The protocol allows anyone to spend a transactions that is over 8 years old.
4. The protocol allows blocks with only half the target difficulty if over 1 hour has elapsed since the last block.

1.4.2 Common assumptions in mining games

In the mining games in Chapters 4–9 we use the following default assumptions. More complicated situations may require these to be relaxed.

Assumption 1 (near-time price stability): We may assume that the value of a bitcoin is stable in the short term. For the long term we can assume that the price today reflects the discounted future value. This means that we do not factor speculation into miners' decisions.

Assumption 2 (near-time hashrate stability): We may assume that the total hashrate and difficulty are stable in the short term (obviously, in situations such as selfish mining where the difficulty becomes part of the game, we cannot make this assumption; see Chapter 8).

Assumption 3 (first-seen rule): Miners will always mine on top of the longest chain that they see first, provided there are no other explicit reasons for them not to.

Assumption 4 (instant broadcasting): All nodes and miners are connected enough so that transactions are broadcast nearly instantaneously.

For the simplest model, consider that there are n mining pools, denoted P_i, and each P_i is capable of producing H_i hashes per second.

At each point in time the state of the game (see §3.3.1) will be the information describing the following:

- all known and broadcast competitive chains, including necessarily the longest chain and forks that may be of comparable length,
- all transactions in the mempool (including their fees), that is, transactions that are valid but may not have been confirmed in the longest chain,
- the current hashrate of all of the pools.

Note that "the" mempool is not well-defined. Each mining node has its own mempool that might not match completely with those of other nodes. However, for our purposes we can assume there is a single mempool.

First, we assign to each pool a hashrate h_i:

$$h_i := \frac{H_i}{H},$$

where H is the total hashrate, given by

$$H := \sum_{i=1}^{n} H_i.$$

Thus $h_i \in (0, 1)$ and

$$\sum_{i=1}^{n} h_i = 1.$$

Each h_i corresponds to the probability of the event that pool i wins the next block.

The expected block reward (the sum of the block subsidy and expected transaction fees in a block) will be given by R. We will denominate all other Bitcoin amounts in terms of this unit. Using R as a base unit allows us to perform computations that are agnostic of the current block reward, which varies over time.

Remark 1.2. Suppose that some miners go offline. In the short term, as long as there is no change in difficulty, which updates every two weeks, there is no greater probability that the miners still online will win blocks in a given time interval. However, there is a greater probability that the remaining miners will win the *next* block. For example, suppose that suddenly 90 % of miners went offline, leaving a single pool of 10 %. This pool's odds of winning a block in the next 10 minutes do not improve. Until the difficulty changes, this pool expects to win one block every 100 minutes. Now, certainly, they will win the next block, and the block after that, provided that other pools stay offline, but these blocks will remain far apart. The single pool will not see an increase in the frequency at which they win blocks, until the difficulty is adjusted, after the 2,016th block period is completed. They may however see an increase in fees paid, as they may select the transactions with the highest fees to include in the blocks. Blocks will be less frequent so the fees may become more competitive.

Remark 1.3. Bitcoin miners and nodes typically run the Bitcoin Core client, downloadable from GitHub. Miners are not required to, and there are other possible ways to run Bitcoin. It is possible that miners or other institutions may be using their own homebrew client. When attacking the problem from a game theory perspective it is easiest to make the assumption that each agent could configure their own software to run a given strategy. This is not completely true; there is some cost to implement one's own version of Bitcoin code. The typical pleb running a Bitcoin node may not have the resources to modify the code for their own purposes. But a larger party, perhaps a large mining pool, a financial institution, or a nation-state, should have the means to hire someone sufficiently familiar with the codebase to modify the software that they are using to interact with the Bitcoin network. For this reason we will treat the cost of switching from default strategy to be negligible.

2 Probability spaces

Probability theory is an important topic in pure and applied mathematics. Here we only give it short shrift, introducing topics that will be necessary for computations in later chapters.

A *probability space*, in the most general sense, consists of the following:

- a set Ω, called the *sample space*,
- a collection $\mathcal{F}$ of subsets of Ω, called *events*, and
- a function $P : \mathcal{F} \to [0, 1]$ which describes the probability of any event.

Example 2.1. Suppose you flip three coins and observe whether each is "heads" or "tails." There are eight possible outcomes which make up the sample space:

$$\Omega = \{\text{HHH}, \text{HHT}, \text{HTH}, \text{HTT}, \text{THH}, \text{THT}, \text{TTH}, \text{TTT}\}.$$

The set of events is much larger, as it consists of all possible subsets of Ω (there are 64 possible subsets). These are often described prosaically, for example, the event $\{\text{HHH}, \text{TTT}\}$ can be described as "all three coins show the same." Similarly, $\{\text{HHH}, \text{HHT}, \text{HTH}, \text{THH}\}$ is the event "at least two heads."

Assuming the coin in Example 2.1 is "fair" we can assign a probability to each event. In this case, the probability of any of the sample outcomes will be the same: HHH is as probable as TTH. The event "at least two heads" has probability 1/2, because this involves four out of the eight equally probable samples, while "exactly two heads" has probability 3/8.

Example 2.2. Consider sampling from the set of real numbers $0 \le x \le 1$ in a uniformly random way. Uniformly random, in this case, means that the probability of the number you sample falling in an interval (a, b) will be precisely $|b - a|$. Here the sample space is quite large (uncountably infinite, in fact). The event space is quite large. The easiest-to-describe and often most useful events will be collections of intervals, for example, $\{x : \frac{1}{2} \le x \le \frac{2}{3}\}$ or $\{x : x < \frac{1}{3} \text{ or } x > \frac{2}{3}\}$. The probability for any singleton in the sample space is precisely zero – this corresponds to the probability of choosing a number from $[0, 1]$ completely at random, and having it be, say, $\pi/\sqrt{19}$. On the other hand, the probability of interval events will be easier to describe. The probability of the number x being larger than 1/3 will be 2/3. The probability of x being in between 1/5 and 4/5 will be 3/5.

We are intentionally avoiding discussing the full set of events in this example, as this would take us well beyond the scope of this text, and would involve a long discussion of "measurability." This will not be necessary for our discussion of Bitcoin.

The probability function $P : \mathcal{F} \to [0, 1]$ should satisfy some basic assumptions in order to be considered a probability function.

https://doi.org/10.1515/9783110772845-002

1. First, the probability that any random sample lies in the set that it has been randomly sampled from must be 1, so we require

$$P(\Omega) = 1$$

(note that Ω is a subset of Ω, albeit improper, so it is itself an event, that is, $\Omega \in \mathcal{F}$).
2. If two events are disjoint, the probability of the two events should add up in the obvious way:

$$P(E_1 \cup E_2) = P(E_1) + P(E_2) \quad \text{whenever } E_1 \cap E_2 = \{\varnothing\}.$$

In words, "the probability of E_1 or E_2 is the probability of E_1 plus the probability of E_2, whenever E_1 and E_2 do not overlap."

As an immediate corollary of these two axioms, one can derive other properties. For example, the probability of an event and its complement must sum to 1:

$$P(E) + P(E^c) = 1,$$

where E^c denotes the complement of E, that is, all elements of Ω that are not in E.

By decomposing $E_1 \cup E_2$, we can derive another formula:

$$P(E_1 \cup E_2) = P(E_1) + P(E_2) - P(E_1 \cap E_2).$$

Also note that because $P(E) \geq 0$ for any event, it follows that

$$P(E_2) \leq P(E_1) \quad \text{whenever } E_2 \subset E_1.$$

2.1 Countably and uncountably infinite

A set X is *finite* if the number of elements in X is a finite number. A set X is called *countably infinite* if the elements of X can be put in one-to-one correspondence with the natural numbers $\mathbb{N} = \{1, 2, 3, 4, \ldots\}$. In other words, it must be possible to start writing down the elements of X as $\{x_1, x_2, x_3, \ldots\}$ in a way that will eventually include every element of X. The term *countable* includes countably infinite and finite. For example, the rational numbers are countably infinite: One can start writing down a sequence $\{0, 1/2, 1, 1/3, 2/3, 4/3, 5/3, 2, 1/4, 3/4, \ldots\}$ of rational numbers in a way such that every rational number will eventually be hit. For example the rational number $q =$ 5,087/17,801,239 will be *somewhere* in the sequence if devised correctly.

A set X is *uncountably infinite* if the elements of X cannot be put in one-to-one correspondence with the natural numbers. Every attempt to describe X as $\{x_1, x_2, x_3, \ldots\}$ must necessarily omit some elements of X. The set of real numbers is uncountable.

An important uncountable set is the set of infinite binary sequences $\mathcal{B}$, that is, the set of all strings of 0's and 1's that start somewhere but do not terminate. A typical example is

$$b = 0110100100100101110010\ldots.$$

Countable sets or countable events are easier to deal with for probability theory. By putting the samples in an order one can proceed through all of them and make statements such as

$$P(E) = \sum_{\omega \in E} P(\omega_i).$$

For an uncountable set, such a sum may not make sense. When dealing with uncountable sets, it is often useful to partition them into countably many subsets.

Exercise 2.1. Prove that the set $\mathcal{B}$ is uncountable. (Hint: Assume it is countable and choose a way to count it. Argue that it is possible to use this counting to construct an element of $\mathcal{B}$ not given in the counting.)

2.2 Partitions

For many purposes we will need to divide a probability space into either finite or countably infinite events, such that

$$\bigcup_i E_i = \Omega$$
$$E_i \cap E_j = \{\varnothing\} \quad \text{for all } i, j.$$

For Example 2.1, one choice of partition is according to the first flip, that is, divide into two events: E_1 = "first flip is heads" and E_2 = "first flip is tails."

We may find it convenient to relax the definition slightly, up to events with probability 0. A more general definition would be

$$P\left(\bigcup_i E_i\right) = 1,$$
$$P(E_i \cap E_j) = 0 \quad \text{for all } i, j,$$

which means that any random outcome ω with probability 1 is found in at least one subset of the partition E_i and will be in more than one subset with probability 0.

For example, for integers $i \geq 1$, let

$$E_i := \left\{x : \frac{1}{i+1} \leq x \leq \frac{1}{i}\right\}.$$

This partitions the unit interval $[0, 1]$:

$$E_1 \cup E_2 \cup E_3 \ldots = \left[\frac{1}{2}, 1\right] \cup \left[\frac{1}{3}, \frac{1}{2}\right] \cup \left[\frac{1}{4}, \frac{1}{3}\right] \ldots .$$

The end point $\{0\}$ does not occur, and the points $\{1/n\}$ occur more than once; however, the probability of any of these being sampled is zero. If a countable set of outcomes each has probability zero, then the event consisting of the countable union is also zero.

2.2.1 Conditional probability

Oftentimes we will be given partial information. We may know that an event has occurred, and we will want to know if this gives us a better idea of the probability of other events. We define the following conditional probability:

$$P(E \mid F) = \frac{P(E \cap F)}{P(F)},$$

which is read as "the probability of E, given F." Given the event F has occurred we use this information to determine the probability of event E occurring. In Example 2.1 let E be "all flips are heads" and let F be "at least two heads." We have

$$\begin{aligned} P(\text{all three heads} \mid \text{at least two heads}) &= \frac{P(E \cap F)}{P(F)} \\ &= \frac{1/8}{3/8} = \frac{1}{3}. \end{aligned}$$

In other words, if you have the information that at least two heads have been flipped (you do not know which ones or the order) you now conclude there is a 1/3 probability that all three are heads.

We say that two events E, F are *independent* if

$$P(E)P(F) = P(E \cap F). \tag{2.1}$$

It follows that for independent events

$$P(E \mid F) = \frac{P(E)P(F)}{P(F)} = P(E),$$

in other words, knowing that F has occurred does not change the probability that E has occurred, justifying the definition.

2.3 Random variables

A random variable is essentially a function on the space Ω. That is, if $f : \Omega \to \mathbb{R}$ determines a real number for each element in the sample space, we say f is a random variable. The value of $f(\omega)$ depends on the randomly sampled event, so is random in this sense.

In Example 2.1, one could let f be "the number of heads". This is a random variable: For each outcome in the sample space, one can determine the number of heads.

In Example 2.2, one could define a random variable $f(x) := x^2$ as follows: given an $x \in [0,1]$ chosen at random, square x.

2.3.1 Expectation

A real-valued random variable will have an *expectation*, also known as "expected value," "mean," "average," or "weighted average." This is denoted $\mathbb{E}[f]$. Heuristically, if many trials were repeated and the resulting values of f were averaged out, one would expect this number to trend towards $\mathbb{E}[f]$.

For a finite probability space, the definition is simple:

$$\mathbb{E}[f] = \sum_{\omega_i \in \Omega} f(\omega_i)P(\omega_i). \tag{2.2}$$

A very similar definition holds for countably infinite spaces, but one has to make the caveat that the sum converges (the St. Petersburg paradox provides a famous example when the expectation is infinite). The "expected number of heads" in Example 2.1 will be

$$\frac{1}{8}(3+2+2+2+1+1+1+0) = \frac{12}{8}.$$

For uncountably infinite spaces, the expectation can often be expressed as an integral. For example, to compute the expected value of x^2 when x is sampled uniformly from $[0,1]$ we compute

$$\mathbb{E}[x^2] = \int_0^1 x^2\,dx = \frac{1}{3}$$

(there is a bit here we are deliberately not unpacking).

The *variance* of a random variable is defined as

$$\mathrm{Var}(X) := \mathbb{E}[(X - \mathbb{E}[X])^2]. \tag{2.3}$$

In other words, variance is the expectation of the difference between the value and its expectation, squared. If a function is usually close to its mean, it will have small variance. The *covariance* for two random variables is

$$\mathrm{Cov}(X,Y) := \mathbb{E}[(X-\mathbb{E}[X])(Y-\mathbb{E}[Y])] = \mathbb{E}[XY] - \mathbb{E}[X]\mathbb{E}[Y].$$

If X and Y tend to be above their means for the same events as one another, the covariance will be positive. The variance satisfies the formula

$$\mathrm{Var}(X+Y) = \mathrm{Var}(X) + \mathrm{Var}(Y) + \mathrm{Cov}(X,Y).$$

Variables X and Y are considered *independent* if information about the value of X does not give you any information about the value of Y. In other words, any event involving knowledge of X sheds no new light on Y. For example,

$$P(X \geq 3 \mid Y \leq a) = P(X \geq 3)$$

regardless of the value a. This implies and is a stronger condition than $\mathrm{Cov}(X,Y) = 0$. For a larger set of variables, say $X_1, X_2, \ldots, X_n$, we say these are independent if information about any combination of them does not give us information about variables not included in that combination.

Importantly for our concerns, whenever variables $X_1, X_2, \ldots X_n$ are independent we have

$$\mathrm{Var}\left(\sum_{i=1}^{n} X_i\right) = \sum_{i=1}^{n} \mathrm{Var}(X_i). \tag{2.4}$$

Note that the operation of computing the expectation is linear in the random variable. Given two random variables f and g and a countable probability space,

$$\begin{aligned}\mathbb{E}[f+g] &= \sum_{\omega_i\in\Omega} (f(\omega_i) + g(\omega_i))P(\omega_i)\\ &= \sum_{\omega_i\in\Omega} (f(\omega_i)P(\omega_i) + g(\omega_i)P(\omega_i))\\ &= \mathbb{E}[f] + \mathbb{E}[g].\end{aligned}$$

Note that this is true even if f and g are not independent.

Exercise 2.2. Show that the variance scales quadratically, that is,

$$\mathrm{Var}(\lambda X) = \lambda^2\,\mathrm{Var}(X).$$

Exercise 2.3 (St. Petersburg paradox). Consider the following game: You flip a coin repeatedly until you obtain tails. If the first flip is tails, you receive \$1 and the game is over; otherwise proceed. If the second flip is tails, you receive \$2 and the game is over. If the first two are heads and the third is tails, you win \$4. If the fourth flip is tails, you will \$8, and so forth. What is the expectation for winnings in this game?

2.3.2 Conditional expectation

Condition expectation is a convenient computational tool that will be frequently used in the upcoming sections. This allows us to split the computation of expectation into pieces that will be more manageable. If we have a nice partition of the sample space, we can compute the conditional expectations over each of the events in the partition, and then sum these up appropriately.

This involves the quantity $\mathbb{E}[f \mid E]$, or "the conditional expectation of f, given event E." In other words, if we restrict ourselves from the whole set Ω to the event E, what is our expectation for f? For countable probability spaces, we use (similar to (2.2))

$$\begin{aligned} \mathbb{E}[f \mid E] &= \sum_{\omega_i \in \Omega} f(\omega_i) P(\omega_i \mid E) \\ &= \sum_{\omega_i \in E} f(\omega_i) \frac{P(\omega_i)}{P(E)}. \end{aligned} \tag{2.5}$$

Here we are using the fact that for each singleton ω_i we have

$$\begin{aligned} &P(\omega_i \mid E) = \frac{P(\omega_i)}{P(E)} \quad \text{if } \omega_i \in E, \\ &P(\omega_i \mid E) = 0 \quad \text{if } \omega_i \notin E. \end{aligned}$$

This allows us to use a partition to compute expectation, via the following formula:

$$\mathbb{E}[f] = \sum_i \mathbb{E}[f \mid E_i] P(E_i). \tag{2.6}$$

The weighted average is simply the weighted average of a set of weighted averages. The proof of this formula follows from (2.5):

$$\begin{aligned} \sum_i \mathbb{E}[f \mid E_i] P(E_i) &= \sum_{E_i} \Big(\sum_{\omega_j \in \Omega} f(\omega_j) P(\omega_j \mid E_i) \Big) P(E_i) \\ &= \sum_{E_i} \Big(\sum_{\omega_j \in E_i} f(\omega_j) \frac{P(\omega_j)}{P(E_i)} \Big) P(E_i) \\ &= \sum_{\omega_j \in \Omega} f(\omega_j) P(\omega_j) = \mathbb{E}[f]. \end{aligned}$$

Returning again to Example 2.1 and letting f be the number of heads we see, if we partition into E_1 := "first flip is heads" and E_2 := "first flip is tails" we have

$$\begin{aligned}\mathbb{E}[f] &= \mathbb{E}[f \mid E_1]P(E_1) + \mathbb{E}[f \mid E_2]P(E_2) \\ &= \frac{1}{2}\mathbb{E}[f \mid E_1] + \frac{1}{2}\mathbb{E}[f \mid E_2].\end{aligned}$$

Intuitively (and as can be verified by writing out the full sum),

$$\mathbb{E}[f \mid E_1] = 2$$

and

$$\mathbb{E}[f \mid E_2] = 1,$$

so not surprisingly, this recovers

$$\mathbb{E}[f] = \frac{1}{2}(2+1) = \frac{3}{2}.$$

A frequently useful partition is according to the value of the random variable itself. For example, $E_c = \{\omega \mid f(\omega) = c\}$ defines an event. If the random variable f has a countable range, the expected value can be written as

$$\mathbb{E}[f] = \sum_{c \in \mathrm{Range}(f)} cP(f = c).$$

In Example 2.1 this gives us

$$\begin{aligned}\mathbb{E}[\text{number of heads}] &= \sum_{c \in \{0,1,2,3\}} c \times P(\text{number of heads} = c) \\ &= 0 \times \frac{1}{8} + 1 \times \frac{3}{8} + 2 \times \frac{3}{8} + 3 \times \frac{1}{8} \\ &= \frac{12}{8}.\end{aligned}$$

2.4 Bernoulli processes and Bernoulli schemes

Most of the probability spaces involved in mining games are best presented in terms of Bernoulli processes. A *Bernoulli process* is a finite or infinite sequence of independent binary-valued random variables $\{X_1, X_2, X_3, \ldots\}$ such that each $X_i \in \{0, 1\}$ and for all values of i we have

$$\begin{aligned}P(\{X_i = 1\}) &= p, \\ P(\{X_i = 0\}) &= 1 - p.\end{aligned}$$

We emphasize that the probability does not depend on i. The probability space involved is the space of binary sequences:

$$\mathcal{B} := \{\text{all binary sequences}\}.$$

Note that with this description of the space, the random variables X_i described above are simply the ith entry in the sequence.

For Bernoulli processes, the probability space is determined by a parameter p describing the probability that each given trial will be 0 or 1. This allows us to compute the probabilities for many types of events. For example:

- The probability that a given $X_i = 1$ is p.
- The probability that the first two outcomes are both 0 is $(1-p)^2$. Since $P(\{X_1 = 0\}) = (1-p)$ and $P(\{X_2 = 0\}) = (1-p)$ and these two events are independent, we conclude

  $$P(\{X_1 = 0 \text{ and } X_2 = 0\}) = (1-p)(1-p)$$

  using (2.1).
- The probability that $X_1 \neq X_2$ is $2p(1-p)$. This is the sum of probabilities of two disjoint events, each with probability $p(1-p)$.

Example 2.3 (random walker). A drunkard lurches forwards or backward in a series of steps. The probability that they step forwards is p, the probability they step backwards is $1-p$. After 10 steps, what can we say about where they are?

When we are only interested in what happens after a fixed number of steps (in this case 10) we do not need to use the full infinite space $\mathcal{B}$. We can deal with $\mathcal{B}^{(10)}$, the set of binary sequences of length 10. This set has $2^{10} = 1{,}024$ elements. The probability that the walker has moved 10 steps ahead is p^{10}, and the probability that the walker has moved 10 steps backward is $(1-p)^{10}$. These probabilities are easy to compute because for each there is exactly one sequence in this event. To compute the probability that the drunkard has moved eight steps ahead, observe that this involves 10 different possibilities. The walker must take nine steps forward and one step back, but because the order in which this occurs does not matter, there are 10 different steps that could have been taken backward. The probability for each individual path is $p^9(1-p)$. So

$$P(\text{walker is eight steps ahead}) = 10 \times p^9(1-p).$$

The probability that the walker is six steps ahead is more interesting. We know that the walker must have taken two steps backwards, and we do not care which of the 10 steps these were. However, we do have to be careful not to count the possibilities twice. This brings us to combinatorics. The notation

$$\binom{10}{2}$$

is read "10 choose 2" and denotes the number of distinct ways to choose two out of a set of 10. We can think of this as follows. If we are given a task of choosing two distinct objects out of an array of 10 possibilities, there will be 10 possibilities for our first choice. Once the first one has been chosen there are nine possibilities to choose the second one. This gives us 90 ways to choose a distinct ordered pair from 10 possibilities. However, choosing four and then eight is no different from choosing eight and then four, at least for our purposes, which means the quantity 90 has over-counted everything by a factor of 2. We conclude that

$$\binom{10}{2} = \frac{10 \times 9}{2} = 45.$$

In general the formula determining ways to select k elements from a set of n is given by the *binomial coefficient*

$$\binom{n}{k} = \frac{n!}{(n-k)!k!}.$$

Using this formula, we see that

$$\begin{aligned} P(\text{walker is six steps ahead}) &= \binom{10}{2} p^8 (1-p)^2, \\ P(\text{walker is four steps ahead}) &= \binom{10}{3} p^7 (1-p)^3, \\ &\dots \\ P(\text{walker is four steps behind}) &= \binom{10}{7} p^3 (1-p)^7, \\ &\text{etc.} \end{aligned}$$

More generally, a *Bernoulli scheme* is similar, but with the possibility of more than two outcomes each trial.

Given a finite collection of mining pools, the sequence of pools that attain the next target hash is a Bernoulli scheme. The random variable X_i will be the answer to the question: "Which pool produced the ith hash meeting the target?" (random variables need not always be real-valued). From the perspective of a single pool, the sequence of "my pool" or "not my pool" is a Bernoulli process. Note that

$$\begin{aligned} P(\{X_i = \text{"my pool"}\}) &= p, \\ P(\{X_i = \text{"not my pool"}\}) &= 1-p \end{aligned}$$

whenever the pool controls a fraction p of the hashrate.

2.4.1 The binomial distribution formula and the De Moivre–Laplace theorem

When $X_i \in \{0, 1\}$ we can compute statistics for the sum of the first n outcomes. Define

$$S_n := X_1 + X_2 + \cdots + X_n.$$

First observe that

$$\mathbb{E}[X_i] = p,$$

and by the linearity of expectation,

$$\mathbb{E}[S_n] = np.$$

Now for a particular k we use the analysis in Example 2.3 to compute $P(\{S_n = k\})$. In order for S_n to be k there must have been exactly k instances among the first n, where $X_i = 1$. There are $\binom{n}{k}$ choices of these, each has probability $p^k(1-p)^{n-k}$. Thus

$$P\{(S_n = k)\} = \binom{n}{k} p^k (1-p)^{n-k}.$$

For n not too large, these numbers can be computed by hand (or by computer). The function

$$\phi(k; n, p) = \binom{n}{k} p^k (1-p)^{n-k}$$

is a *probability mass function*, that is, a function of k that determines the probability that the value of the random variable is k. As n becomes large, we will be less interested in the small possibility that S_n is exactly equal to a particular k and more interested in when S_n is below or above certain thresholds. This requires the *cumulative distribution function*. For a random variable f, the cumulative distribution function is defined by

$$F(s) = P(\{f \leq s\}).$$

Necessarily, we will have

$$\lim_{s \to -\infty} F(s) = 0$$

and

$$\lim_{s \to \infty} F(s) = 1.$$

For the particular case of binomial distributions, for $k \in \mathbb{N}$ we have

$$F(k;n,p) = P(\{S_n \le k\}) = \sum_{i=0}^{k} \binom{n}{i} p^i (1-p)^{n-i}.$$

This can be somewhat involved to compute by hand. It can also be expressed via the incomplete beta function I as

$$F(k;n,p) = I_{1-p}(n-k, k+1) = (n-k)\binom{n}{k} \int_0^{1-p} t^{n-k-1}(1-t)^k \, dt$$

(we are not expecting the average reader to know what the beta function is or why it exists, and we will not offer a proof of this statement; this is just a fun fact). This expression may or may not be easier to compute. We will revisit this in more detail in §4.1.

An approximation formula is given by the De Moivre–Laplace theorem:

$$\phi(k;n,p) \approx \frac{1}{\sqrt{2\pi np(1-p)}} e^{-(k-np)^2/(2np(1-p))},$$

which is closely akin to the central limit theorem. When applied to S_n, the central limit theorem states that as $n \to \infty$, the random variable

$$\tilde{S}_n := \frac{(S_n - np)}{\sqrt{np(1-p)}} \tag{2.7}$$

converges to a random variable with standard normal distribution. Thus for very large values of n, one may use the cumulative distribution function for the standard normal distribution:

$$\Phi(x) := \int_{-\infty}^{x} \frac{1}{\sqrt{2\pi}} e^{-s^2/2} \, ds.$$

Example 2.4. If $p = 0.4$ and $n = 10{,}000{,}000$, what is the probability that $S_n \le 4{,}001{,}000$?
Using (2.7) the event that $S_n \ge 4{,}100{,}000$ is the same as the event that

$$\tilde{S}_n \le \frac{4{,}001{,}000 - 4{,}000{,}000}{\sqrt{10{,}000{,}000 \times (0.4)(0.6)}} = 0.6455.$$

If $\tilde{S}_n$ is a normally distributed random variable, we have

$$P(\{\tilde{S}_n \le 0.6455\}) = \Phi(0.6455) \approx 0.7407.$$

For a full statement and proof of De Moivre's theorem, see [51, Chapter 5, §1.5].

2.4.2 Probability-preserving isomorphisms and comparison of expectations

Occasionally we will need to consider very similar spaces with identical properties. Clearly the space of Bernoulli trials with $p = 1/3$ where the outcomes are $\{0, 1\}$ must be equivalent in some way to the space of Bernoulli trials with $p = 1/3$ and the outcomes are $\{\mathrm{A}, \mathrm{B}\}$. We can make this explicit: Define the map

$$T : \mathcal{B}_{\{0,1\}} \to \mathcal{B}_{\{\mathrm{A},\mathrm{B}\}}$$

via

$$T(00101\ldots) = \mathrm{AABAB}\ldots$$

in the obvious way. Then you can check that T is probability preserving in the sense that

$$P(\{x \in E\}) = P(\{T(x) \in T(E)\}).$$

This will be useful to us via the following: Consider a mining pool with hashrate p. Suppose in one universe they decide to mine a given transaction, and suppose in another universe they decide to not mine that same transaction. Moving forward, the sequence of Bernoulli trials determined by whether or not this pool obtained the block will look different: Because the hashing function behaves chaotically, the fact that the pool obtained the next block in the former situation is not going to guarantee the pool obtained the next block in the latter.

Many of our computations will involve trying to maximize the expected value of a function, given a decision between two or more options. For example, letting R be the rewards a given pool earns over the next 1,000 blocks, we would like to compute $\mathbb{E}(R)$ in the case where one pursues a certain strategy and compare it to the case where one does not pursue that strategy. Naively, the Bernoulli sequences describing who wins the blocks are not the same. However, they are isomorphic: The probability that an event E occurs in one is equal to the probability that a corresponding event $\tilde{E}$ occurs in the other. This allows us to compute the difference of the two values $\mathbb{E}(R)$ by computing them at the same time using the same partition, without worrying about the fact that these are technically different probability spaces.

2.4.2.1 Left-hand right-hand game

We illustrate this with a contrived example. Suppose you have to choose between two games. Game 1 is the following: You flip a coin with your left hand. You win \$10 when you flip heads, you lose \$1 for every tails, and the game is over when you flip heads for the first time. In Game 2 you flip with your right hand. You win \$15 for every heads and you lose \$1 for every tails, and once you flip a heads the game is over. Now intuitively the right-hand game is better, but suppose we want to know by how much. Computing the expectation for either game is possible, but slightly annoying as it involves an infinite sum. A naive approach in comparing the two games is to compute the expectation for one and then the expectation for the other, and compare them side-by-side. However, this computation may end up more complicated. Perhaps you are not interested in knowing exactly how much you will win: You may be more motivated to know how much you would be willing to pay to upgrade from Game 1 to Game 2. It will be an easier computation to analyze the difference of these two games, using the fact that both games occur on the same two isomorphic underlying probability spaces. If you perform a series of coin-flips with your left hand, it will produce a Bernoulli process with identical distribution as if you had flipped the coin with your right hand, even though the sequence sampled will not literally be the same.

We define a function f as follows: f is the money won by playing the left-hand game, subtracted from the money won playing the right-hand game. The game will terminate with probability 1. We may partition the space by defining E_i to be the event that the first H is flipped on the ith flip. Then we compute

$$\mathbb{E}[f] = \sum_{i=1}^{\infty} \mathbb{E}[f \mid E_i] P(E_i).$$

Now we note the following: For event E_i the left-hand game wins $10 - i$, while the right hand game wins $15 - i$. The difference is thus the constant number 5. So we have

$$\mathbb{E}[f] = \sum_{i=1}^{\infty} 5P(E_i) = 5 \sum_{i=1}^{\infty} P(E_i) = 5.$$

This is much easier than computing

$$\begin{aligned}&\mathbb{E}[f_{\mathrm{RH}}] - \mathbb{E}[f_{\mathrm{LH}}]\\ &\quad = \sum_{i=1}^{\infty}(15 - i)\left(\frac{1}{2}\right)^i - \sum_{i=1}^{\infty}(10 - i)\left(\frac{1}{2}\right)^i.\end{aligned}$$

This may appear contrived and pedantic now, but this framework will make computations easier, for example when a mining pool is deciding whether or not to pursue a chain reorg.

Justification

Suppose we have a probability-preserving isomorphism

$$T : \Omega \to \tilde{\Omega}$$

and we have partitions

$$\{E_i\}$$

$$\{\tilde{E}_i = T(E_i)\}$$

of both such that

$$P_\Omega(E_i) = P_{\tilde{\Omega}}(T(E_i)).$$

Certainly,

$$\sum P(E_i) = 1 = \sum P(\tilde{E}_i).$$

We have

$$E_\Omega(f) = \sum P(E_i)E(f \mid E_i),$$
$$E_{\tilde{\Omega}}(\tilde{f}) = \sum P(\tilde{E}_i)E(\tilde{f} \mid \tilde{E}_i),$$

so

$$\begin{aligned} E_\Omega(f) - E_{\tilde{\Omega}}(\tilde{f}) &= \sum P(E_i)E(f \mid E_i) - \sum P(\tilde{E}_i)E(\tilde{f} \mid \tilde{E}_i) \\ &= \sum P(E_i)\{E(f \mid E_i) - E(\tilde{f} \mid \tilde{E}_i)\} \end{aligned}$$

because the probabilities are the same; $P(E_i) = P(\tilde{E}_i)$. Now if we can determine f and $\tilde{f}$ from E_i and the corresponding $\tilde{E}_i$, we can use the simplified formula

$$E_\Omega(f) - E_{\tilde{\Omega}}(\tilde{f}) = \sum P(E_i)E(f - \tilde{f} \mid E_i).$$

2.5 Win–Lose–Pay game

The following game is very simple, and illustrates a type of computation that will come up frequently in the ensuing chapters. We try to offer a thorough explanation here.

The game is as follows. At each round of the game, one of {Win, Lose, Pay} is selected, with probabilities $\{p, q, r\}$, respectively, where $p + q + r = 1$. If "Pay" is selected, you pay a fee κ, and the game continues to the next round. If "Win" is chosen, you win the game and claim a reward λ, and you are also returned any fees you have paid into the game. If "Lose" comes up, the game is over, you win nothing, and you have forfeited anything you have paid into the game. What do you expect to win?

To begin, note that the game will end with probability 1 (provided that $r \neq 1$). The probability that the game ends with a win is

$$w = \frac{p}{p+q}$$

and the probability it ends with a loss is

$$l = \frac{q}{p+q}.$$

The expected amount you will win (only considering positive winnings) is then

$$\frac{p}{p+q}\lambda.$$

The more interesting question is how much you expect to pay. If you win, you get your fees back, so this only comes into play when you lose. Let

$$E := \mathbb{E}[\text{fees lost}].$$

We claim that

$$E = r(E + l\kappa) \tag{2.8}$$

as follows. At the first round of the game, E represent the fees you expect to lose in all future rounds. Applying Bayesian expectation (2.6) to this first round,

$$E = \begin{pmatrix} p \times \mathbb{E}[\text{fees lost} \mid \text{win at round 1}] \\ + q \times \mathbb{E}[\text{fees lost} \mid \text{lose at round 1}] \\ + r \times \mathbb{E}[\text{fees lost} \mid \text{pay at round 1}] \end{pmatrix}.$$

The first two events result in no lost fees, so we have

$$E = r \times \mathbb{E}[\text{fees lost} \mid \text{pay at round 1}].$$

Now if we are back for a round 2, we have paid κ. This particular fee will be forfeited with probability l (if you win, you get it back). Now we examine the probability space in front of us at round 2: It is identical to the space that was in front of us at round 1. So the additional amount we expect to lose will again be given by E. Thus

$$\mathbb{E}[\text{fees lost} \mid \text{pay at round 1}] = E + l\kappa.$$

We conclude (2.8).

Solving for E,

$$E = \frac{rl\kappa}{1-r}.$$

The expected outcome of the game is

$$\frac{p}{p+q}\lambda - \frac{q}{p+q}\frac{r\kappa}{1-r}.$$

2.6 Probability distribution functions

Often the probability space is a continuous spectrum, instead of discrete events. Recall Example 2.2, where points are sampled uniformly from the unit interval. In general, the sample does not have to be uniform, but can be described by a probability distribution function.

Example 2.5. Consider the function

$$\rho(x) = \frac{1}{\sqrt{2\pi}} e^{-x^2/2}$$

and note that

$$\int_{-\infty}^{\infty} \rho(x)\,dx = 1. \tag{2.9}$$

A variable x is *normally distributed* when

$$P(a \leq x \leq b) = \int_a^b \rho(x)\,dx.$$

Again, as in Example 2.2, the probability that x drawn from this distribution is exactly a predetermined number (say 7.19825) is zero, but we can always compute the probability that x lies in an interval.

In general, a probability distribution function will be a non-negative function ρ such that

$$\int_{\Omega} \rho(\omega)\,d\omega = 1.$$

We can compute the expectation of a random variable f via

$$\mathbb{E}[f] = \int_{\Omega} f(\omega)\rho(\omega)\,d\omega.$$

Exercise 2.4. Find the mean and variance of a normally distributed variable. (Hint: After computing the mean, use integration by parts together with (2.9).)

2.6.1 Jensen's inequality

Let f be a random variable, and let φ be a convex real-valued function on real variables. Then

$$\varphi(\mathbb{E}[f]) \leq \mathbb{E}[\varphi(f)]. \tag{2.10}$$

This useful theorem is called *Jensen's inequality* and is true for any probability space; discrete, continuous, or mixed.

Considering Example 2.2, the expected value of x^2 is 1/3, whereas the expected value of x squared is 1/4.

2.7 Poisson processes

A Poisson process is an event whose probability of occurring in the next time interval is independent of how recently the event occurred or did not occur. With a time unit fixed, Poisson processes are determined by a parameter λ, the number of expected occurrences per time unit. The number of occurrences in a time unit is a random variable valued in the non-negatives integers, called a Poisson variable. The probability that the event occurs x times in an interval is given by the Poisson distribution

$$p(x) = \frac{\lambda^x}{x!} e^{-\lambda}. \tag{2.11}$$

Exercise 2.5. Show from the distribution formula (2.11) that the expectation for a Poisson variable is λ.

While the Poisson variable is an integer-valued real variable, the time between any time and the next occurrence is a real-valued random variable with probability distribution

$$\rho(t) = \lambda e^{-\lambda t}.$$

Example 2.6. Block production (ignoring changes in difficulty) is a Poisson process, occurring on average once every 10 minutes. Letting t be the time (in minutes) between now and the next block appearing, t follows the probability distribution

$$\rho(t) = \frac{1}{10} e^{-t/10}, \quad \text{for } t > 0.$$

Note that

$$\int_0^\infty \frac{1}{10} e^{-t/10}\, dt = 1.$$

Note also that

$$\int_0^{10} \frac{1}{10} e^{-t/10}\, dt \approx 0.632,$$

so there is roughly a 63 % chance the next block will occur before 10 minutes elapse.

Example 2.7. While the time between two successive blocks can exhibit a lot of variance, the time for a series of 2,016 blocks is more predictable (relative to the size of the time interval). We can compute this as follows. Noting that the times between the first block and the second block, between the second block and the third block, etc., are all independent random variables, we can think of the space as consisting of 2,016 independent random variables, each with the same distribution function.

If we normalize so that we are using units of 10 minutes, the probability distribution describing the time until the first block is found is

$$p_1(t) = e^{-t}, \quad t > 0.$$

The time between when the first block is found and the second has the same probability distribution

$$p_2(t) = e^{-t}, \quad t > 0,$$

etc. If we would like to compute the time before 2,016 blocks are found, we model the space as $\mathbb{R}_+^{2,016}$ with probability distribution function

$$\prod_{i=1}^{2,016} p_i(t_i).$$

The random variable describing the total time taken for 2,016 blocks is given by

$$T = \sum_{i=1}^{2,016} t_i.$$

With some work one finds that T has a distribution function

$$p(T) = \frac{T^{2,015}}{2,015!} e^{-T}.$$

This is not particularly easy to deal with by hand, but is no problem for computers. Not surprisingly, the expected value of T is 2,016, which follows from the fact that the sum of the expected values of independent variables is the expected value of the sum.

We can compute the probability that the next difficulty period arrives over 1 hour too early:

$$\int_0^{2,010} \frac{T^{2,015}}{2,015!} e^{-T}\, dT \approx 0.4497.$$

The probability that the next difficulty period arrives 12 hours early, however, is

$$\int_0^{1,944} \frac{T^{2,015}}{2,015!} e^{-T}\, dT \approx 0.03088,$$

which is small but not incredibly small.

Note that the difficulty is computed according to time taken between difficulty periods. If one looks at a graph of the difficulty over time (such graphs are readily available online), one observes this is a quite bumpy function, moving up and down, while trending generally upward over time. This variance is due to the randomness in the nature of finding blocks; even after 2,016 blocks there is still some chance that the hashrate can be over- or underestimated by a few percent.

2.7.1 Naturally occurring chain splits

Temporary chain splits happen naturally for network latency reasons: Blocks are created roughly every 600 seconds, but their propagation through the network takes at minimum some fraction of a second. We can model the relationship between splitting and latency by letting $\Delta_{i,j}$ be the delay between miners i and j. Let p_i be the hashrate of miner i. Assuming an omniscient viewpoint which can determine absolutely which miner first finds the next block, miner i finds it first with probability p_i. Recalling the Poisson distribution (§2.7), the probability that miner j does not find a block before learning of the block broadcast by miner i is given by

$$e^{-p_j \Delta_{i,j}},$$

where $\Delta_{i,j}$ is given in units of 10 minutes, which is precisely the probability of miner j mining exactly 0 blocks in the time period $\Delta_{i,j}$. Thus the probability that the block reaches all miners with none having found a block is

$$\prod_j e^{-p_j \Delta_{i,j}} = \exp\left(-\sum_j p_j \Delta_{i,j}\right).$$

Summing over all first finders i, the probability that the next block reaches all miners without any conflicts is

$$\sum_i p_i \exp\left(-\sum_j p_j \Delta_{i,j}\right).$$

To get an upper bound on this probability, we use Jensen's inequality (2.10). Let

$$g_i := \sum_j p_j \Delta_{i,j}.$$

Then

$$\sum_i p_i \exp\Big(-\sum_j p_j \Delta_{i,j}\Big) = \sum_i p_i e^{-g_i} \leq \exp\Big(-\sum_i p_i g_i\Big) = \exp\Big(-\sum_{i,j} p_i p_j \Delta_{i,j}\Big).$$

So the probability of a broadcast block saturating the network without a split is bounded above by the negative exponential of the average propagation time.

This says that if the average delay is 1 second, the probability of a particular block running into a split is at least

$$1 - \exp\left(-\frac{1}{600}\right) = 0.166\,\%$$

and one would expect at least 87 incidental splits per year. This is a lower estimate. Unfortunately, it is difficult to observe how frequently chain splits happen. Bitcoin nodes form a peer-to-peer network with a topology that is somewhat murky by design. Nodes may hear of chain splits, but they do not make this information public: If nodes are in the practice of revealing information that they know, a sophisticated attacker may (in theory with enough information) be able to piece together information about the network. Some nodes do publish information about how many uncle blocks they observe; however, this will only be a small subset and only represents the blocks that they were able to see. Because this information is not generally available, we do not actually know empirically what the average latency is on the Bitcoin network, but we suspect that it is quite small.

Exercise 2.6. Consider the following game. You flip a coin until tails is shown. If you flip an odd number of coins, you win \$1. If you flip an even number of coins, you lose \$1. What is your expected outcome?

Exercise 2.7. Compute the probability that six blocks are found within the next 20 minutes.

Exercise 2.8. Suppose that 30 % of the hashrate is controlled by a single miner. You learn there is an 80 % probability that the miner has suddenly lost power. Find the probability mass function for the number of blocks produced by the entire network in the next hour.

Exercise 2.9. Find a lower bound on the number of forks expected per day if the average delay is 2 seconds.

Exercise 2.10. Suppose that we observe 50 stale blocks in one year. Find a reasonable estimate for the average delay.

3 Game theory basics

3.1 Single-round games

We begin with some very simple motivating examples and then offer some definitions. In the very basic examples in this section, players make moves simultaneously in a single round, without knowledge of the other player's strategy.

Example 3.1 (Rock–Paper–Scissors). Player 1 and Player 2 simultaneously choose Rock, Paper, or Scissors, where Rock beats Scissors, Scissors beats Paper, and Paper beats Rock. This can be summarized in the following table.

Player 2's choice / Player 1's choice	Rock	Paper	Scissors
Rock	Tie	2 wins	1 wins
Paper	1 wins	Tie	2 wins
Scissors	2 wins	1 wins	Tie

Example 3.2 (Prisoners' dilemma). Two partners in a bank robbery are apprehended while jaywalking and brought into questioning separately. Each are given the following choice: (1) confess to the major crime and implicate their associate or (2) refuse to implicate their associate but face charges for jaywalking.

The outcomes are summarized as follows.

Player 2's choice / Player 1's choice	Stay silent	Accuse partner
Stay silent	Both pay fines for jaywalking.	Player 2 goes free, while Player 1 goes to jail for bank robbery and is fined for jaywalking.
Accuse partner	Player 2 goes to jail for bank robbery and is fined for jaywalking. Player 1 goes free.	Both go to jail for bank robbery.

Example 3.3 (Splitting \$4). Two players are supposed to split \$4. They each write down a sum: \$1, \$2, or \$3. If the sum of the numbers is no more than \$4, each player gets the number they wrote down. Otherwise, they get nothing. This can be summarized in the following table.

https://doi.org/10.1515/9783110772845-003

Player 2's choice / Player 1's choice	$1	$2	$3
$1	Each gets $1	Player 1 gets $1, Player 2 gets $2	Player 1 gets $1, Player 2 gets $3
$2	Player 1 gets $2, Player 2 gets $1	Each gets $2	Each gets $0
$3	Player 1 gets $3, Player 2 gets $1	Each gets $0	Each gets $0

3.1.1 Payoff tables

In the sequel, it will be convenient to use a *payoff table* to describe two-person games. The payoff table for the above example (Example 3.3) looks like

	1	2	3
1	$(1,1)$	$(1,2)$	$(1,3)$
2	$(2,1)$	$(2,2)$	$(0,0)$
3	$(3,1)$	$(0,0)$	$(0,0)$

Example 3.1 has the following payoff table:

	Rock	Paper	Scissors
Rock	(T, T)	(L, W)	(W, L)
Paper	(W, L)	(T, T)	(L, W)
Scissors	(L, W)	(W, L)	(T, T)

When we see (a, b) in a payoff table it means that Player 1 wins proceeds a, while Player 2 wins proceeds b.

In general, the games we study will consist of the following:

1. A set of players.
2. A set of moves by each player.
3. The outcomes of the combinations of moves.
4. The preferences of the players for the different outcomes.

3.1.2 Preferences and utility function

In analyzing games, usually what we need to know is the preference between different outcomes. A Rock–Paper–Scissors player, if given a choice, will choose to Win rather than Lose. Similarly, in games involving money, it is evident that players prefer more money rather than less. It will be useful to try to quantify this, with a utility function, U, describing how "good" that outcome is for that player. This function assigns numeric values to different outcomes and will be useful in determining players' strategies, especially when the other players' moves may be undetermined and some sort of expected value is desired. For example, in Rock–Paper–Scissors it may be useful to assign

$$\begin{aligned} U(\text{Win}) &= 2, \\ U(\text{Loss}) &= 0, \\ U(\text{Tie}) &= 1. \end{aligned} \tag{3.1}$$

Note that if we were to use a different utility function, for example

$$\begin{aligned} U(\text{Win}) &= 1, \\ U(\text{Loss}) &= -1, \\ U(\text{Tie}) &= 0, \end{aligned}$$

or even

$$\begin{aligned} U(\text{Win}) &= 10, \\ U(\text{Loss}) &= 0, \\ U(\text{Tie}) &= 1, \end{aligned} \tag{3.2}$$

we get the same preferential ordering.

Whenever the game is deterministic (no probability involved) any utility function that preserves the ordering will result in the same analysis. In this text, we will usually assume that the players have rather large banks, so the concavity of the utility of money is not felt, and players act according to the maxim "money is good and more is better." Fortunately most of the games we discuss will denote currency valued in one fixed denomination, so we will simply be able to use that value as the payoff function in our games.

If we have fixed a utility function, we can then create a payoff table; for example, if we use function (3.1) we get the following payoff table for Rock–Paper–Scissors:

	Rock	Paper	Scissors
Rock	$(1,1)$	$(0,2)$	$(2,0)$
Paper	$(2,0)$	$(1,1)$	$(0,2)$
Scissors	$(0,2)$	$(2,0)$	$(1,1)$

For the prisoners' dilemma game, we could try to quantify it by saying that going to jail for bank robbery is 10,000 times as bad as being fined for jaywalking. The payoff function would look like this:

Player 1's choice \ Player 2's choice	Stay silent	Accuse partner
Stay silent	(−1, −1)	(−10,001, 0)
Accuse partner	(0, −10,001)	(−10,000, −10,000)

However, from a purely preferential perspective, this is equivalent to the following payoff table:

Player 1's choice \ Player 2's choice	Stay silent	Accuse partner
Stay silent	(3, 3)	(0, 5)
Accuse partner	(5, 0)	(1, 1)

All of the preferential orderings in the above two tables are preserved, so there is a sense in which these games are equivalent.

3.1.3 Matrix games

Matrix games are a standard point of exploration for two-player game theory. These are *zero-sum* games, meaning that the loss of one player is equal to the gain of the other player. The payoffs are easy to represent by writing a single number representing the reward to Player 1. For example, Rock–Paper–Scissors can be rewritten as follows:

	Rock	Paper	Scissors
Rock	0	−1	1
Paper	1	0	−1
Scissors	−1	1	0

These can also be described as "row/column" games: Player 1 chooses a row, Player 2 chooses a column, and Player 2 then gives Player 1 the value specified in the matrix, or receives that number if the entry is negative.

Example 3.4. The following matrix determines a matrix game:

$$\begin{pmatrix} 3 & 6 & 1 & -4 \\ -3 & 0 & -4 & 7 \\ -1 & 10 & -3 & -1 \\ 5 & -6 & 1 & 5 \end{pmatrix}.$$

Quickly analyzing this shows the complexity: Player 1 may be tempted to choose Row 3; in this way, if Player 2 chooses Column 2, Player 1 will win a large reward of 10. Of course, Player 2 may anticipate this, and choose Column 3. But if Player 1 believes Player 2 will choose Column 3, Player 1 would prefer Row 1. This reasoning continues;

there is no deterministic strategy. The best strategies are mixed strategies to be discussed in §3.2.

3.1.4 N-player games and more terminology

Many games involve more than two players. When players can communicate and form coalitions this becomes a cooperative game; these are discussed in Chapter 16. When players do not communicate, the games can be set up in a similar way: Each player makes a move and then some outcome is determined. It will not always be possible to draw such nice tables; in a three-player game one would need a 3D array. In general, we set up the following notation. Player p_i has a choice of moves, or strategies, which are given by a set S_i. The Cartesian product space

$$S_1 \times S_2 \times \cdots \times S_n$$

is the set of all possible choices of strategies for all players. For example, in Rock–Paper–Scissors there are nine possible pairs of moves by the players.

$$S_1 \times S_2 = \{(R,R),(R,P),(R,S),(P,R),(P,P),(P,S),(S,R),(S,P),(S,S)\}.$$

There will be an outcome function, which we call Y, which takes any n-tuple of strategies and gives an outcome in the outcome set O, which can be written

$$Y : S_1 \times S_2 \times \cdots \times S_n \to O.$$

For Rock–Paper–Scissors we have the following outcome set:

$$O = \{\text{Player 1 wins, Player 2 wins, Tie}\}.$$

When we have a utility function, we can map the outcomes to utility vectors for each player, namely

$$U : O \to \mathbb{R}^n,$$

that is, an ordered tuple of values describing the utility to each player for each possible outcome:

$$\begin{aligned} U(\text{Player 1 wins}) &= (2,0), \\ U(\text{Player 2 wins}) &= (0,2), \\ U(\text{Tie}) &= (1,1). \end{aligned}$$

Often, we will simply short-circuit the outcome stage and write $\tilde{U}$ as a function on the strategy space, that is,

$$\tilde{U} = U \circ Y : S_1 \times \cdots \times S_n \to \mathbb{R}^n.$$

For example,

$$\tilde{U}(\text{Rock, Paper}) = (2, 0).$$

Because this is a vector-valued function, we can write the utility to each player as

$$\begin{aligned}\tilde{U}_1 &: S_1 \times \cdots \times S_n \to \mathbb{R},\\ \tilde{U}_2 &: S_1 \times \cdots \times S_n \to \mathbb{R},\\ &\text{etc.}\end{aligned}$$

For two-player games with finitely many moves, this is general nonsense that is just as easily encoded by a payoff table. However, when the moves lie on a continuum or there are many players it will make more sense to describe the game using these functions.

So far, we have described *deterministic* games, meaning the outcome is completely determined by knowing the players' moves, by say, reading it off the table or using the function Y.

3.1.5 Dominant strategies

For single-round games, a player's strategy is simply their choice of move. If the game has multiple stages, the strategy involves a response to any possible game state. Let $s, t \in S_i$ be two strategies in Player i's strategy space. We say that strategy s *weakly dominates* strategy t if the strategy s results in at least the utility of t regardless of other players' strategies, and is strictly superior in at least one case. In other words, if s and t are strategies for Player 1,

$$\tilde{U}_1(s, s_2, \ldots, s_n) \geq \tilde{U}_1(t, s_2, \ldots, s_n)$$

for each fixed $(s_2, \ldots, s_n)$ and for at least some $(s_2^*, \ldots, s_n^*)$

$$\tilde{U}_1(s, s_2^*, \ldots, s_n^*) > \tilde{U}_1(t, s_2^*, \ldots, s_n^*).$$

We say that strategy s *strictly dominates* strategy t if strategy s results in superior utility to that of t regardless of any combination of other players' strategies, namely

$$\tilde{U}_1(s, s_2, \ldots, s_n) > \tilde{U}_1(t, s_2, \ldots, s_n)$$

for each fixed $(s_2, \ldots, s_n)$.

A (strictly, weakly) *dominant strategy* is a strategy that (strictly, weakly) dominates all other available strategies.

3.1.6 Examples

- For Rock–Paper–Scissors (Example 3.1), no player has a dominant strategy. If Player 1 chooses Rock, this will be superior or inferior to the choice of Paper, depending on the move of Player 2. Because it must be superior regardless of Player 2's move, it cannot be a dominant strategy.
- For the prisoners' dilemma (Example 3.2), each player has a strictly dominant strategy: accuse the other. Regardless of what Player 2 does, Player 1 can always improve their situation by accusing Player 2, and similarly for Player 2.
- In splitting \$4 (Example 3.3), there is no dominant strategy. If Player 1 writes down \$1, certainly he could have made a better choice by writing down \$2 in the situation where Player 2 wrote down \$1 or \$2. If Player 1 writes down \$2, in the unlikely situation that Player 2 wrote down \$1, Player 1 could have done better by writing down \$3. If Player 1 writes down \$3, this will only be a good choice if Player 2 writes down \$1, as both other choices will result in winning nothing.

The stag hunt is another classic example. As there are variations in how the game is expressed, we specify the following instance for our discussion.

Example 3.5 (Stag hunt). Two hunters set out hunting. They must cooperate to catch a stag, but either can catch a hare on their own. A stag is much preferable to a hare. An example payoff table describing these preferences is given below:

Hunter 1 \ Hunter 2	Hunt stag	Hunt hare
Hunt stag	$(5,5)$	$(0,3)$
Hunt hare	$(3,0)$	$(2,2)$

That is, if both decide to hunt hare, they both get hares, but if one decides to hunt hare while the other hopes to hunt stag, the latter gets nothing, while the hare hunter may bring in slightly more hare.

There is no dominant strategy: If Player 1 hunts hare, they are missing an opportunity if Player 2 has decided to hunt stag. If Player 1 hunts stag, they are left with nothing if Player 2 decides to hunt hare.

Exercise 3.1. Consider a game with the following payoff table:

Player 1 \ Player 2	Left	Right
Up	(a,b)	(c,d)
Down	(e,f)	(g,h)

Show that if both $e > g > a > c$ and $d > h > b > f$, then the game is a *deadlock* game, that is, a game in which the strategy that is mutually most beneficial is also dominant.

3.1.7 Nash equilibria

Clearly, if all players have a dominant strategy, they will play this. However, this usually is not the case. Instead players can look for a strategy that is best, provided they have some idea about the moves by the other players. A Nash equilibrium is an n-tuple of strategies that maximizes each player's outcome among their possibilities, provided that no other players deviate. In mathy notation, this is the following: A Nash equilibrium is an n-tuple of strategies (simply a pair in the case of two-player games) $(s_1, \ldots, s_n)$ such that

$$\begin{aligned} \tilde{U}_1(s_1, s_2, \ldots, s_n) &\geq \tilde{U}_1(t, s_2, \ldots, s_n) \quad \forall t \in S_1, \\ \tilde{U}_2(s_1, s_2, \ldots, s_n) &\geq \tilde{U}_2(s_1, t, \ldots, s_n) \quad \forall t \in S_2, \\ &\text{etc.} \end{aligned} \tag{3.3}$$

Recall that the dominant strategy condition for Player 1 is

$$\tilde{U}_1(s_1, s_2, \ldots, s_n) \geq \tilde{U}_1(t, s_2, \ldots, s_n) \quad \forall t \in S_1 \text{ and } \forall (s_2, \ldots, s_n) \in S_2 \times \cdots \times S_n,$$

which is much more restrictive than (3.3), which only requires the inequality to hold for the specific $(s_2, \ldots, s_n)$.

3.1.8 Examples

- For Rock–Paper–Scissors, there can be no Nash equilibrium in which players use *pure* strategy (see §3.2 for *mixed strategies*, in which players employ some amount of randomness). If Player 1 plays Rock and Player 2 knew this, Player 2 would play Paper. But if Player 1 knew Player 2 was playing Paper, Player 1 would play Scissors.
- For the prisoners' dilemma, the dominant strategy for both players is to accuse the other. If both players have a dominant strategy, this pair is necessarily a Nash equilibrium.
- For splitting \$4, there is a most suggestive Nash equilibrium, where each player writes down \$2. Note that if either player wrote down \$1 while the other wrote down \$2, they would receive only \$1, while if they write down \$3, they receive nothing. So clearly \$2 is the best option when the other player writes down \$2. However, there are actually two more Nash equilibria: If Player 1 writes down \$1 and Player 2 writes down \$3, or vice versa. Each pair satisfies the criteria that either player will have a worse outcome if they modify their move and the opponent does not deviate.
- For stag hunt, there are two Nash equilibria. If both players hunt stag, they maximize their reward. If both hunt hare, they maximize their reward. Given that the other player is hunting hare, switching to stag would result in a worse outcome.

3.1.9 Determining Nash equilibria by looking at a table

Consider the splitting \$4 game. We present an easy method to find Nash equilibria: Player 1 can go through each column, and for each column decide the optimal row, and make a note of this. For example, in the following (splitting \$4 game), each optimal payout for a given column to Player 1 has been starred.

	1	2	3
1	$(1,1)$	$(1,2)$	$(1^*,3)$
2	$(2,1)$	$(2^*,2)$	$(0,0)$
3	$(3^*,1)$	$(0,0)$	$(0,0)$

Player 2 can then do the same thing: for each row, choose the optimal column (note that you may star multiple possibilities if they are all equal). As above, we get:

	1	2	3
1	$(1,1)$	$(1,2)$	$(1^*,3^*)$
2	$(2,1)$	$(2^*,2^*)$	$(0,0)$
3	$(3^*,1^*)$	$(0,0)$	$(0,0)$

We see that there are three entries with two stars. Each time we see two stars for an entry, that entry represents a Nash equilibrium.

Example 3.6. Consider the matrix game determined by

$$\begin{pmatrix} 3 & 6 & 1 & -4 \\ -3 & 0 & -4 & 7 \\ -1 & 10 & -3 & -1 \\ 5 & -6 & 1 & 5 \end{pmatrix}.$$

Each player marking their best strategies (i. e., Player 1 marks the maximum in each column with a $^{(1)}$ while Player 2 marks the minimum in each row with a $^{(2)}$) we get

$$\begin{pmatrix} 3 & 6 & 1^{(1)} & -4^{(2)} \\ -3 & 0 & -4^{(2)} & 7^{(1)} \\ -1 & 10^{(1)} & -3^{(2)} & -1 \\ 5^{(1)} & -6^{(2)} & 1^{(1)} & 5 \end{pmatrix}.$$

We observe that there is no Nash equilibrium: This is perhaps unsurprising for zero-sum games. It is possible, however. Consider

$$\begin{pmatrix} 3 & 6 & 10 & 5 \\ -3 & 0 & 7 & -7 \\ -10 & 10 & -8 & -1 \\ 0 & -6 & 1 & 5 \end{pmatrix}.$$

Note that row 1 and column 1 is a Nash equilibrium: Given Player 1 has chosen row 1, column 1 is the minimum, while given that Player 2 has chosen column 1, row 1 gives the maximum.

Does this mean that (row 1, column 1) will be the result of the game? Perhaps, but not necessarily. Each player will be trying to guess what the other player will play. The only way for Player 1 to do better is to try and correctly guess when and how Player 2 is going to guess how Player 1 is going to deviate. But since Player 1 is coming out ahead by not trying to outsmart Player 2, Player 1 has very little incentive to deviate from row 1. So the most likely outcome is that Player 1 will choose row 1. Thus Player 2's best option is column 1.

Exercise 3.2. Is it possible for a matrix game with no two entries the same to have multiple equilibria? Find a counterexample or prove that this is impossible.

3.2 Mixed strategies

How would one play a game such as Example 3.4 in practice? To fully understand the range of strategies, we need to bring in some randomness. This involves the concept of mixed strategies: Each player chooses a strategy that is non-deterministic. This allows for a wider range of strategies, in a continuum of infinitely many possibilities.

For Rock–Paper–Scissors, instead of choosing a "pure" strategy such as Rock, the player can use a mixed strategy: Roll a six-sided die. If the number shown is 1 or 4, play Rock. If the number is 2 or 5, play Paper. Otherwise, play Scissors. If the die is fair, there is no way for Player 2 to come up with a strategy that will reliably beat Player 1. In fact, each player playing the same $(1/3, 1/3, 1/3)$ random strategy is a Nash equilibrium – even if they both know the other players' strategy, they cannot improve their own odds.

Full exploration of this topic is slightly beyond the scope of this book, but goes back to some of the foundations of modern game theory. A very interesting and noteworthy result of Von Neumann is the following: Every matrix game has a Nash equilibrium if we allow mixed strategies. That is, there is a pair of probability vectors $\vec{p}_1, \vec{p}_2$ such that neither player can improve their expectation, given the other player plays that strategy. A full discussion would take us into the topic of linear programming, also a fascinating and useful area of both pure and applied mathematics.

3.2.1 Mixed strategies in real life

Mixed strategies happen often in real life. Sports is one example. A pitcher in baseball may have a 102 mph fastball, but a batter expecting a fastball can key in on the pitch and get a good swing. So even if a fastball is the best pitch, the pitcher will not throw this 100 % of the time. In soccer, players shooting penalty kicks need to decide where to kick the ball; left, right, or maybe even center. If the goalkeeper anticipates that a player usually kicks right, they can jump right at the instant the ball is kicked, giving themselves a chance to stop the kick. In football, if a team never runs a fake punt, the defenders can leave receivers unguarded while attempting to block the kick.

In international relations, adversaries may attempt to maintain *strategic ambiguity*. While not technically the same thing as a mixed strategy, strategic ambiguity prevents one player from anticipating their opponent's move. If the player is risk-averse, this may serve to their opponent's advantage.

Exercise 3.3. Suppose that you know your opponent's mixed strategy for Rock–Paper–Scissors and suppose that this strategy is not uniformly random. Argue that at least one pure strategy gives you a positive expectation (assume payoffs are 1, 0, and −1).

3.3 Extensive-form games

Single-round games are easy to set up and analyze, but most real-world situations involve a sequence of moves. These are called *extensive-form* or *sequential* games. Some are deterministic like chess, while others like poker are non-deterministic. Chess is extremely difficult to analyze in pure mathematical terms. For blockchain we will see that most games involve some randomness and are extensive-form. Typical two-player games will involve a series of rounds in which players make moves in turn or with turns decided by chance.

The full definition of extensive-form games is somewhat extensive, so we do not offer it here. Extensive-form games can often be analyzed by a *game tree* which allows us to plot the possible paths through the game. Often the "chance player" will have moves represented on the game tree so that all possibilities can be specified.

The strategy set for extensive games can be a bit more difficult to notate, but in the most general form it can be described as the set of functions from the set of game states to the set of moves. That is, a strategy will be a response to each possible game state. Even though the game is played in ordered rounds, one can still describe their responses to each possible scenario before arriving there (although typically to maintain strategic ambiguity, one would not reveal this).

3.3.1 State of a game

Extensive games involve multiple stages, with each stage depending on players' moves at the previous stage. A good general-purpose word that we will use is the *state*. In computer science, the state is often used to describe the current value of all of variables. In a blockchain setting, the state may involve any current variables that are relevant, such as all of the transactions in the longest chain, or other signed and valid transactions that may be competing with transactions that have been confirmed. For a game, the state can be thought of as describing all information needed to reconstruct the game. For example, if a baseball game is delayed due to rain, what is needed to resume the game the next day is the state of the game; the score, the count, which runners are on base, the current lineup, etc. Typically the state is a description of any variables that can change during the course of a game or software program.

We begin with a very simple, two-round example.

Example 3.7 (Ultimatum game). Two players are offered a total of \$100, provided they can split the money as follows: Player 1 proposes a split, and Player 2 either rejects this and both get none, or accepts the split and it is distributed as such. There is no negotiation.

This game has two stages. In Stage 1, Player 1 must decide on the offer. In Stage 2, Player 2 must decide to accept or reject.

The game tree for this game is a rooted tree graph with the start of the game as round zero, and edges coming out of this round zero. In the ultimatum game, each possible dollar amount between 0 and 100 represents a move for Player 1 and a corresponding edge coming out of the root node. From each node in round 2, there are two options for Player 2: Accept or Reject. So if we assume that Player 1 can only use whole dollar amounts, the game tree will have 202 possible end leaves. For this particular game we can determine the payout of the game by the ending leaf. Some games will require we keep track of the game history along the way as value can be won or lost.

Immediately we see that Player 1 has a decided advantage. If Player 2 is an emotionless, rational player who will maximize their winnings from the game, they will certainly accept any positive number. Knowing this basic fact, Player 1 can look at all the possibilities and decide to make a split (\$99, \$1). Since \$1 is preferable to \$0, Player 2 will accept this.

Players 1's strategy set consists of 101 offers, ranging from (\$0, \$100) to (\$100, \$0). Player 2's strategy is more complicated to describe, because Player 2 must have a response to each possible state of the game. There will be 101 such states, so a single strategy for Player 2 will be a function

$$S : \{\text{Player 1's moves}\} \to \{\text{Accept, Reject}\}.$$

In practice, Player 2's strategy may be quite easy to describe more prosaically: "Accept all offers of \$1 or more, reject \$0."

Now because of the distinct ordering of the moves of the game, it does not really matter what Player 2 declares their strategy will be, if they do so in an attempt to persuade Player 1 to offer a better split. Player 2 is forced to make a decision after Player 1 has played, so if acting rationally will probably accept any positive offer.

This is an example of *subgame perfect equilibrium*. A subgame perfect equilibrium is a set of strategies such that these strategies will constitute a Nash equilibrium at every stage of the game. A subgame perfect equilibrium is usually found using backward induction, which describes the reasoning Player 1 would have used to make the offer (\$99, \$1).

A strategy is *sequentially rational* if for each game state, the strategy maximizes the players' expected outcomes, given their beliefs about the game. This can be complicated – in games like chess, the goal is to be as sequentially rational as possible. Doing so is another matter, even for experts. However, in simple examples like the ultimatum game, the sequentially rational strategy should be clear. If Player 1 expects that Player 2 will accept any positive offer and will reject a \$0 offer, the only rational strategy is to offer (\$99, \$1).

3.3.2 Game trees

Example 3.8. Consider a version of the stag hunt (Example 3.5) in which Hunter 1 is able to choose before Hunter 2, who is made aware of the choice of Hunter 1. This game is then sequential. It can be modeled by the following tree.

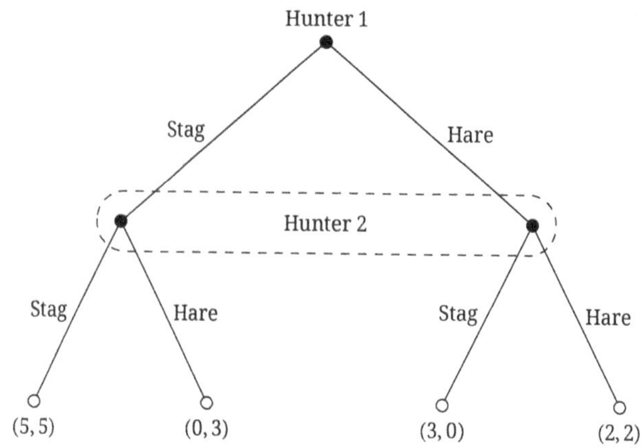

To analyze this game, we work backwards. This two-stage game is easy, but illustrates the mechanics of *backwards induction*. Start by considering the nodes whose child leaves are all terminal (in the current example, this is simply the two situations Hunter 2 may find themselves in). Each of these nodes represents the root of a *subgame*. For example, the event "Hunter 1 chooses stag" sets up a subgame in which only Hunter 2 is the decision maker. The outcome of this subgame can be predicted to be a choice of stag, provided Hunter 2 is rational. For each subgame in which a player gets to choose between terminal outcomes, assume that the decision maker chooses the leaf that is most favorable for them. Then replace that node with the outcome that would result. Work upwards in this way; each step will be deterministic provided the player has a preference between all options. The end result will be a path leading to the best (rational) possible outcome for the decision maker at the top.

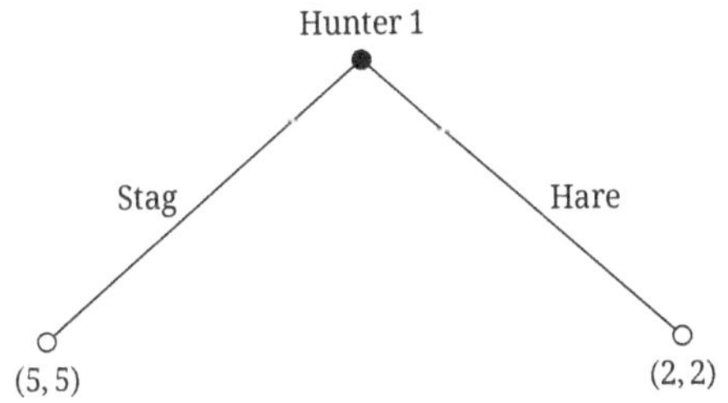

In this case, Hunter 1 knows that if they choose stag, Hunter 2 will choose stag, and if Hunter 1 chooses hare, Hunter 2 will choose hare. Preferring the stag, Hunter 1 chooses stag.

Backward induction works well when terminal states are clear. In the ultimatum game, Player 1 realizes that they essentially get to choose out of all splits up to ($99, $1) (perhaps Player 2 will reject a $0 offer), so they immediately choose this split. We are discounting psychology or bluffing. Player 2 may indeed reject an offer of less than $10, out of spite. This is not rational unless the game is repeated and Player 2 is attempting to signal that they will reject inequitable offers, or if Player 2 derives more utility from rejecting insulting offers. We will not discuss bluffing here; see Chapter 15 for further discussion of the ultimatum game.

The principle of backwards induction has important implications for Bitcoin security. A common argument asserting that a 51 % attack is unlikely cites the belief that any such attack would be followed by users abandoning Bitcoin en masse, leaving the attacker with worthless plunder. But this argument needs to take into consideration the agency of users after the attack: They may or may not abandon Bitcoin. Abandoning Bitcoin en masse would mean huge losses which may be unpalatable and there may be no obvious alternative. A group of miners makes the decision to attack or not attack the network based on their beliefs about what the response would be following the attack. If miners believe that an economic predominance of Bitcoin holders will respond to an attack by continuing to use the network as if the attacked never happened, these miners may be incentivized to attack. We will explore this more in §10.1 and §13.1 and Chapters 14, 16, and 17.

Rationality is an assumption on players' behavior that does not always hold in the strictest sense. For example, a chess player may not possess the mental capacity to work backwards from all possible end-states of a game of chess. In fact the existence of rational strategies to play chess is a mathematical theorem: A famous result in game theory known as Zermelo's theorem says that "in chess either White can force a win, or Black can force a win, or both sides can force at least a draw." However, nobody can compute the winning strategy, so chess remains unsolved, at least for the near future.

Exercise 3.4. Consider the subgame of a tic-tac-toe game in which X moves first that has begun as follows.

O		
O	X	X
X		

Construct a complete subgame tree, with terminal nodes representing points at which the game ends. Does backwards induction determine a winning strategy for either player? (It may be convenient to label the open squares 1, 2, 3, 4.)

3.4 Schelling points

Schelling points, or focal points, are strategies that players will play in absence of communication. Focal points are not acquiescent to a precise mathematical definition. Some may be more clear, for example, in the splitting $4 game, there seems to be a clear focal point – each player writes down $2. This is "fair" and uniquely so, thus intuitively one would imagine the other player would also see this and write down $2.

The original example of a Schelling point is due to Thomas Schelling, who asked the following question: If you are supposed to meet someone in New York City, but have not been able to communicate where or when, where and when would you go to attempt to meet them? In this case, you would attempt to go with what you believed other people would think. In experiments, "Grand Central Station at noon" was the most common answer given. Since this would maximize the probability of meeting the other person, this is probably the best move. Note that the payoff table for "meet up or lose" consists of wins along the diagonal and losses off. Players win if they play the same move and lose if they do not.

Note that there are similarities in the reasoning for finding Schelling points and the *Keynesian beauty contest*. In a Keynesian beauty contest, participants are asked to select faces that will be most attractive to judges. It is an interesting fact that people will often collectively have beliefs about others that are wrong. However, because others share the same collective misconceptions the beliefs turn out to create a Schelling point. For example, players may consider investing in a stock or cryptocurrency. They might have personal opinions about which is stronger on its own fundamentals, but will often make investment decisions based on what they think others will find most compelling. Keynes described this (three quarters of a century before cryptocurrency) as follows: "We have reached the third degree where we devote our intelligence to anticipating what average opinion expects the average opinion to be" [30].

Schelling points are chosen because they are special and stand out for some reason. This general concept underpins cryptocurrency. Technically, the notion of Schelling point requires an absence of communication, so one could argue the concept does not apply. But tomorrow, or the day after, people will continue to use currencies (crypto and otherwise) without holding a collective meeting and deciding to do so. Bitcoin is valuable precisely because people have decided that it is special, and for this reason one could reasonably expect it to maintain a "special" status, at least into the near future. By choosing to transact on the Bitcoin blockchain protocol, you are choosing the Bitcoin network as a "meeting point" where you can meet other users of the same protocol and blockchain. If you vary the protocol or blockchain, you cannot expect the same results; the network is useless without the network effect. This is why the rules of a consensus protocol are crucial. As long we continue to follow the rules of a protocol and prescribe value to coins transferred according to this protocol, we can expect the coin to maintain some status as special.

Open-ended Question 3.1. Where would you meet someone in Seattle? Paris? Boston? San Francisco? Washington, DC? Beijing? London? And Austin, TX?

Open-ended Question 3.2. Two players who do not know one another are brought onto a game show. They are separated and the host shows them two envelopes. One contains \$1,000, the other contains \$10,000. They have to choose between keeping the former themselves or giving the latter to the other player. What do they choose? Is this game different than the prisoners' dilemma?

Open-ended Question 3.3. Suppose there is an outbreak of a contagious pathogen that is transferred primarily on surfaces. You are a guest in a hotel, where guests must enter and leave through one of four front doors, each of which requires contact with a handle. Which door do you use?

4 Monopolizing pool

As a basic introduction to mining games, we consider the case when a single mining pool controls more than 50 % of the hashrate. Suppose this pool would like to control the entire blockchain, by mining their own blocks. There are several reasons to do this: By establishing and then maintaining monopoly power over the mining they can control profit margins – by easing the hashrate down, the difficulty will decrease, while the block reward may or may not decline (unless, of course, such an action causes enough dysphoria in the Bitcoin ecosystem to destroy the benefits; see §10.1). Also, the pool may have direct orders from, or be heavily funded by, a larger entity that would like to censor the network's transactions. To accomplish this end, the pool mines only their own blocks.

To set up the game, consider a point in time at which the monopolizing pool (MP) decides to start mining only their own blocks, ignoring all other blocks that appear on the network. Aware of this, the other resisting miners (RP) instead start mining their own chain. We define a random variable:

$$
\begin{aligned}
S_i := {} & \text{total blocks found by MP when} \\
& i \text{ total blocks have been found by MP and RP combined.}
\end{aligned}
$$

We are assuming that both RP and MP are publishing blocks as they find them. The RP are not colluding and have no motivation to perform a sneak attack on MP, so they are best served by instantly publishing all blocks. The MP is motivated to demoralize the RP, with the goal of attriting all RP. Publishing blocks is likely to be consistent with this goal.

This is different from cases in which one party is trying to perform a stealth double-spend – mining transactions in secret and then foisting these to the network only after they have overtaken the main chain and a certain number of blocks have passed. This requires slightly different analysis; see [22].

Assuming we know each block that has been produced at a given time affords simpler analysis, we do not need to worry about the time variable. If we suspect one pool is mining in secret, the number of blocks they have mined is a random variable itself. By using total blocks mined as the time unit (a unit that increments roughly, but not consistently, at 10 minute intervals) we can obviate this hassle.

4.1 Rough approach: Bernoulli's law of large numbers and Hoeffding's bound

We take this opportunity to introduce more useful results from probability theory, indulging some very simple proofs.

https://doi.org/10.1515/9783110772845-004

Theorem 4.1 (Chebyshev's inequality). *Suppose that X is a non-negative random variable. Then*

$$P(\{X \geq \varepsilon\}) \leq \frac{\mathbb{E}[X]}{\varepsilon}.$$

Proof. Using the conditional expectation formula (2.6),

$$\mathbb{E}[X] = \mathbb{E}[X \mid X \geq \varepsilon]P(\{X \geq \varepsilon\}) + \mathbb{E}[X \mid X < \varepsilon]P(\{X < \varepsilon\}).$$

Certainly the expected value of X, given that $X \geq \varepsilon$, must be at least ε. Because $X \geq 0$, the second term is also non-negative, and we have

$$\mathbb{E}[X] \geq \varepsilon P(\{X \geq \varepsilon\}).$$

Diving by ε gives the result. □

Back to our game, if we define random variables

$$\xi_i := 1 \text{ if MP produces the } i\text{th block,}$$
$$\xi_i := 0 \text{ if RP produce the } i\text{th block,}$$

we have defined a Bernoulli process (§2.4) with $p > 0.5$ being the hashrate of MP. As in §2.4.1,

$$S_n := \sum_{i=1}^{n} \xi_i.$$

Now by applying (2.4) to the independent variables ξ_i,

$$\begin{aligned}\operatorname{Var}(S_n) &= \sum_{i=1}^{n} \operatorname{Var}(\xi_i) \\ &= \mathrm{n}\operatorname{Var}(\xi_i).\end{aligned}$$

We can compute the variation by hand from equation (2.3). We know that

$$\mathbb{E}(\xi_i) = p$$

and that $\xi_i = 1$ with probability p, so

$$(\xi_i - \mathbb{E}[\xi_i])^2 = (1-p)^2 \quad \text{with probability } p,$$
$$(\xi_i - \mathbb{E}[\xi_i])^2 = (0-p)^2 \quad \text{with probability } 1-p,$$

so

$$\begin{aligned}\mathbb{E}(\xi_i - \mathbb{E}[\xi_i])^2 &= p(1-p)^2 + (1-p)p^2 \\ &= p(1-p).\end{aligned}$$

Thus (2.4) gives us

$$\operatorname{Var}(S_n) = np(1-p).$$

Now by definition

$$\begin{aligned}\operatorname{Var}(S_n) &= \mathbb{E}\big[(S_n - \mathbb{E}[S_n])^2\big] \\ &= \mathbb{E}\big[(S_n - pn)^2\big],\end{aligned}$$

that is,

$$np(1-p) = \mathbb{E}\big[(S_n - pn)^2\big].$$

We now apply Chebyshev's inequality to the random variable $(\frac{S_n}{n} - p)^2$:

$$\begin{aligned}P\left(\left\{\left(\frac{S_n}{n} - p\right)^2 \geq \varepsilon\right\}\right) \leq \frac{\mathbb{E}[(\frac{S_n}{n} - p)^2]}{\varepsilon} &= \frac{\mathbb{E}[(S_n - pn)^2]}{n^2\varepsilon} \\ &= \frac{p(1-p)}{n\varepsilon}.\end{aligned}$$

The expression given here shows that as n becomes large, the probability that the fraction of blocks obtained by MP differs from p by more than any fixed $\sqrt{\varepsilon}$ dwindles to 0. This demonstrates a principle known as Bernoulli's law of large numbers. Suppose a mining pool has a hashrate of exactly 51 %. This MP has mined over half of the blocks and controls the longest chain if and only if $\frac{S_n}{n} > 0.5$. In particular, this is true whenever

$$\left(\frac{S_n}{n} - p\right)^2 < (0.01)^2.$$

The probability that this condition fails is bounded above by

$$\frac{0.51 \times 0.49}{n(0.01)^2} = \frac{2{,}499}{n}.$$

This is not that terrific of a bound: If a mining pool has 51 % of the hashrate, we have shown that even after 10,000 blocks are produced, we can only bound (via this method) the probability that MP has not controlled the blockchain by $\frac{1}{4}$.

This is a rough bound and not an exact computation of the probability. We expect this to overestimate the probability by a factor of at least 2 because we are not taking into consideration the fact that $S_n/n > 0.52$ is actually a win for MP.

A more powerful inequality (which we do not prove) is Hoeffding's inequality.

Theorem 4.2 (Hoeffding's inequality). *If $\xi_i \in \{0,1\}$ are independent Bernoulli variables, then*

$$P\left(\left\{\frac{S_n}{n} - p \geq t\right\}\right) \leq e^{-2nt^2}.$$

This statement is much more powerful. To apply it, we switch roles and look at the perspective of the RP pool. In order to maintain control of the chain, we would take $p = 0.49$ and would need

$$\frac{S_n}{n} - p \geq 0.01.$$

So the probability that RP control the chain is bounded above by

$$e^{-0.0002n}.$$

This may seem to decay slowly at first, but this is exponential decay. After 10,000 blocks, this bound becomes

$$e^{-2} \approx 0.13534.$$

Observe that this is close to half of the bound we obtained previously via Bernoulli's law of large numbers. This continues to decay exponentially fast from there: After 20,000 blocks the probability is

$$e^{-4} \approx 0.018316.$$

Of course, the control of a chain by MP would be much more drastic if they had a larger percentage: 51 % means eventually they will control the chain, but not necessarily today or tomorrow. The assurance that this chain will be ahead (from this bound) does not even approach 99 % after 10 weeks.

However, if MP controls 60 % of the hashrate, the computation changes drastically. By Bernoulli's law of large numbers, we get

$$P\left(\left\{\left(\frac{S_n}{n} - p\right)^2 \geq (0.1)^2\right\}\right) \leq \frac{E[(S_n - pn)^2]}{\varepsilon} = \frac{24}{n}.$$

After 144 blocks (roughly one day), the probability that the number of blocks is between 50 % and 52 % is 5/6. Using Hoeffding's inequality,

$$P\left(\left\{\frac{S_n}{n} - p \geq t\right\}\right) \leq e^{-0.02n}.$$

In comparison, after 1,000 blocks, the Hoeffding bound assures us that the probability that MP has failed to control the chain is bounded by 2.06×10^{-9}:

$$e^{-0.02\times1{,}000} \approx 2.06 \times 10^{-9}.$$

4.2 Binomial distribution and the De Moivre–Laplace theorem

To be even more precise, we may apply the theory in §2.4.1.

From the perspective of MP, they are ahead provided that

$$S_n > \frac{n}{2}.$$

Thus

$$\begin{aligned} P\left(\left\{S_n > \frac{n}{2}\right\}\right) &= 1 - P\left(\left\{S_n \le \frac{n}{2}\right\}\right) \\ &= 1 - \sum_{i=0}^{\lfloor n/2 \rfloor} \binom{n}{i} p^i (1-p)^{n-i}. \end{aligned}$$

This can be computed with a computer. Taking $n = 1{,}000$ and $p = 0.51$,

$$1 - \sum_{i=0}^{500} \binom{1{,}000}{i} (0.51)^i (0.49)^{1{,}000-i} \approx 0.72610.$$

Taking $n = 10{,}000$ and $p = 0.51$,

$$1 - \sum_{i=0}^{5{,}000} \binom{1{,}000}{i} (0.51)^i (0.49)^{1{,}000-i} \approx 0.9767. \tag{4.1}$$

Remark 4.1. This computation can be done with readily available software. For example, the python package `scipy.stats` has a function `binom.cdf`, which performed the above computation in a split second. Using

```
from scipy.stats import binom
binom.cdf(5000,10000,0.51)
```

within a python session will result in the probability that a mining pool with a hashrate of 51 % has created more than 5,000 of the first 10,000 blocks.

For practical purposes, this is probably sufficient. A basic installation of python generates a floating point truncation error at about $n = 10^9$, which is more blocks than will be produced in any of our lifetimes.

Without computers, one may use the De Moivre–Laplace theorem as an approximation. Recall that the random variable

$$\tilde{S}_n = \frac{(S_n - np)}{\sqrt{np(1-p)}}$$

converges to a random variable with standard normal distribution. The normal distribution function $\Phi(x)$ has been computed for decades by hand and is available in many reference books. For $p = 0.51$, $n = 10{,}000$ we are interested in,

$$\begin{aligned}
&P(\{S_n - np \le 5{,}000 - 5{,}100\}) \\
&= P\left(\left\{\frac{S_n - np}{\sqrt{np(1-p)}} \le \frac{-100}{\sqrt{10{,}000(0.49)(0.51)}}\right\}\right) \\
&\approx P(\{\tilde{S}_n \le -2.0004\}) \\
&= \Phi(-2.0004) \approx 0.022729.
\end{aligned}$$

Observe that

$$1 - 0.022729 = 0.97727$$

is rather close to the value obtained in (4.1).

The De Moivre–Laplace theorem is useful for following both p and n as either change, or in solving one based on knowledge of the other.

Example 4.1. Suppose we would like to find the smallest value of p that will ensure with probability at least 90 % that the mining pool is ahead by 10 blocks after the first 10,000 blocks have been mined. We want to find p such that

$$P(\{S_n \ge 5{,}005\}) = 0.9.$$

We compute

$$\begin{aligned}
P(\{S_n \ge 5{,}005\}) &= P\left(\left\{\tilde{S}_n \ge \frac{5{,}005 - 10{,}000 \times p}{\sqrt{1{,}000p(1-p)}}\right\}\right) \\
&= 1 - P\left(\left\{\tilde{S}_n < \frac{5{,}005 - 10{,}000 \times p}{\sqrt{1{,}000p(1-p)}}\right\}\right) \\
&= 1 - \Phi\left(\frac{5{,}005 - 10{,}000 \times p}{\sqrt{10{,}000p(1-p)}}\right).
\end{aligned}$$

At this point we solve

$$1 - \Phi(u) = 0.9,$$

or

$$\Phi(u) = 0.1,$$

which can be done by consulting any table of the standard normal cumulative distribution function and determine that

$$u \approx -1.29.$$

Then we solve

$$\frac{5{,}005 - 10{,}000p}{\sqrt{10{,}000p(1-p)}} = -1.29$$

and get

$$p = 0.50695.$$

Exercise 4.1. What is the probability that a miner with 35 % of the hashrate obtains at least 672 blocks during a difficulty adjustment epoch (2,016 blocks)?

Exercise 4.2. Suppose a miner has 45 % of the hashrate. Give a number of blocks N so that there is at least a 90 % probability that the miner has mined less than $N/2$ of the first N blocks.

5 Basic double-spend game

Suppose that Alice controls a transaction output, which she double-spends by creating a valid transaction to Bob and then creating another valid transaction to Charlie. While these transactions might both be valid, they cannot go on the same blockchain. Miners are left to decide, perhaps arbitrarily, which transaction to include in the next block. This need not be random; they can choose the transaction that has the higher fee or choose the transaction which came first to their attention.

Let us consider the simplest possible scenario: Bob operates mining pool P_1 while Charlie operates mining pool P_2. In this case each will certainly try to confirm their own transaction, while the remaining pools may go either way. In a few minutes, one of the transactions will be confirmed. Without loss of generality, suppose that this is Bob's transaction. Thus at the latest block, Bob is in possession of the transaction output sent by Alice. What should Charlie do?

Most miners, by default, will simply continue to append to the chaintip. Charlie, who controls h_2 of the total hashrate, needs to decide if it is worthwhile to try and split the blockchain at the previous block, in hopes to overtake and surpass the current chain. The default rule followed by miners is that they continue to mine on the first-seen longest chain. So not only does Charlie need to catch up; Charlie must create a longer blockchain in order for miners to begin building on Charlie's block.

We can analyze Charlie's probability of accomplishing this and decide if it is worth the effort. Even though it appears somewhat simple, the question is slightly involved. This is an extensive-form game. Charlie can choose from a variety of strategies. The simplest strategy is to forget the transaction. More risky strategies are to continue to mine even when many blocks behind.

We assume that the payment to Charlie that Charlie wants to claim is λR. In our first computation we will ignore not only Bob's motivations, but also the accumulated block rewards of different mining pools, and focus only on the difference in length. (The situation where Bob has the hashpower to fight back is more complicated. This involves an extensive-form game fought in multiple rounds. Note that if Bob's hashrate is small, it is fair to ignore this case.) The other miners may be thought of as non-playable characters, essentially following the first-seen longest-chain rule at all times, even if that means supplanting a block they have mined in the past. Thus the instant that Charlie's chain becomes longer, Charlie wins. For this game, it will be convenient to think in terms of stages, each new stage occurring when a new block is produced on either chain. For this game setup, the relevant game state information is just the number describing the difference between the lengths of the chains. We do not need to keep track of who mined the block first, as this only matters when the two chains are equal in length, and because the opposing chains would not have mined a chain after Charlie took the lead, we can assume that equal lengths means that Charlie's is not first-seen. If Charlie's was first-seen, the game would be over and Charlie would have won. The game state should also include information about how many blocks Charlie has produced on his own chain.

https://doi.org/10.1515/9783110772845-005

We define the state space as follows:

$$\left\{(i,j) \mid \begin{pmatrix} i \in \{-1,0,1,2,\ldots\} \text{ denotes how far behind Charlie's chain is} \\ j \in \{0,1,2,\ldots\} \text{ denotes how many blocks Charlie has mined} \end{pmatrix}\right\}. \tag{5.1}$$

To illustrate this, the game starts with state $(1,0)$, denoting that Charlie is one block behind and has produced zero blocks since the game started. Any state of the form $(-1,j)$ is terminal: This means Charlie has overcome the main chain and taken the lead.

Now we denote $p := h_2$. Hereafter, the probability that the state moves from $(1,0)$ to $(0,1)$ is p, while the probability that the state moves from $(1,0)$ to $(2,0)$ is $1-p$.

This can be generalized: The probability that the first state value decreases by 1 will always be p. The probability that it increases by 1 is $1-p$. The second state variable increases by 1 every time the first state vector decreases by 1 and stays the same otherwise.

Charlie has an array of strategies available to him. For each possible state of the game, Charlie can either continue to pursue mining his own blockchain or give up and go back to mining the main chain. One basic strategy available to Charlie is "mine only when not behind by more than k blocks." This strategy only considers the first state variable and might not necessarily allow Charlie the optimal strategy: The second variable is not purely a "sunk cost," because it is recoverable if (and only if) Charlie proceeds to win.

The probability space we work on will be the Bernoulli space $\mathcal{B}$, which consists of all sequences chosen from

$$\{\text{Charlie finds next hash, Someone else finds next hash}\}.$$

Recall the remarks in §2.4.2, in particular the left-hand right-hand game comparison. There are two universes, with equivalent probability spaces $\mathcal{B}$: one in which Charlie employs a divergent strategy and one in which Charlie simply mines the longest chain. Because these probability spaces are equivalent we may ignore the fact that the particular outcomes in $\mathcal{B}$ are different in the two different universes (recall §2.4.2.1) and define random variables as follows:

$$\begin{aligned} f_{\text{strategic}} &:= \text{how many block rewards Charlie wins} \\ &\quad\ \text{using the strategy when the game ends,} \\ f_{\text{null}} &:= \text{how many block rewards Charlie would} \\ &\quad\ \text{have won when the game ends using a null strategy.} \end{aligned}$$

Then define

$$f := f_{\text{strategic}} - f_{\text{null}}. \tag{5.2}$$

Again the "would have won" is a slight technical abuse: It does not quite make sense because the game would never have been played. But due to the remarks in §2.4.2 we may ignore this technicality.

Clearly, if

$$E(f) > 0,$$

then Charlie has chosen a better strategy over the null strategy. Thus we seek to maximize, if possible, the value $E(f)$ over all possible strategies. It is not so simple to develop a formula for each possible strategy. Short of maximizing over all strategies, we are happy to find at least one strategy that yields a positive expectation. We start with the simple strategy "mine only when not behind by more than k blocks."

Remark 5.1. We are presenting the analysis from Charlie's perspective, which is omniscient. It is possible in general that a double-spend attempt may involve a secret fork (see §5.3). Suppose you want perform a double-spend by sending an output to yourself and also to an exchange. You keep the output to yourself secret and mine this output onto a chain in secret, hoping to spring your chain onto the network once you have been credited by the exchange and are able to withdraw elsewhere or take possession of whatever you purchased. The analysis from the viewpoint of the exchange (how many blocks should pass before we credit the transaction in order to minimize risk?) is slightly more complicated as it involves the time variable as well. A much more general discussion of double-spends, taking other angles and considerations into account, is offered in a series of works by Cyril Grunspan and Ricardo Pérez-Marco [20, 22, 23].

5.1 Simplest strategy: Charlie elects to mine his own chain when less than k blocks behind

This game starts at state $(1, 0)$, meaning that Charlie is one block behind and has yet to mine any blocks on his forked chain. Charlie commits to a strategy of mining his own chain until the moment that his chain falls k blocks behind, at which point Charlie will switch back to the main chain.

First of all, we would like to know the probability that Charlie will win, given the current state. The probability of winning depends only on the first state variable, so we define

$$w_i := \text{probability that Charlie will win, given that Charlie is } i \text{ blocks behind.}$$

5.1.1 Recursion formulas

The analysis depends on the conditional probability formula (2.6). Note that by letting W denote the random variable that is 1 if Charlie wins and 0 if he loses, we have

$$w_i = \mathbb{E}[W \mid \text{Charlie is } i \text{ blocks behind}].$$

If Charlie is i blocks behind, there are two events that could happen immediately next, and we know these probabilities. We write

$$\begin{aligned}&\mathbb{E}[W \mid \text{Charlie is } i \text{ blocks behind}]\\&\quad = \mathbb{E}\left[W \mid \begin{matrix}\text{Charlie is } i \text{ blocks behind}\\ \text{and wins the next block}\end{matrix}\right] P(\text{Charlie wins the next block})\\&\qquad + \mathbb{E}\left[W \mid \begin{matrix}\text{Charlie is } i \text{ blocks behind}\\ \text{and loses the next block}\end{matrix}\right] P(\text{Charlie loses the next block})\end{aligned}$$

by (2.6). But

$$\mathbb{E}\left[W \mid \begin{matrix}\text{Charlie is } i \text{ blocks behind}\\ \text{and loses the next block}\end{matrix}\right] = \mathbb{E}[W \mid \text{Charlie is } i+1 \text{ blocks behind}],$$

so we can write in simpler terms

$$w_i = pw_{i-1} + (1-p)w_{i+1}. \tag{5.3}$$

This is true for any i in between -1 and k, and it becomes most useful if we start working from the end points. We know that

$$w_{-1} = 1,$$

which simply states that if Charlie is -1 blocks behind, he has pulled ahead, and all miners now mine his chain; he has won. At the other end,

$$w_k = 0, \tag{5.4}$$

stating that Charlie gives up the fight and concedes defeat when he is k blocks behind. There are multiple ways to proceed in solving for the values w_i. We will offer two. The first method is somewhat straightforward and gives nice-looking formulas for smaller values of k. The second method uses difference equations and is somewhat more high-tech. The formulas are not as pretty but this method will lend itself to more complicated computations in later sections and will be easy to program using linear algebra.

5.1.2 A method for computing winning probability

Begin by noting that

$$\begin{aligned}w_{k-1} &= pw_{k-2} + (1-p)w_k\\&= pw_{k-2}\end{aligned}$$

from (5.4). Proceed with the goal to continue to define w_i in terms of w_{i-1}, implicitly defining q_i by writing

$$w_i = q_i w_{i-1}. \tag{5.5}$$

We have determined that

$$q_{k-1} = p.$$

With this in mind, we may write, using (5.3) and (5.5),

$$\begin{aligned} w_{i-1} &= p w_{i-2} + (1-p) w_i \\ &= p w_{i-2} + (1-p) q_i w_{i-1} \end{aligned}$$

or

$$w_{i-1}(1 - (1-p) q_i) = p w_{i-2},$$

which can be solved as

$$w_{i-1} = \frac{p w_{i-2}}{(1 - (1-p) q_i)},$$

that is,

$$q_{i-1} = \frac{p}{(1 - (1-p) q_i)}.$$

We can iterate this, until we get to

$$\begin{aligned} w_0 &= q_0 w_{-1} \\ &= q_0, \end{aligned}$$

where

$$\begin{aligned} q_0 &= \frac{p}{(1 - (1-p) q_1)}, \\ q_1 &= \frac{p}{(1 - (1-p) q_2)}, \\ &\dots \\ q_{k-1} &= p, \\ q_k &= 0. \end{aligned} \tag{5.6}$$

For example, when $k = 2$, we have

$$q_1 = p,$$

$$q_0 = \frac{p}{(1-(1-p)p)}.$$

Hence

$$w_0 = \frac{p}{(1-(1-p)p)},$$
$$w_1 = \frac{p^2}{(1-(1-p)p)}.$$

Exercise 5.1. Let $m = -(1-p)p$. Use (5.6) to show that there is a sequence of polynomials $\rho_i(m)$ such that for a given k, we have

$$q_{k-i} = p\frac{\rho_{i-2}(m)}{\rho_{i-1}(m)}$$

and

$$w_0 = p\frac{\rho_{k-2}(m)}{\rho_{k-1}(m)},$$
$$w_1 = p^2\frac{\rho_{k-3}(m)}{\rho_{k-1}(m)},$$
$$\dots$$
$$w_{k-1} = p^k\frac{\rho_{-1}(m)}{\rho_{k-1}(m)},$$

where

$$\rho_{-1}(m) = 1,$$
$$\rho_0(m) = 1,$$

and the polynomials satisfy the recursion relation for $i \geq 1$

$$\rho_i(m) = \rho_{i-1}(m) + m\rho_{i-2}(m).$$

5.1.3 Another method: difference equations

While the above by-hand method will yield a formula for w_i in any given situation, there is another observation that offers us a streamlined computation. Consider the equation

$$w(x) = pw(x-1) + (1-p)w(x+1), \tag{5.7}$$

thinking here of w as a function of the continuous variable x on the interval $[-1, k]$. The equation appears to be a second-order finite difference equation. Drawing from intuition from second-order homogeneous ordinary differential equations (ODEs) (cf. [32, Section 2.1]) we take a guess of the solution of the form

$$w(x) = e^{\tau x},$$

in which case (5.7) becomes

$$e^{\tau x} = pe^{\tau x - \tau} + (1-p)e^{\tau x + \tau}.$$

Dividing by $e^{\tau x}$,

$$1 = pe^{-\tau} + (1-p)e^{\tau}$$

which can be solved as

$$e^{\tau} = \frac{p}{(1-p)}.$$

Hence,

$$w(x) = \left(\frac{p}{1-p}\right)^x$$

is a solution to (5.7). Also note that if w is constant, (5.7) will be solved. So we define two solutions,

$$w_1(x) = \left(\frac{p}{1-p}\right)^x,$$
$$w_2(x) = 1,$$

which are independent solutions (two functions are independent if one function is not simply a real multiple of the other; more than two functions are independent if no function is a linear combination of multiples of the other functions). Noting that equation (5.7) can be written as a linear homogeneous equation,

$$w(x) - pw(x-1) - (1-p)w(x+1) = 0,$$

we may easily check that two solutions added together will also be a solution. By general theory for homogeneous equations, as in the case of linear second-order ODEs (cf. [32, Section 2.1]), we can write the general solution as

$$w(x) = c_1\left(\frac{p}{1-p}\right)^x + c_2.$$

By prescribing boundary values

$$w(-1) = 1,$$
$$w(k) = 0,$$

we may solve for the coefficients

$$c_1\left(\frac{p}{1-p}\right)^{-1} + c_2 = 1, \tag{5.8}$$

$$c_1\left(\frac{p}{1-p}\right)^{k} + c_2 = 0 \tag{5.9}$$

and obtain

$$c_1 = \frac{p}{1-p}\frac{1}{(1-(\frac{p}{1-p})^{k+1})}, \tag{5.10}$$

$$\begin{aligned} c_2 &= -\left(\frac{p}{1-p}\right)^{k+1}\frac{1}{(1-(\frac{p}{1-p})^{k+1})} \\ &= -\left(\frac{p}{1-p}\right)^{k} c_1. \end{aligned} \tag{5.11}$$

Hence,

$$w(x) = \frac{(\frac{p}{1-p})^{x+1} - (\frac{p}{1-p})^{k+1}}{1-(\frac{p}{1-p})^{k+1}}. \tag{5.12}$$

For simplicity of computation in the future, we will let

$$m := \frac{p}{1-p}$$

and note that the coefficients are determined by the solution to the following linear system, which uses a matrix to express equations (5.8) and (5.9):

$$\begin{pmatrix} m^{-1} & 1 \\ m^k & 1 \end{pmatrix}\begin{pmatrix} c_1 \\ c_2 \end{pmatrix} = \begin{pmatrix} 1 \\ 0 \end{pmatrix}. \tag{5.13}$$

Remark 5.2. The above expression (5.12) is reminiscent of the formula for computing sums of geometric series. This is not a coincidence. Most of the results in this section and the next could be derived using geometric series.

When $k = 2$ we have

$$w(0) = \frac{(\frac{p}{1-p})^{1} - (\frac{p}{1-p})^{3}}{1-(\frac{p}{1-p})^{3}},$$

$$w(1) = \frac{(\frac{p}{1-p})^2 - (\frac{p}{1-p})^3}{1 - (\frac{p}{1-p})^3}.$$

5.1.4 Computing expected values

We now compute the expected value to Charlie of the function f, defined by (5.2). We are most interested in

$$\mathbb{E}[f \mid (1,0)],$$

that is, the expected benefit of mining the reorg strategy when starting from one block behind.

From state $(1,0)$ there are two possible next states, $(2,0)$ and $(0,1)$, so we write, again using (2.6),

$$\mathbb{E}[f \mid (1,0)] = p\mathbb{E}[f \mid (0,1)] + (1-p)\mathbb{E}[f \mid (2,0)].$$

Introducing a less cumbersome notation,

$$e(x,s) := \mathbb{E}[f \mid (x,s)],$$

this relation becomes

$$e(1,0) = pe(0,1) + (1-p)e(2,0).$$

For brevity, we only present the solution method using difference equations. Unlike in the previous example, when there was only one state variable, in this case, we have two. Observe the following relation:

$$e(x,s) = pe(x-1,s+1) + (1-p)e(x+1,s). \tag{5.14}$$

Claim 5.1. *We have*

$$e(x,s+1) = e(x,s) + w(x) - 1. \tag{5.15}$$

Proof. Given we are at state (x,s) and following a given strategy that only depends on x, the probability space of possible paths starting at (x,s) and going until the game ends is identical to the probability space of possible paths starting at $(x,s+1)$. For each path that ends in success, the reward to Charlie is the same in both cases, because all of the block rewards that would have been won along the way are granted when he takes the longest chain. But if the path ends in a loss for Charlie, he loses all block rewards he would have won had he pursued the null strategy. He loses one more block reward if the path started from state $(x, s+1)$ than if the same path had started from path (x, s). The difference will be the probability of a loss, which is $1 - w(x)$. □

Exercise 5.2. Using a Bernoulli space and arguments from §2.4.2.1, offer a more rigorous proof of Claim 5.1.

Now equation (5.14) becomes

$$\begin{aligned} e(x,s) &= p\big[e(x-1,s) + w(x-1) - 1\big] + (1-p)e(x+1,s) \\ &= pe(x-1,s) + (1-p)e(x+1,s) + p\big(w(x-1) - 1\big). \end{aligned}$$

This now only involves a single value in the second state variable. So if we are interested in computing $e(x, 0)$, we may define

$$e(x) := e(x, 0),$$

which must satisfy

$$\begin{aligned} e(x) - pe(x-1) - (1-p)e(x+1) &= p\big(w(x-1) - 1\big) &\quad (5.16) \\ &= p\left(c_1\left(\frac{p}{1-p}\right)^{x-1} + c_2 - 1\right) &\quad (5.17) \end{aligned}$$

for the c_1 and c_2 obtained above in (5.10). This is similar to the finite difference equation (5.7) in x, but with extra terms on the right-hand side, making this a non-homogeneous equation. As with second-order linear non-homogeneous ODEs, we first find a particular solution and then add this to the solutions w_1 and w_2 obtained previously to obtain the general solution.

Consider the non-homogeneous finite difference equation

$$y(x) - py(x-1) - (1-p)y(x+1) = p(c_2 - 1). \quad (5.18)$$

We try a very basic guess (this is similar to the method of undetermined coefficients from ODE theory)

$$y_1(x) := \beta x$$

for some constant β to be determined. Plugging y_1 into (5.18) and solving for β, we get

$$\beta = \frac{p(c_2 - 1)}{2p - 1}.$$

Thus

$$y_1 = p\frac{1 - c_2}{1 - 2p}x$$

solves (5.18). Next we solve

$$y(x) - py(x-1) - (1-p)y(x+1) = pc_1\left(\frac{p}{1-p}\right)^{x-1}. \tag{5.19}$$

For this we try a solution of the form

$$y_2(x) := ax\left(\frac{p}{1-p}\right)^x.$$

Plugging y_2 into (5.19) and doing some straightforward computations leads to

$$a = \frac{c_1(1-p)}{1-2p}. \tag{5.20}$$

We conclude that

$$y_2(x) = \frac{(1-p)}{(1-2p)}c_1x\left(\frac{p}{1-p}\right)^x$$

is a solution to (5.19).

Thus we arrive at the general solution to (5.16):

$$y(x) = \tilde{c}_1\left(\frac{p}{(1-p)}\right)^x + \tilde{c}_2 + p\frac{1-c_2}{(1-2p)}x + \frac{(1-p)}{(1-2p)}c_1x\left(\frac{p}{1-p}\right)^x.$$

To be clear, c_1 and c_2 are the coefficients that were previously determined by (5.10), while $\tilde{c}_1$ and $\tilde{c}_2$ are coefficients that need to be solved when finding a solution to a boundary value problem. We can now solve for the boundary conditions:

$$e(-1) = \lambda,$$
$$e(k) = 0.$$

Recall that when Charlie takes the lead, he wins the blocks he would have won, plus the extra λ times the block reward, which is the value of the transaction. If Charlie gives up mining without having won a single block, he is no better or worse off than if he mined the null strategy. This analysis is available because we have reduced the equation to one where the second state variable is 0.

Solving with the boundary conditions determines $\tilde{c}_1$ and $\tilde{c}_2$ by

$$\tilde{c}_1\left(\frac{p}{(1-p)}\right)^{-1} + \tilde{c}_2 - \frac{p(1-c_2)}{(1-2p)} - \frac{(1-p)}{(1-2p)}c_1\left(\frac{p}{1-p}\right)^{-1} = \lambda, \tag{5.21}$$

$$\tilde{c}_1\left(\frac{p}{(1-p)}\right)^k + \tilde{c}_2 + \frac{p(1-c_2)}{(1-2p)}k + \frac{(1-p)}{(1-2p)}c_1k\left(\frac{p}{1-p}\right)^k = 0. \tag{5.22}$$

It is convenient to write this as a 2×2 system of equations,

$$\begin{pmatrix} m^{-1} & 1 \\ m^k & 1 \end{pmatrix}\begin{pmatrix} \tilde{c}_1 \\ \tilde{c}_2 \end{pmatrix} = \begin{pmatrix} \lambda + mr(1-c_2) + rc_1 m^{-1} \\ -mr(1-c_2)k - rc_1 km^k \end{pmatrix}$$
$$= \lambda \begin{pmatrix} 1 \\ 0 \end{pmatrix} + r \begin{pmatrix} m^{-1} & m \\ -km^k & -km \end{pmatrix}\begin{pmatrix} c_1 \\ 1-c_2 \end{pmatrix},$$

once again using less cumbersome notation

$$m = \frac{p}{1-p},$$
$$r = \frac{1-p}{1-2p},$$

with

$$\begin{pmatrix} c_1 \\ c_2 \end{pmatrix} = \begin{pmatrix} m^{-1} & 1 \\ m^k & 1 \end{pmatrix}^{-1} \begin{pmatrix} 1 \\ 0 \end{pmatrix}. \tag{5.23}$$

Here

$$\begin{pmatrix} m^{-1} & 1 \\ m^k & 1 \end{pmatrix}^{-1} = \frac{1}{m^{-1} - m^k} \begin{pmatrix} 1 & -1 \\ -m^k & m^{-1} \end{pmatrix}$$

is the inverse matrix from standard linear algebra. Now

$$\begin{pmatrix} \tilde{c}_1 \\ \tilde{c}_2 \end{pmatrix} = \begin{pmatrix} m^{-1} & 1 \\ m^k & 1 \end{pmatrix}^{-1} \begin{pmatrix} \lambda + mr(1-c_2) + rc_1 m^{-1} \\ -mr(1-c_2)k - rc_1 km^k \end{pmatrix}. \tag{5.24}$$

At this point we may plug the constants into the expression to obtain the function:

$$e(x,0) = \tilde{c}_1 m^x + \tilde{c}_2 + mr(1-c_2)x + rc_1 x m^x. \tag{5.25}$$

The expansion of this expression is messy and not particularly enlightening, so we leave it as is.

Example 5.1 ($p = 0.1$, k small). Suppose a miner with 10 % of the hashrate considers a strategy to mine a reorg chain if two or less blocks behind and give up if three blocks behind. Taking $p = 0.1$, $k = 3$ gives us

$$m = \frac{1}{9},$$
$$r = \frac{9}{8},$$
$$c_1 = \frac{1}{9}\frac{1}{1 - \frac{1}{9}^4} = \frac{729}{6{,}560}, \tag{5.26}$$

$$c_2 = -\frac{1}{9}^4 \frac{1}{1-\frac{1}{9}^4} = -\frac{1}{6{,}560},$$

$$\begin{pmatrix} \tilde{c}_1 \\ \tilde{c}_2 \end{pmatrix} = \begin{pmatrix} \frac{729}{6{,}560}\lambda + \frac{1{,}554{,}957}{8{,}606{,}720} \\ -\frac{1}{6{,}560}\lambda - \frac{3{,}234{,}573}{8{,}606{,}720} \end{pmatrix},$$

and

$$\begin{aligned} e(x) = {} & \left(\frac{729}{6{,}560}\lambda + \frac{1{,}554{,}957}{8{,}606{,}720}\right)\left(\frac{1}{9}\right)^x - \frac{1}{6{,}560}\lambda \\ & - \frac{3{,}234{,}573}{8{,}606{,}720} + \frac{1}{8}\left(1 + \frac{1}{6{,}560}\right)x + \frac{9}{8} \times \frac{729}{6{,}560}x\left(\frac{1}{9}\right)^x. \end{aligned}$$

We can plug in the values for x:

$$\begin{aligned} e(-1) &= \lambda, \\ e(0) &= \frac{91}{820}\lambda - \frac{6{,}561}{33{,}620}, \\ e(1) &= \frac{1}{82}\lambda - \frac{729}{3{,}362}, \\ e(2) &= \frac{1}{820}\lambda - \frac{405}{3{,}362}, \\ e(3) &= 0. \end{aligned}$$

We conclude that in order to mine with these odds, a miner starting with a one-block deficit is only motivated to attempt a reorg if

$$e(1) = \frac{1}{82}\lambda - \frac{729}{3{,}362} > 0, \tag{5.27}$$

that is,

$$\lambda > \frac{729}{41} \approx 17.78,$$

and for a two-block deficit,

$$\lambda > \frac{4{,}050}{41} \approx 98.78.$$

In other words, for a pool with 10 % of the hashrate, in order for it to be prudent to attempt to perform a reorg with a “mine to three blocks behind” strategy, when starting from two blocks behind, the reward would need to be roughly 100 times the block reward.

On the other hand, if a mining pool mines a transaction, only to discover that all the other pools have discovered another block seconds earlier, in order to push forward, this pool is wise to do so if

$$\lambda > \frac{6{,}561}{3{,}731} \approx 1.7585.$$

We can ask the same question, but for $k = 2$, in which case similar calculations yield

$$\begin{aligned} e(-1) &= \lambda, \\ e(0) &= \frac{10}{91}\lambda - \frac{729}{8{,}281}, \\ e(1) &= \frac{1}{91}\lambda - \frac{810}{8{,}281}, \\ e(2) &= 0. \end{aligned}$$

So if the pool adopts a strategy to mine until two blocks behind, starting from one block behind, this makes sense if

$$\lambda > \frac{810}{91} \approx 8.901.$$

The following table shows the "break-even" multiple λ of the block reward required to mine the reorg strategy when starting from one block behind:

	$k = 2$	$k = 3$	$k = 4$	$k = 5$	$k = 6$
$p = 0.1$	8.9011	17.780	27.546	37.507	47.501
$p = 0.2$	3.8095	7.5294	11.976	16.769	21.696
$p = 0.3$	2.0675	3.9425	6.3118	9.0544	12.048
$p = 0.4$	1.1842	2.0769	3.1822	4.4973	6.0114
$p = 0.49$	0.70766	1.0824	1.4714	1.8749	2.2929

One may observe monotonicity in the chart: Break-even numbers increase as k becomes larger. However, this only tells part of the story. It is also interesting to look at the expected value for a fixed λ. The following table shows the expected advantage of mining the reorg chain when we take a fixed $\lambda = 10$:

	$k = 2$	$k = 3$	$k = 4$	$k = 5$
$p = 0.1$	0.01208	−0.0949	−0.2163	−0.3395
$p = 0.2$	0.29478	0.14533	−0.12169	−0.42150
$p = 0.3$	0.90370	0.93995	0.63324	0.16885
$p = 0.4$	1.8560	2.4378	2.4557	2.1514

This is more instructive as to how Charlie would behave in practice. Given the extra reward λ, Charlie plugs p into the formula for several values of k. The value of k that maximizes the expected surplus will be the strategy. It appears (we make no attempt to prove this here) that this function exhibits concave behavior with respect to k, in which case there is a unique maximizer k which can be found by increasing k until the function begins to decrease.

There is an improvement in this strategy that will be difficult to optimize precisely, but can be described. Suppose Charlie is sent a transaction worth 10 times the block reward but fails to process it in time, and he must perform a reorg starting from one block behind. The above analysis says that Charlie is wise to pursue a strategy, starting from time 0, to mine until two blocks behind, and then give up. However, there is a (somewhat low) possibility that Charlie will trade a string of alternating blocks with the rest of the chain. If this continues for long enough, Charlie's difference between winning and losing will be not only the reward λ, but $\lambda + s$, where s is the number of blocks that Charlie has won. So there may be an opportunity to refine the strategy as the game continues to get a slightly better strategy. So the estimates proved here are lower bounds that are slightly but strictly below the optimal strategy. For p closer to $\frac{1}{2}$ and k larger, this becomes more pronounced. The computation of this optimized strategy is left as an exercise for the very ambitious reader.

Exercise 5.3 (Involved). Compute a formula for $e(1,0)$ for $k = 2$ and p using convergent geometric sequences. (Hint 1: Partition the probability space of Bernoulli processes into events described by which round the game ends. The game ends with probability 1, so this is possible. Find the pattern and write it as a geometric series.) (Hint 2: Compute and compare the Taylor series for the functions $\frac{1}{1-x}$ and $(\frac{1}{1-x})^2$ and use this to rewrite the expression you see.)

Exercise 5.4. Find values of a, b, c so that both

$$y(x) = e^{2x}$$

and

$$y(x) = e^{-x}$$

are solutions of the third-order finite difference equation

$$y(x) - ay(x-1) - by(x-2) - cy(x-3) = 0.$$

5.2 Longer-duration attacks

If we look only at the chance of winning when a mining pool commits to mining forever, even if they are very far behind, their odds are given by the function $w(x)$ by taking the limit

$$W_\infty(x) = \lim_{k\to\infty} \frac{(\frac{p}{1-p})^{x+1} - (\frac{p}{1-p})^{k+1}}{[1-(\frac{p}{1-p})^{k+1}]} = \left(\frac{p}{1-p}\right)^{x+1}. \tag{5.28}$$

In particular, if the pool with hashrate p starts k blocks behind and commits to trying to catch up and win, never cutting losses, their chances of catching up are

$$\left(\frac{p}{1-p}\right)^{k+1}.$$

Remark 5.3. Often the probability of simply catching up ([20, Lemma 3.1]) is computed; this is $(\frac{p}{1-p})^k$. We require *catching up and winning.*

For example, if $p = 0.3$, the chance of overcoming a six-block deficit is

$$\left(\frac{0.3}{0.7}\right)^7 \approx 0.0026556.$$

For $p = 0.15$ the chances are much lower,

$$\left(\frac{0.15}{0.85}\right)^7 \approx 5.3297 \times 10^{-6},$$

but better than 1 in a million!

We also see that as $p \to \frac{1}{2}$, the odds of overcoming a large deficit do not decrease quite as rapidly; for example, if we let $p = 0.4$ and $k = 20$,

$$\left(\frac{0.4}{0.6}\right)^{21} \approx 2.0049 \times 10^{-4}.$$

This may make sense when compared to familiar sports events: A basketball team that should otherwise have a 0.400 winning probability may surge 21 games above 0.500. This is rare, but not quite a black swan.

In fact, we observe that (5.28) tells us the following: If the probability is exactly $p = 1/2$, the probability of winning is 1 regardless of starting position. This follows from the statement that an unbiased random walk on the set of integers will eventually hit every integer, excepting certain perversely chosen sequences that happen with probability 0.

Example 5.2 ($k = 7$). Six blocks is often to referred to as a "safe" depth, in the sense that transactions are "confirmed" once six blocks have been mined starting with the block mining the transaction. Again, part of the issue that complicates this is that we may not know if someone is attacking the network in secret or not; we take the omniscient perspective of the potential attacker who will be making the decision to attack or not from already six blocks behind.

The follow table shows the values of λ required to break even when starting from six blocks behind:

	$k = 7$	$k = 8$	$k = 9$	$k = 10$	$k = 11$
$p = 0.1$	672,596	1,210,680	1,796,068	2,391,839	2,989,396
$p = 0.2$	7,270.2	11,638.7	16,632	21,918	27,320
$p = 0.3$	477.27	673.71	900.35	1,148.87	1,411.7
$p = 0.4$	69.892	87.617	107.59	129.64	153.58
$p = 0.45$	30.986	36.502	42.559	49.157	56.293
$p = 0.49$	16.423	18.138	19.910	21.740	23.628

So unless the reward for the double-spend is huge or the attacker has nearly 50 % of the hashpower, the attacker rarely finds it wise to attempt starting from already six blocks behind.

5.3 Stealth six-block double-spend

The attack described in Example 5.2 is unlikely to succeed, because it starts from a position significantly behind. A more feasible attack would be to immediately begin mining a secret chain, keeping this from the network until the appropriate time.

For example, some exchanges will credit an account bitcoin after six blocks have been found confirming the transaction. This is a common practice, but as pointed out in [44] and [22] this does not accurately account for the probability of success of the attackers without considering the time as well. An exploit that one may try is the following. Send a large transaction to an exchange, say 100 bitcoins. Once the credit has been applied to your account on the exchange, immediately trade the currency (say for Ethereum) and then withdraw Ethereum (note that one must be careful withdrawing Bitcoin from the exchange when planning to mess with outputs the exchange believes they control; if they send you a transaction downstream of the one you intend to invalidate, your efforts are all for naught). Then attempt to annul the original large transaction. If the exchange has policies in place that pause larger withdrawals, there may be ways around this. For example, one may set up other ringer accounts with sky-high sell orders in a very thinly traded currency. One may purchase the thinly traded currency with bitcoins ("fat-finger" orders are not uncommon) and then withdraw the Bitcoin from the ringer account, perhaps days or months later.

The best way to pull this off is to attempt to mine a competing transaction and have the exchange transaction ready to go. To begin, you must control a large output. You compute but do not publish a transaction directing this output to the exchange. Meanwhile, you have a secret transaction that directs this same output to an address you control. You mine this transaction in secret, until you have a block. Once you have obtained a block that includes the self-directed transaction, you immediately publish the exchange-directed transaction, but not the block you have mined. At this point, you have a one-block lead against the main chain. Your goal is to secretly maintain that lead until the main chain has been extended by six blocks. After the main chain has mined six blocks, the exchange considers your deposit good. If your secret chain is seven blocks long, you can go forwards with whatever shenanigans you had in mind. Of course, you may always abort. If you have not mined any more secret blocks, you can forget the whole operation and consider it just a waste of an hour's worth of mining power.

The analysis is performed by computing the probability of each of the different states that the game can be in when the main chain adds six blocks. Then for each of these states, we can apply the computation we have already done. There are six events to measure. One event is that the attacker has added six or more blocks (seven in total) by the time that the main chain gets to six. This outcome is clear; the attacker wins by immediately publishing their seven-block extension. If the attacker has produced five more secret blocks, this sets off a tie situation as in the computations above; we must compute $w(0)$ to determine the probability that the attacker will win and $e(0, 5)$ for the expected net winnings. In turn, we perform computations for each of the other possible states.

As before, assume that the loss-cutting strategy is determined by a constant k. This is probably a simplification of the optimal strategy: In practice an attacker who has built 10 secret blocks and is eight blocks behind may be more likely to proceed than, say, the attacker who is simply eight blocks behind having only produced a single secret block. To keep things simple, assume that $k = 6$.

The computation involves what is called the *negative binomial distribution* formula (cf. [20, Section 5]).

5.3.1 Negative binomial distribution

The negative binomial distribution describes probabilities for the number of successes in a sequence of Bernoulli trials before a specified number of failures occur. From the attacker's perspective, "success" is the attacker obtaining a block. This happens with probability p. A "failure" is the main chain adding a block. This happens with probability $1 - p$. We are interested in knowing the probabilities of the events that there are zero, one, two, three, four, five, or more than six successes at the moment that the sixth failure occurs.

Computing the negative binomial distribution involves the probability that in the first $x + r - 1$ trials, there have been $r - 1$ failures and x successes, and then the $(x + r)$th trial results in a failure. The distribution is given by

$$\phi(x; r, p) = \binom{x + r - 1}{r - 1} p^x (1 - p)^r,$$

where

$$\phi(x; r, p) = P(\text{number of success} = x \text{ when the } r\text{th failure happens}).$$

Remark 5.4. Different versions of this formula exist and sometimes the roles of success and failure have been swapped. Also note that

$$\binom{x+r-1}{r-1} = \binom{x+r-1}{x},$$

which may make some formulas found elsewhere appear different.

Thus the probability that the attacker has mined x blocks when the main chain reaches its sixth is given by (note that here we are assuming that one block has already been credited to the attacker)

$$\phi(x; 6, p) = \binom{x-1+5}{5} p^{x-1}(1-p)^6.$$

As before, we let f be the net winnings for the attack. This will involve the reward λ, which is the amount of the double-spend that has been recovered in a successful attack. We want to compute the probability of attaining this and account for the expected number of mined blocks that will be lost in a failed attempt.

We define the following events:

$$\begin{aligned} Q_i &:= \text{attacker has mined } i \text{ blocks when the main chain reaches six} \\ Q_{7+} &:= \text{attacker has mined seven or more blocks when the main chain reaches six.} \end{aligned}$$

Now

$$\begin{aligned} \mathbb{E}[f] = &\sum_{i=1}^{6} \mathbb{E}[f \mid Q_i] P(Q_i) \\ &+ \lambda P(Q_{7+}). \end{aligned}$$

The values $P(Q_i)$ are given precisely (note that we are starting at one block):

$$\begin{aligned} P(Q_i) &= \binom{i+4}{5} p^{i-1}(1-p)^6 \\ &= \phi(i-1; 6, p), \\ P(Q_{7+}) &= 1 - \sum_{i=1}^{6} P(Q_i). \end{aligned}$$

Now we compute $\mathbb{E}[f \mid Q_i]$. Note that Q_i means, in the convention set up in (5.1), that we are at state $(6-i, i)$. Fortunately, we have already begun to compute the function $e(x, s)$. Applying (5.15) iteratively,

$$e(x, i) = e(x, 0) + i(w(x) - 1).$$

So

$$\begin{aligned}\mathbb{E}[f \mid Q_i] &= e(6 - i, i) \\ &= e(6 - i, 0) + i(w(6 - i) - 1).\end{aligned}$$

We can combine this all into

$$\begin{aligned}\mathbb{E}[f] = &\sum_{i=1}^{6}[e(6 - i, 0) + i(w(6 - i) - 1)]\phi(i - 1; 6, p) \\ &+ \lambda\left(1 - \sum_{i=1}^{6}\phi(i - 1; 6, p)\right)\end{aligned} \tag{5.29}$$

where the expressions for $e(6 - i, 0)$ are obtained from (5.25), using (5.23) and (5.24).

5.3.1.1 Computations

We would like to know when the double-spend attack described above pays off. Consider when an attacker with hashpower p seeks a double-spend with a value of λ and plans to cut losses by giving up at six blocks behind. The computation proceeds as follows:

1. Compute c_1, c_2, and $\tilde{c}_1, \tilde{c}_2$, by (5.23), (5.24).
2. Compute values $w(0), \ldots, w(6)$ and then $e(0, 0), e(1, 0), \ldots, e(6, 0)$ using (5.25).
3. Plug into (5.29) to get an expression of the form

$$\mathbb{E}[f] = a(p, 6)\lambda - b(p, 6).$$

4. The break-even value is

$$\lambda = \frac{b(p, 6)}{a(p, 6)}.$$

The negative binomials are easily computed with the `scipy.stats.nbinom` python package.

The following table describes the break-even multiple of the block reward:

Hashrate	Break-even
0.025	5,849,594.63919
0.05	119,708.096744
0.075	13,694.1085119
0.1	3,163.39696518
0.15	462.049924291
0.2	133.404850221
0.25	54.4772052334
0.3	26.8071499673
0.35	14.4129053512
0.4	7.85014186382
0.45	4.03873239636
0.475	2.7768062571

For example, with 30 % of the hashrate, a pool would be incentivized to try this strategy if the reward is 27 times the block reward.

6 Censorship attacks

Censorship attacks can happen in a number of ways. We are going to explore what portion of the hashrate is necessary to perform an attack and how that influences the fee that must be paid in order to overcome the incentives against including the transaction on the blockchain.

The least effective way for a pool or collection of pools to censor is to simply to decline to process a transaction. Even if the pool has a majority of the hashrate, anything less than 100 % hashrate will allow the transaction on the chain. This is why Bitcoin is often described as *censorship-resistant*. Effective censorship requires a more aggressive strategy. To effectively censor, the censoring parties should commit to supplanting any blocks that process forbidden transactions. If the censoring parties have more than 51 % they should be able to accomplish this in the long run, with probabilities as described in Chapter 4. If the censoring pool has less than 51 % of the hashrate, they should not expect to be able to overwhelm the rest of the hashrate on a consistent basis, at least by brute force. Even if they accomplish this feat once, or twice, the transaction will stay in the mempool and should eventually be confirmed.

The most effective approach for the censoring party is to announce their intention to attempt to supplant any blocks containing undesirable transactions and then allow incentives to influence other rational miners. At this point, presuming the censors can be taken seriously with this commitment, this promise will be taken into consideration by other miners. Before getting into the details, we offer a quick example. Suppose that a pool with 25 % of the hashrate announces that they will attempt to uncle any blocks which contain transactions which spend a certain output. Assume this threat is received as credible by other miners on the network. Now suppose a transaction appears in the mempool. The non-censoring miners have two options: leave it alone or attempt to mine it. If the transaction fee is small or non-existent, the miners have little or nothing to gain by including the transaction when attempting to construct a block. On the other hand, if they do include the transaction in their next block, there will be a risk: There is at least a 1/16 chance that the censoring pool will immediately write two blocks on top of the latest compliant block, before anyone appends to the offending block. So the expected value for including such a transaction in a block is at least 1/16 of a block reward less than leaving the transaction alone. In order to compensate for this the transaction should include a fee of at least 1/16 of a block reward or expect that miners are either ignorant of or resistant to the censorial pool's threats to supplant the block.

In practice, an entity with a bit of cash on hand who wanted to censor transactions may try the approach of committing to rewarding miners who supplant offending blocks. It is easy to choose the reward large enough so that rational miners will be induced to pursue a reorg whenever an offending transaction is included in a block. By backwards induction, rational miners will simply choose not to include that transaction, unless the fees are large enough to merit the risk. Of course the censor is likely to be a party who can simply bid higher, to the point that the fee required to spend

https://doi.org/10.1515/9783110772845-006

the transaction may be more than the unspent output. Another practical way to censor transactions is to have major financial institutions agree that the outputs from certain UTXOs are *tainted*. If such institutions refuse to treat these UTXOs and their progeny as valid Bitcoin, the effect could cascade throughout the ecosystem: Parties are unlikely to accept an output if they do not believe they will be able to trade it for something of value.

We model a case where a subset of miners tries to practice censorship by making a commitment to mining only on certain chaintips. This can happen in practice if a group of mining corporations resides in a jurisdiction that can exercise control over the miners, in which case the authority can declare it illegal to participate in the settlement of certain transactions. In this case "settlement" could mean mining a chaintip in which such transactions are less than three blocks deep. We assume that the miners will continue to mine the old chain, instead of turning off machines (it is possible that a miner who wishes to remain law-abiding without actively participating in censorship could just refuse to participate in reorgs, satisfying authorities by idling machines until the transactions are considered settled).

We consider three sets of mining pools. The censorial (C) mining pool will have hashrate p and is not interested in minimizing costs. They will follow a fixed predetermined strategy, with the primary objective of keeping offending transactions off of the blockchain. There may also be a non-compliant (NC) pool, which has hashrate q, which is not interested in censorship for ideological or technical reasons and hence will always follow the longest-chain rule and will mine any transactions. Finally there will be a rational (R) set of pools. These may have different sizes $\varepsilon_1 + \cdots + \varepsilon_n = r$ and are considered economic free agents: They will join the compliant pool when more profitable and proceed with non-compliant mining when profitable.

The NC pool will be assumed to not use any strategy: simply mine the longest chain. The C pool will declare their strategy at the beginning, which will be simple, and other players in the game will take this to be credible. For example, the C pool can declare that they will attempt to supplant all non-compliant blocks that are no more than k blocks deep, but will not pursue reorgs beyond a certain depth. It is then left to the remaining rational pools to determine their respective strategies.

If the C pool owns 51 %, they control the network. They can uncle any block, eventually, as is their prerogative. If the C pool has less than 50 % it makes sense to use a loss-cutting strategy. There is no sense pursuing an increasingly unlikely attempt to censor a specific transaction indefinitely.

6.1 Working example: two rational pools

Suppose we have only two rational pools of sizes $\varepsilon_1, \varepsilon_2$ so that

$$p + q + \varepsilon_1 + \varepsilon_2 = 1.$$

We assume that p is less than 50 %. Suppose that C declares that their strategy is to attempt to uncle any offending blocks that are less than k blocks deep. If k blocks behind, they will drop the fight and look for a more recent forking point or mine the chaintip with compliant transactions. With this declared strategy we will attempt to begin to build a payoff table for a few possible basic strategies for the two rational pools. For simplicity, assume that the distasteful transactions are happening somewhat infrequently, so that we do not expect to see a series of consecutive blocks containing offending transactions that may demand a more sophisticated censoring strategy.

The game starts when the discountenanced output holder attempts to send a transaction. The unspent output (fee) will be given by λ in units of the block reward. We will assume that NC has positive hashrate q. This assumption means that eventually the NC will win k consecutive blocks and the transaction will be accepted. In practice, depending on the strategies of the rational pools and the size of NC, this terminal event may occur within an hour, or it may take centuries. However, it is a convenient mathematical assumption as we can then model the game as a game that terminates in finite time with probability 1.

Remark 6.1. As a side note, we point out that there is a huge mathematical difference between the words "finite" and "bounded." Unfortunately these words are often used interchangeably by non-mathematicians. For example, saying the supply of a currency is always finite is usually a true statement, whereas saying the supply is "bounded" (as specified in Bitcoin's whitepaper) is a completely different statement. "Bounded" requires a number that provides the upper bound. Similarly, saying that a game terminates in finite time with probability 1 can be true, even if the expectation of the time it takes is infinite. The payoff in any play of the St. Petersburg paradox game (Exercise 2.3) is finite, but not bounded.

From a rational pool's perspective, there are two questions to be asked:

1. What is the expected number of block rewards that a rational pool would waste in fighting the censors?
2. What is the probability that they receive the extra reward from the fee?

Multiplying the answer to the latter by λ and subtracting the former serves as a useful metric in comparing different strategies.

We can think of the game as occurring in a series of rounds. Each round begins when the offending transaction is mined and terminates in one of two ways. The round may terminate with the compliant chain pulling a full block ahead, in which case the game continues. The NC pool attempts to mine the offending transaction on the chaintip, instead of making a come-from-behind attempt. The next round of the game starts when the transaction is again mined on the chaintip. The round may also terminate with the NC chain pulling ahead k blocks. In this case, the round terminates, and so does the game. So to compute the expected outcome for a given pool, we divide the analysis into these two cases.

In each of the scenarios determined by the pools' strategies, there will be a function $w(x)$, which must be computed, that determines the probability that the censorial pool will win the round. When the transaction is mined and the round begins, the probability that the game will continue beyond that round is $w(1)$. The probability that the game terminates is $1 - w(1)$. We can compute the expected value that a pool will have won in the latter case. In the former case we will need to set up some recursion relationships to complete the computation.

In particular, after fixing a strategy regime and a Pool i, define a random variable $W_i \in \{0, 1\}$ via

$$W_i := 1 \text{ if Pool } i \text{ claims the fee } \lambda \text{ when the game ends,}$$
$$W_i := 0 \text{ otherwise}$$

and a random variable $L_i \in \mathbb{Z}_+$,

$$L_i := \text{blocks mined by Pool } i \text{ that have become stale when the game ends.}$$

Then we define the random variable (dropping subscripts when clear)

$$f := \lambda W - L.$$

The goal then is to compute

$$\mathbb{E}[f] = \lambda \mathbb{E}[W] - \mathbb{E}[L].$$

Now

$$\mathbb{E}[f] = \left\{ \begin{array}{c} (1 - w(1))\mathbb{E}[f \mid \text{C loses round 1}] \\ +w(1)\mathbb{E}[f \mid \text{C wins round 1}] \end{array} \right\}, \tag{6.1}$$

and then

$$\begin{aligned} \mathbb{E}[f \mid \text{C wins round 1}] = (1 - w(1))&\mathbb{E}[f \mid \text{C loses round 2}] \\ &+ w(1)\mathbb{E}[f \mid \text{C wins round 2}]. \end{aligned} \tag{6.2}$$

Now define the following quantity:

$$\tau := \mathbb{E}[\text{blocks lost by Pool } i \text{ in any given round} \mid \text{C wins that same round}],$$

which is a constant that does not depend on the particular round. Noting that looking forward, the game looks the same at the start of every round, we can write

$$\mathbb{E}[f \mid \text{C wins round 1}] = \mathbb{E}[f] - \tau. \tag{6.3}$$

We also use notation

$$\sigma = \mathbb{E}[f \mid \text{C loses round 1}]$$

and combine (6.1) with (6.3) to get

$$\mathbb{E}[f] = (1 - w(1))\sigma + w(1)(\mathbb{E}[f] - \tau),$$

which reduces to

$$\mathbb{E}[f] = \sigma - \tau\frac{w(1)}{1 - w(1)}. \tag{6.4}$$

In the following sections, our goal will be to compute the values σ, τ and the function $w(x)$.

6.1.1 Consideration for this choice of function

One difficulty in modeling a game like this is coming up with an appropriate quantity to measure. We have made the decision to essentially model the game as a one-shot affair. The miners in question will mine until the game is over and then return to mine as they had been before the game started. In reality, this may or may not be optimal. In particular, we are not considering situations in which new offending transactions will pop up while a previous transaction is being fought over.

The goal of the computations will be to demonstrate that, given a fee and a hashing power of the censorial pool, the rational pools can compare different strategies. The results will be meaningful with the assumption that the game will be ending in a reasonably short period of time (this is not an unreasonable assumption).

For this reason, the values that appear in forthcoming payoff tables involve only the expected gain or loss from what would be mined in that time period, and do not consider how much time is taken to complete the game. When discussing incentives for mining games, especially those that may be repeated, the *revenue ratio* (expected revenue per expected time, cf. [21, §3]) can be more instructive, but the computation can become more intricate.

6.1.2 There are many strategies

We focus on strategies that are easier to describe and implement.

1. (Non-compliant) First, there is the non-compliant-at-all-times strategy. The pool simply ignores the censors and includes the transaction, and always mines the longest chain using the first-seen rule.

2. (Mildly compliant) A second strategy would be to attempt to build on the censoring fork when the difference is zero blocks but not when it is one block deep. In particular, the pool would never mine the offending transaction, always leaving it be while in the mempool, and will break ties for the censors.
3. (More compliant) A third strategy would be to not only exclude the transaction, but also attempt to build on the censoring fork when the difference is zero or one block and go back to the longest chain if two blocks deep.
4. (Neutral/selective) The pool may decide to abstain from mining when the difference is one block deep, not wanting to risk the event this block would become an uncle. If the energy cost is significant, these sorts of strategies may come into play.

There are many more variations. Computing outcomes with each of these would give us a 4×4 payoff table with 16 outcomes. Even with symmetries, this would already lead to a somewhat lengthy computation, so we restrict ourselves to the first two strategies. This will result in a 2×2 payoff table with essentially four computations, due to symmetries. Our attainable goal is to populate the following table:

Pool 2 strategy / Pool 1 strategy	Non-compliant	Mildly compliant
Non-compliant	?	?
Mildly compliant	?	?

6.1.3 Case 1: Pool 1 and Pool 2 are both non-compliant

We consider the game from the perspective of Pool 1; the payoff to Pool 2 computation would be identical. From Pool 1's perspective, there is no difference between Pool 2 and NC, as both employ the same behavior.

Our goal is to compute σ and τ from the perspective of Pool 1, given this strategy regime. Each round of the game has two state variables. The first state variable, i, will refer to how far behind the compliant chain is. Then $i = -1$ denotes the fact that the compliant chain has won the round, while $i = 0$ means the NC chain has been equalized by the compliant chain, so the first-seen rule favors the NC chain. At block k, C will concede the transaction. The second state variable, s, will refer to the number of block rewards Pool 1 has contributed to the NC chain.

A third variable, which is determined when the game begins and does not change after the first block is mined, will be the amount that Pool 1 has won by mining the offending transaction, taking values in $\{0, \lambda\}$ (either they mined the block or some other NC mined the block). This does not need to be dragged through computations along the game, but is necessary information when the game ends in favor of the NC chain.

While the game begins when the objectionable transaction is posted, the round begins (as with all subsequent rounds) when the transaction is mined and the longest chain becomes non-compliant. The block will have been mined by Pool 1, Pool 2, or NC. With probability $\varepsilon_1/(1-p)$ the round begins in state $(1,1)$ and with $1-\varepsilon_1/(1-p)$ probability the game begins in state $(1,0)$. We consider these cases to begin computing the expected value of L_1:

$$\mathbb{E}[L_1] = \frac{\varepsilon_1}{1-p}\mathbb{E}[L_1 \mid (1,1)] + \left(1 - \frac{\varepsilon_1}{1-p}\right)\mathbb{E}[L_1 \mid (1,0)]. \tag{6.5}$$

Using conditional expectations,

$$\mathbb{E}[L_1 \mid (x,s)] = \left\{ \begin{array}{l} p\mathbb{E}[L_1 \mid (x-1,s)] \\ \quad + (q+\varepsilon_2)\mathbb{E}[L_1 \mid (x+1,s)] \\ \quad + \varepsilon_1\mathbb{E}[L_1 \mid (x+1,s+1)] \end{array} \right\}. \tag{6.6}$$

This incorporates the three possibilities; the next block is won by C, NC/Pool 2, or Pool 1. We use a fact that can be concluded from an argument similar to the one leading to Claim 5.1:

$$\mathbb{E}[L_1 \mid (x,s+1)] = \mathbb{E}[L_1 \mid (x,s)] + w(x). \tag{6.7}$$

Letting

$$e_1(x) = \mathbb{E}[L_1 \mid (x,0)],$$

we have from (6.6) and (6.7) with $s = 0$

$$\begin{aligned} e_1(x) &= pe_1(x-1) + (q+\varepsilon_2)e_1(x+1) + \varepsilon_1(e_1(x+1) + w(x+1)) \\ &= pe_1(x-1) + (1-p)e_1(x+1) + \varepsilon_1 w(x+1). \end{aligned}$$

Because for the entirety of the round, C is playing catch-up, the probability that C wins w can be computed exactly as in §5.1.3:

$$w(x) = c_1\left(\frac{p}{1-p}\right)^{x+1} + c_2,$$

where c_1, c_2 are determined by (5.13). So we have the equation

$$e_1(x) - pe_1(x-1) - (1-p)e_1(x+1) = \varepsilon_1\left(c_1\left(\frac{p}{1-p}\right)^{x+1} + c_2\right), \tag{6.8}$$

which is quite similar to equation (5.16).

As before,

$$y_1 = -\varepsilon_1\frac{c_2}{1-2p}x,$$

$$y_2 = \frac{\varepsilon_1 c_1 x}{1-2p}\left(\frac{p}{1-p}\right)^{x+1}$$

will solve the equations

$$y(x) - py(x-1) - (1-p)y(x+1) = \varepsilon_1 c_2,$$
$$y(x) - py(x-1) - (1-p)y(x+1) = \varepsilon_1 c_1\left(\frac{p}{1-p}\right)^{x+1}.$$

So the general solution is

$$y(x) = \tilde{c}_1\left(\frac{p}{1-p}\right)^{x} + \tilde{c}_2 - \varepsilon_1\frac{c_2}{1-2p}x + \frac{\varepsilon_1 c_1 x}{1-2p}\left(\frac{p}{1-p}\right)^{x+1}. \tag{6.9}$$

We can solve this with boundary conditions

$$e_1(-1) = 0,$$
$$e_1(k) = 0.$$

Again recall why the zero boundary conditions are there: Because we are solving for $s = 0$ in the second state variable, there will be nothing gained nor lost by the mining pool.

Solving,

$$\tilde{c}_1\left(\frac{p}{1-p}\right)^{-1} + \tilde{c}_2 + \varepsilon_1\frac{c_2}{1-2p} - \frac{\varepsilon_1 c_1}{1-2p} = 0,$$
$$\tilde{c}_1\left(\frac{p}{1-p}\right)^{k} + \tilde{c}_2 - \varepsilon_1\frac{c_2}{1-2p}k + \frac{\varepsilon_1 c_1}{1-2p}k\left(\frac{p}{1-p}\right)^{k+1} = 0,$$

or

$$\begin{pmatrix} m^{-1} & 1 \\ m^k & 1 \end{pmatrix}\begin{pmatrix} \tilde{c}_1 \\ \tilde{c}_2 \end{pmatrix} = \frac{\varepsilon_1}{1-2p}\begin{pmatrix} 1 & -1 \\ -km^{k+1} & k \end{pmatrix}\begin{pmatrix} c_1 \\ c_2 \end{pmatrix}.$$

As before, we get

$$\begin{pmatrix} \tilde{c}_1 \\ \tilde{c}_2 \end{pmatrix} = \frac{m}{1-m^{k+1}}\frac{\varepsilon_1}{1-2p}\begin{pmatrix} 1 & -1 \\ -m^k & m^{-1} \end{pmatrix}\begin{pmatrix} 1 & -1 \\ -km^{k+1} & k \end{pmatrix}\begin{pmatrix} c_1 \\ c_2 \end{pmatrix}. \tag{6.10}$$

With these values evaluated we have the solution:

$$e_1(x) = \tilde{c}_1 m^x + \tilde{c}_2 - \varepsilon_1\frac{c_2}{(1-2p)}x + \varepsilon_1\frac{c_1}{1-2p}xm^{x+1}.$$

Now we are interested in the values

$$\mathbb{E}[L_1 \mid (1,0)] = e_1(1) = \tilde{c}_1 m + \tilde{c}_2 - \varepsilon_1 \frac{c_2}{(1-2p)} + \varepsilon_1 \frac{c_1}{1-2p} m^2,$$
$$\mathbb{E}[L_1 \mid (1,1)] = e_1(1) + w(1). \tag{6.11}$$

We have determined that the expected number of blocks that are lost in a given round, recalling (6.5), is

$$\begin{aligned}\mathbb{E}[L_1] &= \frac{\varepsilon_1}{1-p}(e_1(1) + w(1)) + \left(1 - \frac{\varepsilon_1}{1-p}\right)e_1(1) \\ &= e_1(1) + \frac{\varepsilon_1}{1-p} w(1).\end{aligned}$$

We are not done yet – we wanted to know the expected number of blocks that are lost, given the fact that C won the round. To compute this use

$$\mathbb{E}[L_1] = w(1)\mathbb{E}[L_1 \mid \text{C wins}] + (1 - w(1))\mathbb{E}[L_1 \mid \text{C loses}].$$

Because

$$\mathbb{E}[L_1 \mid \text{C loses}] = 0,$$

we can determine that

$$\begin{aligned}\mathbb{E}[L_1 \mid \text{C wins}] &= \frac{\mathbb{E}[L_1]}{w(1)} \\ &= \frac{e_1(1)}{w(1)} + \frac{\varepsilon_1}{1-p}.\end{aligned} \tag{6.12}$$

Thus we determine

$$\tau = \frac{e_1(1)}{w(1)} + \frac{\varepsilon_1}{1-p}. \tag{6.13}$$

Now σ is easier; it is the expected share of the additional reward. This is easily seen to be

$$\sigma = \frac{\varepsilon_1}{1-p}\lambda$$

as once the round starts the odds of winning do not depend on which pool mined the offending block. So the payoff to Pool 1 is

$$\begin{aligned}&\frac{\varepsilon_1}{1-p}\lambda - \left(\frac{e_1(1)}{w(1)} + \frac{\varepsilon_1}{1-p}\right)\frac{w(1)}{1-w(1)} \\ &\quad = \frac{\varepsilon_1}{1-p}\left(\lambda - \frac{w(1)}{1-w(1)}\right) - \frac{e(1)}{1-w(1)},\end{aligned} \tag{6.14}$$

where $e_1(1)$ is determined by (6.11) and (6.10).

Summary

To summarize:
1. Solve c_1, c_2 via (5.13).
2. Solve $\tilde{c}_1, \tilde{c}_2$ via (6.10).
3. Solve $e_1(1)$ using (6.11).
4. Note that $w(1) = c_1 m + c_2$.
5. Harvest the result in (6.14). This is $\mathbb{E}[\lambda W_1 - L_1]$ in the (non-compliant, non-compliant) strategy regime.

The computation for Pool 2 is identical. One only needs to modify (3) to incorporate the different value ε_2.

6.1.4 Case 2: Pool 1 is non-compliant and Pool 2 is mildly compliant

We will begin by attempting to compute $w(x)$. However, there is a wrinkle that prohibits us from continuing to use exactly the same calculus leading up to (5.13). These finite difference equations to be solved are "broken" by the fact that the strategy and transition probabilities change at intermediate states. So we need to perform some slightly more involved computations in order to solve the problem.

6.1.4.1 Broken difference equations

We begin with the most basic computation, the probability $w(x)$ that C wins the round, given we are at state x. The equations are different from previously. In particular, letting

$$p^* = p + \varepsilon_2,$$

we have

$$w(0) = p^* w(-1) + (1 - p^*) w(1). \tag{6.15}$$

This encodes the fact that when $i = 0$, Pool 2's hashrate of ε_2 is joining the censors (hence p^* and not p). However, for integers $x \geq 1$, Pool 2 returns to mine on the true chaintip, and we have

$$w(x) = pw(x - 1) + (1 - p)w(x + 1). \tag{6.16}$$

We solve the broken difference equation using somewhat of a cheat: Look at equation (6.16). This holds for $x > 0$ and only excludes the point 0, so we may think of (6.15) as a single relation that needs to be satisfied and solve the problem on $[0, k]$ considering this single relation and one degree of flexibility built into the boundary value problem. Note that this works when there is only one relation in the set of relations (6.15).

To begin, because $w(-1) = 1$, from (6.15) we have

$$w(0) = p^* + (1 - p^*)w(1). \tag{6.17}$$

Now if we solve the boundary value problem for equation (6.16) with

$$w(0) = b,$$
$$w(k) = 0$$

(for b to be determined), then necessarily from (6.17) we see that

$$w(1) = \frac{b - p^*}{1 - p^*}$$

(we have skipped the step of solving the first equation (6.15) as an equation; knowing $w(-1) = 1$, specifying $w(0) = b$, we know what $w(1)$ must be). So now, we may solve equation (6.16) with three values,

$$\begin{aligned} w(0) &= b, \\ w(1) &= \frac{b - p^*}{1 - p^*}, \\ w(k) &= 0. \end{aligned}$$

This extra parameter b allows us room to put in the third condition, and we expect to get a unique solution.

Equation (6.16) is the same equation we have seen before, with general solution given by

$$w(x) = c_1 m^x + c_2.$$

Thus the boundary conditions above become

$$\begin{aligned} c_1 + c_2 &= b, \\ c_1 m + c_2 &= \frac{b - p^*}{1 - p^*}, \\ c_1 m^k + c_2 &= 0. \end{aligned}$$

As a matrix,

$$\begin{pmatrix} 1 & 1 \\ m & 1 \\ m^k & 1 \end{pmatrix} \begin{pmatrix} c_1 \\ c_2 \end{pmatrix} = \begin{pmatrix} b \\ \frac{b-p^*}{(1-p^*)} \\ 0 \end{pmatrix}. \tag{6.18}$$

Now we augment the matrix in order to solve the problem completely using linear algebra. Using that

$$\begin{pmatrix} b \\ \frac{b-p^*}{1-p^*} \\ 0 \end{pmatrix} = \begin{pmatrix} 1 \\ \frac{1}{1-p^*} \\ 0 \end{pmatrix} b + \begin{pmatrix} 0 \\ \frac{-p^*}{1-p^*} \\ 0 \end{pmatrix},$$

we see that (6.18) is equivalent to

$$\begin{pmatrix} 1 & 1 \\ m & 1 \\ m^k & 1 \end{pmatrix} \begin{pmatrix} c_1 \\ c_2 \end{pmatrix} - \begin{pmatrix} 1 \\ \frac{1}{1-p^*} \\ 0 \end{pmatrix} b = \begin{pmatrix} 0 \\ \frac{-p^*}{1-p^*} \\ 0 \end{pmatrix}$$

or

$$\begin{pmatrix} 1 & 1 & -1 \\ m & 1 & \frac{-1}{1-p^*} \\ m^k & 1 & 0 \end{pmatrix} \begin{pmatrix} c_1 \\ c_2 \\ b \end{pmatrix} = \begin{pmatrix} 0 \\ \frac{-p^*}{1-p^*} \\ 0 \end{pmatrix}.$$

Now, in order to solve for c_1 and c_2 (and b incidentally) we need to invert this 3×3 matrix,

$$\begin{pmatrix} c_1 \\ c_2 \\ b \end{pmatrix} = \begin{pmatrix} 1 & 1 & -1 \\ m & 1 & \frac{-1}{1-p^*} \\ m^k & 1 & 0 \end{pmatrix}^{-1} \begin{pmatrix} 0 \\ \frac{-p^*}{1-p^*} \\ 0 \end{pmatrix}. \tag{6.19}$$

Now that we have c_1 and c_2, we can compute our new function $w(x)$, now given as a multi-part function:

$$\begin{aligned} w(-1) &= 1, \\ w(x) &= c_1 m^x + c_2 \quad \text{for } x \geq 0. \end{aligned}$$

Remark 6.2. If we had cutoff at, say, $i = 2$, the computation would have been more complicated. Another method would be to solve both equations, but with "buckled" boundary conditions that overlap at two points and a free parameter (such a b above) specified for one of the boundary conditions.

6.1.4.2 Expected value for Pool 1

This same strategy works to compute expected values. To begin, we have

$$\mathbb{E}[L_1] = \frac{\varepsilon_1}{q+\varepsilon_1}\mathbb{E}[L_1 \mid (1,1)] + \frac{q}{q+\varepsilon_1}\mathbb{E}[L_1 \mid (1,0)]. \tag{6.20}$$

But then, using $e_1(x,s)$ for short-hand we have a single relation describing the transitions at state $(0,s)$:

$$e_1(0,s) = p^* e_1(-1,s) + \varepsilon_1 e_1(1,s+1) + qe(1,s).$$

As in (6.7),

$$e_1(1,s+1) = e_1(1,s) + w(1), \tag{6.21}$$

so

$$e_1(0,s) = p^* e_1(-1,s) + \varepsilon_1(e_1(1,s) + w(1)) + qe_1(1,s),$$

or

$$e_1(0,s) - p^* e_1(-1,s) - (1-p^*)e_1(1,s) = \varepsilon_1 w(1). \tag{6.22}$$

As above when we solved the broken equation for $w(x)$, this relation will be plugged in directly to create a three-parameter boundary value problem when solving for $e_1(x)$.

For $x \geq 1$,

$$e_1(x,s) = pe(x-1,s) + \varepsilon_1 e_1(x+1,s+1) + (q+\varepsilon_2)e_1(x+1,s),$$

which gives a familiar equation:

$$e_1(x,s) - e_1(x-1,s) - (1-p)e_1(x+1,s) = \varepsilon_1 w(x+1). \tag{6.23}$$

The general strategy for solving this type of broken system of non-homogeneous equations is similar to the strategy for solving homogeneous equations. We will set up the general solution for (6.23) and then use (6.22) to set up a third condition.

Solving (6.8) as before to get the general solution (6.9),

$$y(x) = \tilde{c}_1\left(\frac{p}{1-p}\right)^x + \tilde{c}_2 - \varepsilon_1\frac{c_2}{1-2p}x + \frac{\varepsilon_1 c_1 x}{1-2p}\left(\frac{p}{1-p}\right)^{x+1}, \tag{6.24}$$

where c_1, c_2 are obtained by (6.19). Set boundary conditions

$$e_1(0) = b_1,$$
$$e_1(k) = 0$$

and then bring in (6.22):

$$e_1(1) = \frac{b_1 - \varepsilon_1 w(1)}{1-p^*}. \tag{6.25}$$

We are trying to solve for $\tilde{c}_1, \tilde{c}_2$, so we plug the values for e_1 in the three previous expressions to get

$$
\begin{aligned}
\tilde{c}_1 + \tilde{c}_2 &= b_1, \\
\tilde{c}_1 m + \tilde{c}_2 - \varepsilon_1 \frac{c_2}{1-2p} + \frac{\varepsilon_1 c_1}{1-2p} m^2 &= \frac{b_1 - \varepsilon_1 w(1)}{1-p^*}, \\
\tilde{c}_1 m^k + \tilde{c}_2 - \varepsilon_1 \frac{c_2}{1-2p} k + \frac{\varepsilon_1 c_1 k}{1-2p} m^{k+1} &= 0,
\end{aligned}
$$

which tidies up into matrix notation as

$$
\begin{aligned}
\begin{pmatrix} 1 & 1 & -1 \\ m & 1 & -\frac{1}{1-p^*} \\ m^k & 1 & 0 \end{pmatrix} \begin{pmatrix} \tilde{c}_1 \\ \tilde{c}_2 \\ b_1 \end{pmatrix} &= \begin{pmatrix} 0 \\ \frac{-\varepsilon_1 w(1)}{1-p^*} + \varepsilon_1 \frac{c_2}{1-2p} - \frac{\varepsilon_1 c_1}{1-2p} m^2 \\ \varepsilon_1 \frac{c_2}{1-2p} k - \frac{\varepsilon_1 c_1 k}{1-2p} m^{k+1} \end{pmatrix} \\
&= \varepsilon_1 \begin{pmatrix} 0 & 0 & 0 \\ \frac{-w(1)}{1-p^*} & -\frac{m^2}{1-2p} & \frac{1}{1-2p} \\ 0 & -\frac{k}{1-2p} m^{k+1} & \frac{k}{1-2p} \end{pmatrix} \begin{pmatrix} 1 \\ c_1 \\ c_2 \end{pmatrix}.
\end{aligned} \tag{6.26}
$$

We can solve for the coefficients $\tilde{c}_1, \tilde{c}_2$ using linear algebra. Once we have done this, we plug back into (6.20) and (6.21) and compute

$$
\begin{aligned}
\mathbb{E}[L_1] &= \frac{\varepsilon_1}{q+\varepsilon_1}(e_1(1) + w(1)) + \frac{q}{q+\varepsilon_1} e(1) \\
&= e_1(1) + \frac{\varepsilon_1}{q+\varepsilon_1} w(1).
\end{aligned}
$$

As before (equations (6.12), (6.13)), we can then determine that

$$
\begin{aligned}
\tau &= \frac{e_1(1)}{w(1)} + \frac{\varepsilon_1}{q+\varepsilon_1}, \\
\sigma &= \frac{\varepsilon_1}{q+\varepsilon_1} \lambda.
\end{aligned}
$$

Once the round starts, the odds of winning do not depend on which pool mined the offending block. So the payoff to Pool 1 is

$$
\begin{aligned}
&\frac{\varepsilon_1}{q+\varepsilon_1} \lambda - \left(\frac{e_1(1)}{w(1)} + \frac{\varepsilon_1}{q+\varepsilon_1} \right) \frac{w(1)}{1-w(1)} \\
&= \frac{\varepsilon_1}{q+\varepsilon_1} \left(\lambda - \frac{w(1)}{1-w(1)} \right) - \frac{e_1(1)}{1-w(1)}.
\end{aligned} \tag{6.27}
$$

This gives the expected value for Pool 1 in the second case.

Summary

1. Solve for w, that is, find coefficients c_1, c_2 using (6.19).
2. Solve for $\tilde{c}_1, \tilde{c}_2, b_1$ using (6.26) and $w(1) = c_1 m + c_2$.

3. Solve for $e_1(1)$ using (6.25).
4. Harvest $\mathbb{E}[\lambda W_1 - L_1]$ using (6.27).

We note that because we are using b_1 to solve for the value $e_1(1)$, we end up not needing the information given by $\tilde{c}_1$ and $\tilde{c}_2$ in this particular computation.

6.1.4.3 Expected value for Pool 2

We have already computed $w(x)$ in this strategy regime. In setting up the computation of the outcome for Pool 2, we now let the second state variable denote the number of blocks that Pool 2 has mined. Note that because Pool 2 is mining with the censorial pool only when the chains are of equal length, Pool 2 has no risk of censorial blocks being lost. The risk to Pool 2 is realized only if Pool 2 adds to an NC chain that is already one block ahead and this chain ends up losing (the given strategy allows for the unlikely but still possible event that the chain will mine a block on the NC chain and then end up breaking a tie in favor of C, making an uncle of their own block, so the strategy would probably be tweaked in practice).

Because the pool is not interested in mining the objectionable transaction itself, we start with

$$\mathbb{E}[L_2] = \mathbb{E}[L_2 \mid (1,0)] \tag{6.28}$$

as the initial state necessarily involves either Pool 1 or NC having just mined the block. Now

$$\mathbb{E}[L_2 \mid (0,s)] = (p+\varepsilon_2)\mathbb{E}[L_2 \mid (-1,s)] + (1-p-\varepsilon_2)\mathbb{E}[L_2 \mid (1,s)],$$

and for $x \geq 1$,

$$\mathbb{E}[L_2 \mid (x,s)] = p\mathbb{E}[L_2 \mid (x-1,s)] + (1-p-\varepsilon_2)\mathbb{E}[L_2 \mid (x+1,s)] + \varepsilon_2\mathbb{E}[L_2 \mid (x+1,s+1)]. \tag{6.29}$$

Also, as before,

$$\mathbb{E}[L_2 \mid (x,s+1)] = \mathbb{E}[L_2 \mid (x,s)] + w(x).$$

Using $e_2(x) = \mathbb{E}[L_2 \mid (x,0)]$, (6.29) becomes

$$e_2(x) - pe_2(x-1) - (1-p)e_2(x+1) = \varepsilon_2 w(x+1) \quad \text{for } x > 0.$$

The function $w(x)$ has been computed with coefficients determined by (6.19). The analysis to determine $e_2(x)$ is very similar to the analysis line following (6.24). With general solution

$$y(x) = \tilde{c}_1\left(\frac{p}{1-p}\right)^x + \tilde{c}_2 - \varepsilon_2\frac{c_2}{1-2p}x + \frac{\varepsilon_2 c_1 x}{1-2p}\left(\frac{p}{1-p}\right)^{x+1}, \tag{6.30}$$

we solve for boundary conditions

$$\begin{aligned} e_2(0) &= b_2, \\ e_2(k) &= 0 \end{aligned} \tag{6.31}$$

and the additional condition with free parameter b_2

$$e_2(1) = \frac{b_2}{1-p^*}. \tag{6.32}$$

This follows from (6.31) and

$$e_2(-1) = 0$$

and

$$e_2(0) - p^* e_2(-1) - (1-p^*)e_2(1) = 0, \tag{6.33}$$

which differs from (6.22) because there is no $w(1)$ term on the right-hand side.

We are trying to solve for $\tilde{c}_1, \tilde{c}_2$, so we plug the boundary values in and convert to a matrix:

$$\begin{aligned} \begin{pmatrix} 1 & 1 & -1 \\ m & 1 & -\frac{1}{1-p^*} \\ m^k & 1 & 0 \end{pmatrix} \begin{pmatrix} \tilde{c}_1 \\ \tilde{c}_2 \\ b_2 \end{pmatrix} &= \begin{pmatrix} 0 \\ \varepsilon_2 \frac{c_2}{1-2p} - \frac{\varepsilon_2 c_1}{1-2p} m^2 \\ \varepsilon_2 \frac{c_2}{1-2p} k - \frac{\varepsilon_2 c_1 k}{1-2p} m^{k+1} \end{pmatrix} \\ &= \frac{\varepsilon_2}{1-2p} \begin{pmatrix} 0 & 0 \\ -m^2 & 1 \\ -km^{k+1} & k \end{pmatrix} \begin{pmatrix} c_1 \\ c_2 \end{pmatrix}. \end{aligned} \tag{6.34}$$

Inverting the first matrix, we can solve for the coefficients $\tilde{c}_1$, $\tilde{c}_2$, and b_2. Once we have done this, we plug back into

$$\mathbb{E}[L_2] = e_2(1) = \frac{b_2}{1-p^*}. \tag{6.35}$$

As before we can then determine that

$$\begin{aligned} \tau &= \frac{e_2(1)}{w(1)}, \\ \sigma &= 0. \end{aligned}$$

So the total expectation for Pool 2 is

$$-\frac{e_2(1)}{1-w(1)} = -\frac{b_2}{(1-w(1))(1-p^*)}. \tag{6.36}$$

Summary

1. Solve for w, that is, find coefficients c_1, c_2 using (6.19).
2. Solve for $\tilde{c}_1, \tilde{c}_2, b_2$ using (6.34) and $w(1) = c_1 m + c_2$.
3. Solve for $e_2(1)$ using (6.32).
4. Harvest $\mathbb{E}[\lambda W_2 - L_2]$ using (6.36).

6.1.5 Case 3: both pools are mildly compliant

This will be almost identical to the computation in §6.1.4.3. This time we take

$$p^* = p + \varepsilon_1 + \varepsilon_2.$$

Because neither pool is interested in mining the objectionable transaction itself, we start with

$$\mathbb{E}[L_i] = \mathbb{E}[L_i \mid (1,0)]$$

as the initial state is that only NC has mined the block. Now we have

$$\mathbb{E}[L_i \mid (0,s)] = p^* \mathbb{E}[L_i \mid (-1,s)] + (1-p^*)\mathbb{E}[L_i \mid (1,s)]$$

for $x = 0$ and

$$\mathbb{E}[L_i \mid (x,s)] = p\mathbb{E}[L_i \mid (x-1,s)] + (1-p-\varepsilon_i)\mathbb{E}[L_i \mid (x+1,s)] + \varepsilon_i \mathbb{E}[L_i \mid (x+1,s+1)]$$

for $x \geq 1$. Also, as before,

$$\mathbb{E}[L_i \mid (x,s+1)] = \mathbb{E}[L_i \mid (x,s)] + w(x),$$

leading to

$$e_i(x) - pe_i(x-1) - (1-p)e_i(x+1) = \varepsilon_i w(x+1) \quad \text{for } x \geq 1.$$

The function $w(x)$ is determined from coefficients given by (6.19). Setting boundary conditions

$$\begin{aligned} e_i(0) &= b_i^*, \\ e_i(1) &= \frac{b_i^*}{1-p^*}, \\ e_i(k) &= 0 \end{aligned} \tag{6.37}$$

gives us a matrix equation:

$$\begin{pmatrix} 1 & 1 & -1 \\ m & 1 & -\frac{1}{1-p^*} \\ m^k & 1 & 0 \end{pmatrix} \begin{pmatrix} \tilde{c}_1 \\ \tilde{c}_2 \\ b_i^* \end{pmatrix} = \varepsilon_i \begin{pmatrix} 0 & 0 \\ -\frac{m^2}{1-2p} & \frac{1}{1-2p} \\ -\frac{k}{1-2p} m^{k+1} & \frac{k}{1-2p} \end{pmatrix} \begin{pmatrix} c_1 \\ c_2 \end{pmatrix}. \tag{6.38}$$

Inverting the matrix, we can solve for the coefficients $\tilde{c}_1, \tilde{c}_2$. Once we have done this,

$$\mathbb{E}[L_i] = e_i(1). \tag{6.39}$$

So the total expectation for Pool 2 is

$$-\frac{e_i(1)}{1 - w(1)} \tag{6.40}$$

$$= -\frac{b_i^*}{(1 - w(1))(1 - p^*)}. \tag{6.41}$$

Inspecting (6.38) we see that

$$\frac{b_1^*}{b_2^*} = \frac{\varepsilon_1}{\varepsilon_2}.$$

Summary

1. Solve for w, that is, find coefficients c_1, c_2 using (6.19).
2. For either ε_1 or ε_2, solve for $\tilde{c}_1$, $\tilde{c}_2$, and b using (6.38) and $w(1) = c_1 m + c_2$.
3. Solve for $e_i(1)$ using (6.37).
4. Harvest $\mathbb{E}[\lambda W_2 - L_2]$ using (6.40).

6.1.6 Analysis: low fee regime

We begin by looking at situations in which the fee paid by the censored party is relatively small. As of September 2022, the block reward is about \$125,000. So if we assume the censored party is willing to part with \$125 in order to process the transaction, this will amount to $\lambda = 0.001$.

6.1.6.1 Normalized tables

We will use "normalized tables," which divide the expectations of a pool by the hashrate of that same pool. Re-normalization of any agent's payoff will not affect strategic considerations, because it does not change preferential ordering. Dominant strategies will remain dominant, and so forth. Agents are attempting to maximize their own outcomes, not their outcomes relative to someone else's outcomes. Note that the hashing fraction of a pool represents the expected block rewards per 10 minutes of mining. So the payoff number given in the normalized table is given in units of expected payoff for 10 minutes of mining for that given pool.

To illustrate this, we start with $p = 0.2$, $\varepsilon_1 = 0.15$, $\varepsilon_2 = 0.1$. A censoring pool has 20 % of hashing power, and two other rational pools with hashing powers of 15 % and 10 %

are deciding between non-compliant and mildly compliant strategies. We start with the unnormalized table with $\lambda = 0.001$ and $k = 4$:

	Non-compliant	Mildly compliant
Non-compliant	$(-0.01944, -0.01296)$	$(-0.03154, -0.00416)$
Mildly compliant	$(-0.00719, -0.0256)$	$(-0.00897, -0.00598)$

Unnormalized payoff: $p = 0.2, \varepsilon_1 = 0.15, \varepsilon_2 = 0.1, \lambda = 0.001, k = 4$.

For comparison, we also show the normalized payoff table:

	Non-compliant	Mildly compliant
Non-compliant	$(-0.1296, -0.1296)$	$(-0.2103, -0.0416)$
Mildly compliant	$(-0.0479, -0.256)$	$(-0.0598, -0.0598)$

Normalized payoff: $p = 0.2, \varepsilon_1 = 0.15, \varepsilon_2 = 0.1, \lambda = 0.001, k = 4$.

Observe this normalized table has some symmetry along the diagonal: It represents that for a censor with censoring power 20 % and a transaction only offering 0.1 % of a block reward as compensation, if both miners use a non-compliant strategy, they are each expected to lose about 13 % of their expected reward over 10 minutes of typical mining (the game itself could last several blocks). By comparison, if both mining pools adopt a mildly compliant strategy they expect to lose about 6 % of their expectations over 10 minutes.

We observe quite clearly the type of game this represents. Both players will play their dominant strategy, and this will maximize the payoff to both players collectively. This is called a *deadlock game*. A deadlock game is a game in which each player has a clearly dominant strategy, and this is mutually beneficial. Typically such games are less interesting, because players choose the strategy that benefits others represented in the payoff table out of their own self-interest. Of course, the party whose interest is not represented by this table is the party sending the transaction. However, once they have sent the transaction and offered a fee, they have no agency in the mining game.

Both pools are motivated to use a mildly compliant strategy, at least when given the choice between mildly compliant and non-compliant. The strategy space is quite large, so we cannot conclude that this strategy is dominant among all possible strategies; it is only dominant when compared against a non-compliant strategy. We should be careful to put this into perspective offered by the normalized table; a total of 7 % of a pool's block reward is about 4 minutes of mining time.

A pool of size 20 % may be considered a large censoring pool, as it represents already 40 % of the way towards full control of the network. How effective is censorship with lower combinations of hashpower? Reducing this censor's power to 5 % shows a marked difference:

	Non-compliant	Mildly compliant
Non-compliant	(−0.00219, −0.00219)	(−0.0096, −0.0005)
Mildly compliant	(−0.00067, −0.0139)	(−0.0010, −0.0010)

Normalized payoff: $p = 0.05, \varepsilon_1 = 0.15, \varepsilon_2 = 0.1, \lambda = 0.001, k = 4$.

Note that while the dynamics of the game are still essentially the same, for low hashrate of the censor the motivations towards rational pools may be approaching indifference: lose 0.2 % of your expected rewards for 10 minutes of mining versus lose 0.1 %. This is equivalent to a difference of about 6 seconds of mining.

What if we consider larger rational pools?

	Non-compliant	Mildly compliant
Non-compliant	(−0.00219, −0.00219)	(−0.00287, −0.000203)
Mildly compliant	(−0.00182, −0.0691)	(−0.00185, −0.00185)

Normalized payoff: $p = 0.05, \varepsilon_1 = 0.5, \varepsilon_2 = 0.01, \lambda = 0.001, k = 4$.

Still, the incentive to comply is still rather small, approaching indifference.

Now what about when the C pool is larger, say 35 %? First we consider small-sized rational pools:

	Non-compliant	Mildly compliant
Non-compliant	(−1.204, −1.204)	(−1.204, −0.423)
Mildly compliant	(−0.423, −1.204)	(−0.423, −0.423)

Normalized payoff: $p = 0.35, \varepsilon_1 = 0.0001, \varepsilon_2 = 0.0001, \lambda = 0.001, k = 4$.

Perhaps not surprisingly, when the rational pools are small, the payoff tables are nearly independent of the other pool's behavior (they actually are different, but the difference is smaller than the precision offered in the tables). For either pool there is a stronger incentive (about 8 minutes of mining) to switch to a mildly compliant strategy.

For comparison:

	Non-compliant	Mildly compliant
Non-compliant	(−1.204, −1.204)	(−1.612, −0.5065)
Mildly compliant	(−0.5795, −2.126)	(−0.6438, −0.6438)

Normalized payoff: $p = 0.35, \varepsilon_1 = 0.2, \varepsilon_2 = 0.1, \lambda = 0.001, k = 4$.

Unsurprisingly, the incentive becomes more certain as the censorial pool approaches 50 %:

	Non-compliant	Mildly compliant
Non-compliant	(−5.565, −5.565)	(−6.196, −2.321)
Mildly compliant	(−2.321, −6.196)	(−2.416, −2.416)

Normalized payoff: $p = 0.47, \varepsilon_1 = 0.05, \varepsilon_2 = 0.05, \lambda = 0.001, k = 4$.

Here a decision to resist the censors will result in an additional expected loss of over 20 minutes of mining. This may still seem small. When the censorial pool extends k (which will make more sense as $p \to 0.5$) there is a significant increase in expected loss:

	Non-compliant	Mildly compliant
Non-compliant	(−29.224, −29.224)	(−30.952, −21.571)
Mildly compliant	(−21.571, −30.952)	(−22.096, −22.096)

Normalized payoff: $p = 0.47, \varepsilon_1 = 0.05, \varepsilon_2 = 0.05, \lambda = 0.001, k = 10$.

6.1.7 Higher-fee regime

The game changes if the party sending the offending transaction is willing to pay. In the following, we take

$$\lambda = 0.16,$$

which represents one bitcoin when the block reward is 6.25 bitcoins.

	Non-compliant	Mildly compliant
Non-compliant	(0.06912, 0.06912)	(0.04383, −0.03525)
Mildly compliant	(−0.03525, 0.04383)	(−0.04167, −0.04167)

Normalized payoff: $p = 0.2, \varepsilon_1 = 0.05, \varepsilon_2 = 0.05, \lambda = 0.16, k = 4$.

The tables have turned. The non-compliant strategy is now dominant, at least when $p = 0.2$. Increasing to $p = 0.25$, we have:

	Non-compliant	Mildly compliant
Non-compliant	(−0.07725, −0.07725)	(−0.1306, −0.09205)
Mildly compliant	(−0.09205, −0.1306)	(−0.1052, −0.1052)

Normalized payoff: $p = 0.25, \varepsilon_1 = 0.05, \varepsilon_2 = 0.05, \lambda = 0.16, k = 4$.

Note that this game has no dominant strategies available, and it has two Nash equilibria, similar to the stag hunt.

6.1.8 Conclusion

In terms of pure game-theoretical considerations, it is possible for a pool with censorial objectives to change the incentives landscape so that rational pools are motivated to join the censors. However, from the tables above, if the pool controls a smaller portion, say 5 % of the hashrate, the motivations are likely to not come into play, unless there are extremely low profit margins for the miners. A censorial pool cannot expect to win all their battles, but the goal would be to cause enough losses over time so the rational miners would join them in increasing numbers. The effect is to grind down the non-compliant pools making non-compliant mining less profitable on average.

There are many other possible strategies that can be explored, and this becomes much more interesting when profitability of the miners is taken into consideration. We are not considering the possibility that non-compliant pools will punish those that join compliant efforts by breaking consensus protocols to mine non-compliant blocks, even at a loss.

Neither are we considering competition for blockspace. A computation (more favorable to the censors) would subtract from λ the average fee obtainable for the blockspace, according to the market rates. In 2022, this might not be a large difference, but as the block reward decreases over the next 40 years, this modified computation may become more significant.

7 Economics of mining

This chapter attempts to give a brief background to the fundamental market concepts that apply to Bitcoin mining. We start by introducing some classical models, and then apply these directly to the naturally occurring economics in Bitcoin mining. The full picture depends on at least four markets, and indirectly on many more markets beyond that. Perhaps the most significant factor is the price of Bitcoin, which is driven by independent forces (this separation is by design, some would argue, and serves to distinguish proof of work from proof of stake). The central market is the production of the hashes, which are converted to Bitcoin via proof of work and the difficulty adjustment. But because Bitcoin requires sophisticated ASICs to mine, the mining market depends heavily on the market for the ASICs themselves. In 2022, the capacity to produce ASICs is heavily competitive and depends on the ability of ASIC producers to gain access to chip-making facilities. These foundries are in hot demand: ASIC production is in competition with worldwide markets for other technologies, such as iPhones. In this chapter we present models that may be useful for analyzing mining as an economic market. The market for hash production is fairly simple and straightforward, thanks to Bitcoin's simple rules and the difficulty adjustment. This will be covered in §7.1.

The market for special equipment is a bit more complicated. A short discussion is offered in §7.2. It is exceedingly difficult to present a grand unifying theory, and also somewhat pointless as the reality is complicated. More theoretical analysis of the ASIC market is presented in [53].

7.1 Hashing as a market

7.1.1 Cournot oligopoly model

We begin by introducing a market model describing the mining production. As has been noted [35], the Bitcoin mining market is one of the most perfectly competitive markets that occur in nature. There are no transportation costs for hashes, the barriers to entry are relatively small, and the products (proofs of work) are identical. These dynamics will lead to a competitive market with many participants.

Rather than simply observing that Bitcoin mining is a perfectly competitive market, we will go ahead and derive this fact by taking a limit of a classical model, the Cournot oligopoly model. That is, we are going to consider the model for oligopolistic markets, but take the number of firms to be quite large. As the number of firms becomes large, the dynamics approach those of perfectly competitive markets. This will give us an opportunity to explore the dynamics and also the motivations for and against collusion.

A discussion of mining as an oligopoly market is similar to the standard discussion found in many economic textbooks. Mining has the feature that the price is not determined by the supply of hashes, but the share of revenue is. Thus the revenue is inversely

https://doi.org/10.1515/9783110772845-007

proportional to the supply. We will introduce the classical model, and then describe how this should be tweaked to apply to mining.

7.1.1.1 Classical model

In a classical oligopolistic market, there will be a number of firms $\{F_i\}$ producing a certain good. Each producer $\{F_i\}$ incurs a per-unit cost c_i of producing the good and sells to consumers at price P. Each producer produces output quantity q_i. For this particular model, we assume that price is determined by where supply and demand meet, that is, there is a function $P(q)$ which describes the price when the total quantity of goods (over all the firms) is q. The function $P(q)$ should be decreasing: More quantity produced means lower prices. We also assume that there is one price at which the good is sold across the market (no price differentiation).

For each firm, we define a profit function

$$\pi_i(q_1, \dots, q_n) := P(q)q_i - c_i q_i.$$

This represents an n-player game: Each player chooses a quantity q_i and receives a payoff π_i which depends on the choices $\{q_j\}$.

To determine the Nash equilibrium, we use multi-variable calculus. The space of all possible choices of $\{q_j\}$ will be

$$\mathbb{R}_n^+ := \{(q_1, \dots, q_n) \mid q_i \geq 0 \text{ for all } i\}.$$

If we assume that the costs $c_i > 0$ and the price $P(q) \to 0$ when $q \to \infty$, the positive values for the profit function π_i must occur inside a compact region and must occur in the interior, provided each firm is making a profit (if that firm will not make a profit, we ignore this firm and reassign the number n to a smaller number). Suppose that $P(q)$ is a differentiable function. Then if q_i is an interior value that maximizes π_i while holding each $\{q_j, j \neq i\}$ constant, we will need

$$\frac{\partial \pi_i}{\partial q_i}(q_1, \dots, q_n) = 0.$$

Computing, we check that this is

$$\frac{dP}{dq}(q)q_i + P(q) - c_i = 0. \tag{7.1}$$

So for a Nash equilibrium, we need to find $(q_1, q_2, \dots, q_n)$ and $q = q_1 + \dots + q_n$ such that (7.1) is satisfied.

Exercise 7.1. Let

$$P(q) := 100 - \frac{q}{3}$$

and assume there are two firms producing goods, with costs 4 and 5 per unit, respectively. Find the Nash equilibrium. (Hint: Equation (7.1) is a 2×2 system of linear equations.)

7.1.2 A model for the mining market

In order to apply the general model to mining, we need a reasonable description of how price (meaning the amount paid out per hash, not the price of Bitcoin) depends on the output. The price of Bitcoin itself is not determined by the supply of hashes; rather, the price is determined by the independent market of supply and demand for Bitcoin. Bitcoin's price P_B will be considered to be a variable that does not depend on the number of hash created by the miners. Nonetheless, the price paid per hash has a very specific expression, thanks to the difficulty adjustment:

$$P(q) = \chi \frac{P_B}{q},$$

where:

$$\begin{aligned} P_B &:= \text{price of a bitcoin in dollars,} \\ q &:= \text{total hashes produced by all miners in a 10-minute period,} \\ \chi &:= \text{block reward, in units of bitcoins.} \end{aligned}$$

7.1.3 Linear cost model

In our simplest model, we will assume that the miners are paying a fixed per-unit cost. This is unlikely for large demand, but reasonable for smaller demand, as we will discuss, but it will provide the basis for a toy model.

Each producer produces q_i hashes at a cost $c_i q_i$. The profit is determined by

$$\pi_i := \chi \frac{q_i}{q} P_B - c_i q_i.$$

We are assuming a difficulty-adjusted view. This model is not accurate in the very short term, but holds when zooming out beyond two weeks. For this reason we also assume that P_B nor q is going to change drastically over a two-week period.

Applying (7.1),

$$-\chi \frac{P_B}{q^2} q_i + \chi \frac{P_B}{q} - c_i = 0. \tag{7.2}$$

Now at a Nash equilibrium, this will be true for each i and we may sum:

$$\sum_i \left(-\chi \frac{P_B}{q^2} q_i\right) + \sum_i \chi \frac{P_B}{q} - \sum_i c_i = 0,$$

that is,

$$(n-1)\chi \frac{P_B}{q} = \sum_i c_i.$$

So

$$q = \frac{\chi P_B (n-1)}{\sum_i c_i}.$$

The total quantity q uniquely determined, we can then solve for each q_i using (7.2):

$$\begin{aligned} q_i &= \frac{\chi \frac{P_B}{q} - c_i}{\chi \frac{P_B}{q^2}} \\ &= \frac{\chi P_B (n-1)}{\sum_j c_j} \left(1 - \frac{(n-1)c_i}{\sum_j c_j}\right). \end{aligned} \tag{7.3}$$

Note that if all costs $c_i = c$ are equal, this reduces to

$$q_i = \frac{\chi P_B (n-1)}{n^2 c}$$

and profits will be

$$\pi_i = \frac{1}{n^2} \chi P_B. \tag{7.4}$$

Conclusion 7.1. *If we assume that each producer has the same fixed cost per hash, profits decrease on the order of $\frac{1}{n^2}$, where n is the number of miners.*

Inspecting (7.3) we see that the condition for profit to be positive for miner i can be formulated as

$$\sum_{i=1}^{n} c_j \geq c_i (n-1),$$

that is,

$$c_i \leq \frac{n}{n-1} \text{average}\{c_j\}.$$

This suggests that if this model holds true, for n large, there is a narrow band of per-unit costs that will remain profitable.

Note here we are assuming that producers are able to increase their output to maximize profit. In reality this may not hold as producers are constrained by their access to

ASICs. Mining costs may change depending on how many hashes are produced: A firm may have more efficient machines that are turned on first and less efficient machines turned on to meet demand.

Exercise 7.2. Suppose only three firms are mining. They can produce as many hashes as they like, at costs of \$3, \$4, and \$5 per unit of hash, respectively. Determine the percentage of the hashrate obtained by each in the Nash equilibrium.

7.1.3.1 Uniqueness of *n*

The quantities determined by (7.3) are well-defined provided the values $\{c_j\}$ are known; necessarily this requires knowledge of how many and which firms are mining profitably. The computation above assumed that all miners are mining profitably: The q_i value must not be negative. In the situation where one or more of the producers are predicted to produce a negative quantity, this firm should be excluded, and the computation should be redone, with only the firms that can produce profitably.

It is a reasonable question to ask: Are there different combinations of mining firms that will be able to mine profitably to the exclusion of other firms? Excluding one firm from the computation should change the profitability for all firms.

It turns out that for linear costs, there is a unique maximal subset of firms that can mine profitably (the proof is left as an exercise). That is, given a set of firms S who are interested in mining, there will be a unique subset $S' \subset S$ such that: (1) the values determined by (7.3) when considering only firms from S' provide non-negative profits for all members of S'; and (2) for any larger set S'' strictly containing S', the Nash equilibrium equation (7.3) will prescribe negative quantities to some miners.

Exercise 7.3. Show that if the firms are ordered by costs $c_1, \ldots, c_n$, then if there is a Nash equilibrium which is profitable for the kth firm, it must be profitable for all firms $i \leq k$.

Exercise 7.4. If the firms are ordered by costs $c_1, \ldots, c_n$ and the solutions to (7.3) when only the firms $1, \ldots, k$ participate prescribe a negative value for the kth firm, show that these equations will also prescribe a negative value for the $(k+1)$st firm if the first $k+1$ firms are included.

Exercise 7.5. Use induction on Exercise 7.4 together with the results of Exercise 7.3 to conclude that there is a unique maximal profitable subset of mining firms.

7.1.4 Monopolist pricing and collusion

The monopolist hashing model is drastically different. If there is a single miner producing hashes, they will maximize the function

$$\pi = \frac{q}{q}\chi P_B - cq.$$

The upper bound on values that can be obtained here is χP, which will be approached as the hash output approaches 0. So a monopolist who is the only player in a hash market would expend as little positive energy as possible and still claim 100 % of the rewards. In practice, this is unlikely; even the slightest bit of competition for the mining market would change this dynamic.

Note that the reward for collusion in the mining market is extremely high, but the reward for defecting is also extremely high: If the set of miners is fixed and cartelized, they could in theory agree to reduce hashrates all the way to near zero and split the proceeds. Costs would go to near zero, but the total revenues would be the same and profits would be large. However, the nature of Nakamoto consensus allows anyone to sweep in and claim a windfall of profit by defecting from (or simply never agreeing to) such an arrangement.

Example 7.1. Suppose that there are five miners, each with identical costs to produce hashes. Each will make a profit of $\chi P_B/25$. If all five where to collude and produce a very small number of hashes, each could claim profits of $\chi P_B/5$. If three collude to exclude the other two and proceed to produce very few hashes, they could claim profits of $\chi P_B/3$. This latter scenario requires that the other two have given up and dropped out.

7.1.5 Perfectly competitive markets

Bitcoin mining, by design, is supposed to approach a perfectly competitive market. In contrast to perfectly competitive markets, oligopolist markets are characterized by situations in which a few firms earn substantial profits over the long run because barriers to entry make it more difficult for new firms to enter. This is a typical market structure in the real world: perfectly competitive markets rarely exist in practice.

It is worth noting that (short of selfish mining, see Chapter 8) there is a decreasing return to scale. This prevents, as the solution to Exercise 7.2 shows, firms with cost advantages to monopolize the market – they will be eating into their own share of the pie beyond a certain point.

7.1.5.1 Barriers to entry

There are some small barriers to enter the mining game. In order to be competitive one must buy ASIC equipment that costs more than most computers, and one must also have a relatively cheap source of energy and a secure location where a mining apparatus can make noise. If one is planning to mine without joining a pool, the barrier to entry is more significant: One needs enough capital to overcome the risk involved: Since the minimum attainable block reward (as of September 2022) is about $125,000, this means that one could easily spend this much money and still not have received a single block reward. Someone spending $5,000 per month on mining could go several years without obtaining a single block reward.

We can be precise: Winning the block lottery is a Poisson process (recall §2.7). If a miner produces a fraction $p \ll 1$ of the total hashrate, then for any 10-minute time period, their expected number of blocks produced will be a Poisson variable with $\lambda = p$. Thus the probability that they produce k blocks in T 10-minute time windows is given by the Poisson distribution:

$$P(\text{miner produces } k \text{ blocks in } 10T \text{ minutes}) = \frac{(pT)^k e^{-pT}}{k!}. \tag{7.5}$$

In particular, the probability that they have not received a single block after $10T$ minutes have elapsed is given by the probability

$$P(\text{miner produces no blocks}) = e^{-pT}.$$

For example, if a miner produces enough hashrate to expect 1/10 of block reward every four weeks (about $12,500 as of September 2022), they are producing a fraction $1/40{,}320$ of the entire hashrate. So even after four weeks, there is still a probability

$$e^{-1/10} \approx 0.90484$$

that they will have received nothing for their efforts. After 52 weeks, the probability of still receiving nothing is slightly larger than 25 %. Especially as Bitcoin price grows, the barrier to entry caused by the granularity of the block reward becomes more severe. However, this is overcome by pooling. By joining a large enough pool, miners can drastically bring down the variance in their expected winnings.

Exercise 7.6. Suppose that a miner with p of the hashrate joins a pool with Kp of the hashrate. Using statistics for the Poisson distribution, show that joining the pool reduces the variance by a factor of K. Argue that this does not depend on the time interval. (Hint: For a Poisson variable with parameter λ, the variance and the mean are both λ, but the variance is a "quadratic" quantity. Exploit this.)

With pooling taken into consideration, the largest barrier to entry is the acquisition of ASIC machines.

7.1.6 Nash equilibrium in the mining market

We recall a theorem due to Rosen in 1965 [43].

Theorem 7.1. *Suppose that the strategies of n players are described by a convex and compact (closed and bounded) subset of Euclidean space $(q_1, \ldots, q_n) \in \mathbb{R}^n$. If the payoff functions π_i are continuous and strictly concave in the variable q_i, a unique Nash equilibrium will exist.*

This theorem, combined with Exercise 7.7, tells us that there is a unique Nash equilibrium for miners with increasing marginal cost functions.

Exercise 7.7. Show that if the marginal cost of producing hash to a miner is increasing, then the profit function is concave as a function of hash output.

Remark 7.1. The theorem guarantees uniqueness for a fixed n-player game with concave payoff: It applies even when we include miners who will be non-profitable as their outputs will be zero and we can take our compact convex region to be a cube

$$\{(q_1, \dots, q_n) \mid 0 \leq q_i \leq R\}$$

for some R exceeding a maximum feasible hashrate. This allows some values of q_i to take the value 0.

7.1.6.1 Profitability

Allowing each mining firm to produce until production capacity runs out (equivalently, $c_i(q) = +\infty$), there will be a Nash equilibrium, that is, a point $(q_1, q_2, \dots, q_n)$ satisfying

$$\chi \frac{q - q_i}{q^2} P_B - c_i(q_i) = 0 \quad \text{for all } i. \tag{7.6}$$

We can solve for the value of a unit of hash,

$$\chi \frac{P_B}{q} = \chi \frac{q_i}{q^2} P_B + c_i(q_i),$$

and compute the profit,

$$\begin{aligned}\pi_i &= q_i \left(\chi \frac{q_i}{q^2} P_B + c_i(q_i) \right) - \int_0^{q_i} c_i(s) ds \\ &= \chi \frac{q_i^2}{q^2} P_B + \int_0^{q_i} [c_i(q_i) - c_i(s)] ds.\end{aligned}$$

The first term, involving the square of the firm's proportion of the pool, is reminiscent of (7.4), but allows for more general distribution of hashrates. The next term represents the surplus between the maximum that they pay for hashrate and what they pay for less expensive hashes.

Observe that solving for $c_i(q_i)$ in (7.6) will tell us the maximum marginal cost $c_i^{\max}(q_i)$ paid by each firm. Summing as before we conclude that

$$q_i = q \left(1 - \frac{(n-1) c_i^{\max}(q_i)}{\sum_j c_j^{\max}(q_i)} \right). \tag{7.7}$$

So we conclude the following: A miner will be able to mine profitably, provided that

$$c_i^{\max} \leq \frac{n}{n-1}\text{average}\{c_j^{\max}\}.$$

7.2 Market for ASICs

In the Cournot model, it is assumed that the price depends on the total quantity of goods produced. Firms that produce the goods decide how much to produce, and price is a dependent variable that will respond to the choice of quantities.

Another model, called Bertrand competition, does not assume that price is determined precisely by output, but rather that firms will choose a price, and the quantity of output will increase or decrease to meet the demand which happens at that price. In short, the quantity sold by each firm is a function of the prices charged by each of the suppliers.

The profit function is similar, but now depends on a vector of prices $p_1, \dots, p_n$:

$$\pi_i(p_1, \dots, p_n) := D_i(p_1, \dots, p_n)p_i - \tilde{c}_i q_i.$$

Here D_i represents the demand met by firm i given the firms have specified these prices, and $\tilde{c}_i$ is the production cost. In the most competitive settings, the model requires that consumers will always choose the producer that sells at the lowest price. This is a bit of an extreme assumption: It assumes that the buyer of a \$129.99 product will switch to a competitor when the competitor offers the same product for \$129.98. In practice, methods of buyer or product differentiation deal with this discontinuity. However, the basic Bertrand model makes this assumption. Necessarily, this also requires the assumption that the capacity to produce is unconstrained: The firm can produce billions of dollars worth of goods on a moment's notice.

7.2.1 Nash equilibrium with Bertrand assumptions

Suppose that multiple firms are competing to sell the same good. Is there a Nash equilibrium in the choice of prices?

First we observe that each firm sells only at price $p_i \geq \tilde{c}_i$; otherwise they would be producing and selling at a loss. Observe next that any firm hoping to make a sale must sell goods at costs no higher than any other firm. So $p_i = p_j$ for all i, j unless the firm is already resigned to not making any sales. Now if all firms are selling at the same price, they should be splitting the quantity sold. But if $p_i = p_j > \tilde{c}_i$, firm i can charge price $p_i - \varepsilon > \tilde{c}_i$ (this is called epsilon-undercutting), and steal all of the customers from the other firms. An infinitesimal decrease in per-unit revenue combined with a significant non-infinitesimal increase in quantity yields a much better profit. We conclude that there is only one possible Nash equilibrium: Each firm with cost $\tilde{c} := \min_i\{\tilde{c}_i\}$ charges $\tilde{c}$ and each other firm either drops out completely or charges a higher price, knowing they will expect no sales and zero profits.

The Bertrand model says that with such competitive assumptions in place, products will be sold at cost, for zero profit, and only firms that produce at the minimum cost will engage the market. In practice this is a bit extreme. Product and consumer differentiation allow for a less severe cutoff function happening near $p_1 = p_2$, allowing for Nash equilibria with positive profits. Reality also dictates that production cannot instantly toggle from 0 to millions, so this model is somewhat imperfect. However, in many cases, it is approximately correct.

Clearly, because hashrate cannot be increased arbitrarily, the Bertrand model does not apply to the market for mining. However, it may apply, with some modifications and constraints, to the market for mining equipment.

7.2.2 Four regimes

We split the market regimes for ASICs in three categories, according to the supply capacity of ASICs. A fourth classification would be an inhomogeneous mixture of the three.

1. *Bertrand regime.* In this regime, the capacity for ASIC producers to produce is essentially unlimited or is at least larger than any reasonably expected demand. The economics behave like the Bertrand model predicts. This regime is where there is maximum competition among ASIC producers and machines will be offered at near cost. This is probably not a realistic model in 2022.
2. *Edgeworth regime.* When the supply is constrained, but not too tightly, we get a situation where market dynamics allow for no Nash equilibrium, due to the Edgeworth paradox. This is characterized by moderate constraints on supply. The Edgeworth paradox occurs when suppliers can always undercut each other, but cannot fully meet the demand of market, so they are torn when hunting for an equilibrium price. Lowering price without enough capacity to steal market share from competitors allows competing suppliers to raise their price and increase profits. Suppliers will then raise prices to almost meet the prices of their competitors, at which point undercutting becomes profitable again. As a result, no equilibrium can be found.
3. *Strongly constrained regime.* When the supply is strongly constrained, Nash equilibria will be predictable and often computable. The incentive for competing suppliers to raise prices so that their sales are below capacity which occurs in the Edgeworth paradox is not present: The best pricing options will always ensure that demand saturates supply.
4. *Mixed.* When there are differentiated miners and suppliers, there can be an abundance of one type of ASIC and dearth of another. The market can be predictable in some segment of the curve and not in others. For example, there may be an abun-

dance of inefficient mining equipment, a lack of efficient mining equipment, and many potential miners with limited access to very cheap electricity.

In [53] the author explores the strongly constrained regime and argues that there is a unique Nash equilibrium when certain criteria are in place. In this Nash equilibrium, the least efficient mining equipment is paired with the cheapest electricity.

8 Selfish mining

In 2013, in a seminal paper of Eyal and Sirer [15], it was demonstrated that if one entity controls more a third of the total mining power, Nakamoto consensus fails to be *incentive-compatible*.

In this chapter we unpack this concept and its ramifications. If the reader would like to skip the details, we offer the following summary.

Summary 1. Selfish mining is an exploit that is available under Nakamoto consensus, made possible by the difficulty adjustment. Selfish mining allows an entity who controls $p > 1/3$ of the hashrate to allocate to themselves a fraction $S(p) > p$ of the block rewards, thus unfairly increasing their revenue. Because the portion $S(p)$ grows non-linearly with p above $p = 1/3$, the selfish miner is incentivized to be excessively greedy in acquiring hashrate, with increasing return on investment all the way up to the point where the selfish miner controls 50.1 % of the miner power, at which point selfish mining becomes monopolistic control.

The notion of incentive-compatibility is from mechanism design. A mechanism is considered to be *incentive-compatible* if every participant achieves their best outcome by following the rules as they were designed to be followed. This concept arises in many systems such as voting; many voting methods such as plurality, approval voting, ranked choice, score-then-automatic-runoff, etc., are designed so that the voters will choose honestly rather than strategically (a famous theorem, Arrow's impossibility theorem, states that there is no voting mechanism that is free from the possibility of manipulation by strategic voting).

Definition 8.1. The *default compliant* strategy for Bitcoin miners is the following:

- Each miner works to append the heaviest chain. If there are two, they work on the first one that came to their attention.
- Each miner publishes each block the instant they have discovered it. This allows other miners to immediately start working on the next block with transactions from the mempool.
- Miners do not intentionally omit fee-paying transactions from the mempool for which there is sufficient blockspace.

Remark 8.1. The last condition is not completely deterministic. Typically miners would order available transactions by fee per blockspace and process as many as possible. There are small wrinkles in this; for example, they may or may not prioritize replace-by-fee transactions, and they may group child-pays-for-parents transactions together. Different miners may use different block-packing algorithms, and some may choose to impose minimum transaction fees.

When miners follow these rules, the network should function continuously as expected. However, Nakamoto consensus allows a miner with enough hashrate to employ

https://doi.org/10.1515/9783110772845-008

what is called selfish mining. Selfish mining works by taking advantage of the difficulty adjustment. By withholding mined blocks from the public, selfish miners can force other miners to waste resources, making the network as a whole less efficient at producing blocks. A careful calculation reveals that the selfish miner can increase their proportion of the blocks produced over a longer time period, at the cost of creating a few stale blocks before the difficulty adjusts. Because the difficulty adjustment rewards miners according to their relative block production, not their hashrate, miners with 34 % of the hashrate can increase their rewards over time.

This may seem counterintuitive at first: By keeping a block secret, there is a risk that someone else will discover a block first. In this case the selfish miner still publishes the block, but it may be too late: Other miners will have seen the other block first and will ignore the block. The selfish miner then needs to find another block on top of their previously hidden block or lose it forever. However, on the other side of this is the possibility that they will mine another block, which they also keep secret. Once they have built a two-block lead, they may continue to add to the lead in secret without any fear of losing the block. The instant that a competing miner publishes a block that comes within one block of catching up, the selfish miner reveals their secret chain to the network and this supplants the block of the competing miner or miners.

Remark 8.2. We use p (as we have been through much of this text) to denote the hashrate of the selfish miner, but much of the literature (including the original paper [15]) on selfish mining uses α.

8.1 Exploiting the difficulty adjustment

The difficulty adjustment is the only major real-time "parameter tweak" provided by the protocol. Because the difficulty parameter is something that responds to miners' behavior, it is not surprising that this is the wrinkle that allows for the most significant exploit.

If the difficulty adjustment were not present, selfish mining would not work; this has been proven in [21, Theorem 4.4]. In particular, selfish mining is a net loss for the miner in the short term, where the effect of selfish mining is to drastically slow down the block production. After block production has been slowed down, the next difficulty adjustment compensates for this, and the rate of block production returns to normal. In this chapter we present basic selfish mining and assume this is a long-term strategy used by a miner who plans on using it for several subsequent difficulty periods.

Exercise 8.1. Using chainwork, explain how difficulty adjustment allows a party with extremely large hashpower to catch up and become the chain with the most work while producing only a fraction of the total blocks.

Exercise 8.2. Suppose that a miner with hashrate p waits until N blocks have been mined in a difficulty period, and then mines the 2,016th block of the previous difficulty period. The miner then mines this chain in secret, producing blocks with lower difficulty. Suppose that instead of the heaviest-chain rule measuring most chainwork, Bitcoin used the longest-chain rule. What hashrate p would the miner need in order to be likely to mine 2,016 blocks in the current period before the other miners do? For this exercise, you may replace the Poisson process with a steady deterministic process, for simplicity.

8.2 Basic selfish mining

In order to analyze this game, we need to realize it as a series of repeated rounds. Each round ends with either the selfish miner capitulating and returning to mine the longest chain or the selfish miner publishing their last two secret blocks to win the round.

We can be more precise: As in Chapter 5 we set up the game states. Instead of being behind in the block race, the selfish miner may be secretly ahead in the block race. The game states will be

$$\{(i,j) \mid \text{ selfish miner has mined } i \text{ blocks total, others have mined } j \text{ blocks}\}.$$

The game starts at state $(0,0)$, at which point the selfish miner attempts to mine a block. If the others find a block first, the selfish miner starts again mining on top of the longest chain; another round begins and the state reverts to $(0,0)$.

However, if the selfish miner finds a block, they keep this secret, and the state moves to $(1,0)$. From $(1,0)$ there are two options:

1. If the others find the next block, the selfish miners typically will go ahead and publish their block, hoping that perhaps some fraction of miners will see this block first and mine on top. This puts the game at state $(1,1)$. Regardless of who sees the block, the selfish miners continue to attempt to mine on top of their own block; if they are lucky, they will find a second block and supplant the block the others produced – the round ends with the miners winning two blocks and the next round restarts at $(0,0)$. Similarly, if the others win a block from state $(1,1)$ they permanently claim for themselves two blocks and a new round starts.
2. The selfish miner finds a block and the state moves from $(1,0)$ to $(2,0)$. Now at this point the selfish miner has the upper hand and uses the following strategy: If they are at least three blocks ahead and they see the others publish a block, they publish an equalizing block. If they are two blocks ahead and the others publish a block, they publish two blocks, exhausting their secret supply. (This may or may not be the optimal strategy; the selfish miners may wish to push their luck and keep these blocks hidden. However, the strategy being used here is straightforward enough to model.)

In our model, we assume:

- A single selfish miner has hashrate p.
- The other miners follow the first-seen, longest-chain rule.
- Network propagation is near instantaneous.
- Fees are not important.

We are going to compute first the expected value of the number of blocks won by the selfish miners in each round and then the expected number of blocks won by the others in each round. This will determine what proportion of blocks is expected to be obtained by the selfish miners as the number of rounds becomes large.

Let

$$\begin{aligned} f_{\mathrm{SM}} &:= \text{number of blocks won by selfish miners,} \\ f_O &:= \text{number of blocks won by other miners.} \end{aligned}$$

We would like to find

$$\mathbb{E}[f_{\mathrm{SM}} \mid (0,0)]$$

and

$$\mathbb{E}[f_O \mid (0,0)].$$

To begin, note that

$$\mathbb{E}[f_{\mathrm{SM}} \mid (0,0)] = p\mathbb{E}[f_{\mathrm{SM}} \mid (1,0)] + (1-p)\mathbb{E}[f_{\mathrm{SM}} \mid (0,1)],$$

where the rightmost term is 0 as the selfish miners win nothing and the round restarts if the others take the first block (we continue to use the conditional expectation formula to compute expectations; recall (2.6)).

Then we compute

$$\mathbb{E}[f_{\mathrm{SM}} \mid (1,0)] = p\mathbb{E}[f_{\mathrm{SM}} \mid (2,0)] + (1-p)\mathbb{E}[f_{\mathrm{SM}} \mid (1,1)], \tag{8.1}$$

$$\mathbb{E}[f_{\mathrm{SM}} \mid (2,0)] = p\mathbb{E}[f_{\mathrm{SM}} \mid (3,0)] + (1-p)\mathbb{E}[f_{\mathrm{SM}} \mid (2,1)]. \tag{8.2}$$

Now

$$\mathbb{E}[f_{\mathrm{SM}} \mid (1,1)] = 2p.$$

This follows because in this tie-breaking situation, one team or the other must win, taking two blocks.

Also,

$$\mathbb{E}[f_{\mathrm{SM}} \mid (2,1)] = 2, \tag{8.3}$$

as this state results in the selfish team immediately broadcasting their two blocks (the state would only exist for a small amount of time depending on network latency, which we assume is 0).

We claim also that for $k \geq 3$, $l \geq 2$,

$$\mathbb{E}[f_{SM} \mid (k, k-l)] = \mathbb{E}[f_{SM} \mid (k-1, k-l-1)] + 1.$$

To make this argument, recall that we are working on a Bernoulli space. The set of paths starting from state $(k, k-l)$ and the set of paths starting from $(k-1, k-l-1)$ are identical. Each path ends with probability 1, with the selfish miners winning the round. In the former case when they win they will have one more block reward to their credit, compared to the same path in the latter case.

Now for $k \geq 3$,

$$\begin{aligned}\mathbb{E}[f_{SM} \mid (k, 0)] &= p\mathbb{E}[f_{SM} \mid (k+1, 0)] + (1-p)\mathbb{E}[f_{SM} \mid (k, 1)] \\ &= p\mathbb{E}[f_{SM} \mid (k+1, 0)] + (1-p)(\mathbb{E}[f_{SM} \mid (k-1, 0)] + 1).\end{aligned}$$

Now we define

$$e(x) = \mathbb{E}[f_{SM} \mid (x, 0)],$$

and we see that for $x \geq 3$ we have

$$p(e(x) - e(x+1)) + (1-p)(e(x) - e(x-1)) = (1-p).$$

To make this equation work for $x = 2$, we declare that $e(1) = 1$. Then, combining equation (8.2) with (8.3) we can check that with $e(1) = 1$,

$$p(e(2) - e(3)) + (1-p)(e(2) - e(1)) = (1-p).$$

Thus the function $e(x)$ satisfies

$$e(x) - pe(x+1) - (1-p)(e(x-1)) = 1 - p. \tag{8.4}$$

We can set this up as a boundary problem, declaring $e(1) = 1$ as the left boundary value. There is not an obvious natural choice for the right boundary value, as there is no right boundary. However, what we can do is solve the problem for some reasonable boundary values with boundary points far to the right (and hence improbable) and then take the limit.

Specifically, recalling discussions following (5.7) and (5.18) with p and $(1-p)$ swapped, we know that

$$e(x) := c_1\left(\frac{1-p}{p}\right)^x + c_2 + \frac{1-p}{1-2p}x$$

is the general solution to the non-homogeneous finite difference equation (8.4).

We can deal with the lack of a right boundary value by imposing a harmless condition on the strategy: Suppose we assume that for some large value N, if the selfish miner is N values ahead they will broadcast their transactions and start the round over. This is an increasingly improbable event for larger and larger values of X. In this case we can set the right boundary value to be $e(N) = N$. Then solving for the boundary conditions

$$c_1\left(\frac{1-p}{p}\right) + c_2 + \frac{1-p}{1-2p} = 1,$$
$$c_1\left(\frac{1-p}{p}\right)^N + c_2 + \frac{1-p}{1-2p}N = N,$$

we get

$$c_1\left[\left(\frac{1-p}{p}\right) - \left(\frac{1-p}{p}\right)^N\right] = \frac{p}{1-2p}(N-1),$$

so

$$c_1 = \frac{\frac{p}{1-2p}(N-1)}{(\frac{1-p}{p}) - (\frac{1-p}{p})^N}.$$

Taking $N \to \infty$, this goes to zero, as the geometric term in the denominator grows faster than the linear term on the top. So we take $c_1 = 0$, and we can immediately solve

$$c_2 = -\frac{p}{1-2p},$$

so

$$e(2,0) = -\frac{p}{1-2p} + \frac{2-2p}{1-2p} = \frac{2-3p}{1-2p}.$$

Now

$$\begin{aligned}\mathbb{E}[f_{SM} \mid (0,0)] &= p^2\mathbb{E}[f_{SM} \mid (2,0)] + 2p^2(1-p)\\ &= p^2\frac{2-3p}{1-2p} + 2p^2(1-p).\end{aligned}$$

Computing the expected number of blocks recorded by the others is a bit easier:

$$\begin{aligned}\mathbb{E}[f_O \mid (0,0)] &= (1-p) + p\mathbb{E}[f_O \mid (1,0)]\\ &= (1-p) + 2p(1-p)^2.\end{aligned}$$

So now we may compute the ratio of the expected number of blocks won by the selfish miner to the expected total of the blocks mined on the longest chain over a long time period, namely,

$$\frac{\mathbb{E}[f_{SM} \mid (0,0)]}{\mathbb{E}[f_{SM} \mid (0,0)] + \mathbb{E}[f_O \mid (0,0)]} = \frac{p^2\frac{2-3p}{1-2p} + 2p^2(1-p)}{p^2\frac{2-3p}{1-2p} + 2p^2(1-p) + (1-p) + 2p(1-p)^2} = \frac{4p^4 - 9p^3 + 4p^2}{p^3 - 2p^2 - p + 1}.$$

Define

$$S(p) := \frac{4p^4 - 9p^3 + 4p^2}{p^3 - 2p^2 - p + 1}. \tag{8.5}$$

The following is a graph of $\frac{S(p)}{p}$ (see Fig. 8.1). When p crosses the value $\frac{1}{3}$, $\frac{S(p)}{p}$ becomes larger than 1 and increases towards 2.

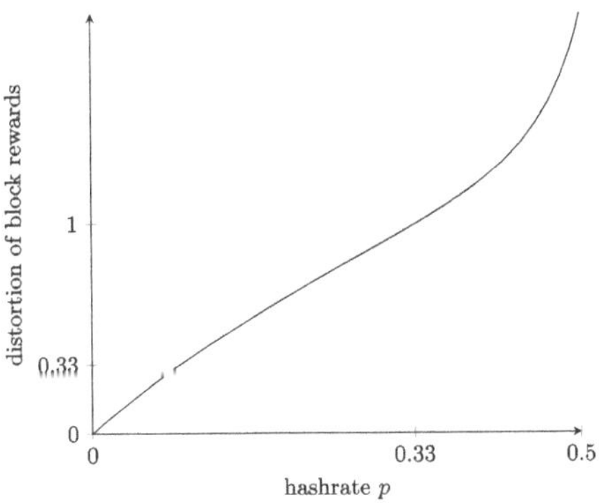

Figure 8.1: Distortion of hashrate to selfish miners $\frac{S(p)}{p}$.

For example, a miner with 35 % of the hashrate can increase their fraction of block rewards to 36.651 %, while a miner with 40 % can increase their portion to 48.372 %.

We may also consider the graph of

$$\frac{1 - S(p)}{1 - p},$$

which describes the distortion of the returns of the other miners (see Fig. 8.2). For example, at $p = 0.4$ the others will notice an 86.047 % return of what they would expect were selfish mining not occurring.

Open-ended Question 8.1. What logistical problems would a pool attempting to selfishly mine need to overcome? Are there ways to overcome this?

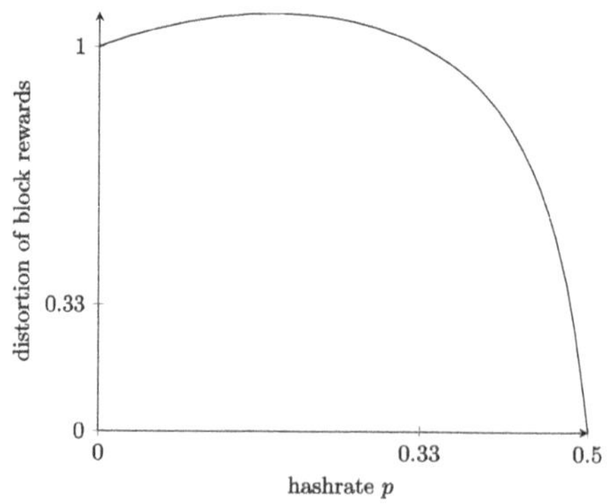

Figure 8.2: Distortion of hashrate to non-selfish miners $\frac{1-S(p)}{1-p}$.

8.3 Market distortions

Selfish mining creates obvious market distortions. Selfish miners enjoy an increase in revenue while the others suffer a decrease in revenue. For an individual miner who is given the option to join a selfish mining cartel, the difference can be quite significant. A selfish miner will always be willing to expand their cartel: Each increase in hashrate increases the distortion itself, further increasing the return of selfish mining. Thus the moment a miner with 33.33 % of the hashrate begins selfish mining and welcomes others to join, this creates a prisoners' dilemma among the remaining miners, possibly leading to *snowballing*, as the rewards for joining the cartel become increasingly steeper as the cartel grows. Defecting to the selfish mining pool will always be a dominant strategy. The distortion will also be passed onto the ASIC market, countering the anti-centralization tendencies described in Chapter 7.

In fact, if we consider examples such as Exercise 7.2 in the regime of selfish mining, we see much different dynamics. A miner with the lowest energy costs, instead of mining to the point where returns are diminished, will aggressively take over the entire block creation game: Selfish mining with 50 % of the hashrate should return 100 % of the rewards with high probability.

8.4 Markov chains

The original selfish mining paper used Markov chains to analyze the expected benefits of selfish mining. In this section we give a quick overview of Markov chains. These will be useful in later chapters as well.

A Markov chain is a process in which the state changes between a set of possible states, such that the probability of changing from one state to another does not depend on the current time or history. These processes are often analyzed using the transition matrix, which consists of the probabilities of moving between states. The matrix $\{a_{ij}\}$ is determined by

$$a_{ij} := P(\text{next state is } j \mid \text{current state is } i).$$

Such a matrix is called a *stochastic matrix* (note that we are using left matrix multiplication, which is non-traditional).

Example 8.1. Suppose that a flea is randomly hopping around the vertices of a triangle. Each second, with probability p, the flea hops counterclockwise, and with probability $1-p$ the flea hops clockwise. If we order the vertices moving counterclockwise N_1, N_2, and N_3, the system has three possible states. The transition matrix is

$$\begin{pmatrix} 0 & p & 1-p \\ 1-p & 0 & p \\ p & 1-p & 0 \end{pmatrix}.$$

In general the transition matrix will be *right-stochastic*, that is, the rows must sum to 1. This simply means that the conditional probabilities a_{ij} for a fixed current state i must sum to 1.

Example 8.2. Suppose a miner with hashrate p attempts to mine from behind, starting from one block behind, giving up only if they fall three blocks behind. This can be set up as a Markov chain. There are five states: (1) The miner has given up, (2, 3) the miner is two blocks or one block behind, (4) the miner has tied but is not first-seen, and (5) the miner pulls ahead and has won. With this ordering, the transition matrix is

$$\begin{pmatrix} 1 & 0 & 0 & 0 & 0 \\ 1-p & 0 & p & 0 & 0 \\ 0 & 1-p & 0 & p & 0 \\ 0 & 0 & 1-p & 0 & p \\ 0 & 0 & 0 & 0 & 1 \end{pmatrix}.$$

Observe that this is not *left-stochastic*: The columns do not sum to 1. Also observe that the first and last rows contain only a single non-zero entry, which must necessarily be a 1. This means that these states are fixed, or terminal. Once the miner has given up, the game is over. Once the miner has won, the game is over, and the game remains in the state "miner won." If the game is over, there is 0 probability of moving to any other state.

The utility of a transition matrix becomes clear when iterating: One can take powers of the matrix to determine the probabilities of the various game states at future times. This is because of the following fact:

$$P\{\text{state after next is } j \mid \text{current state is } i\}$$
$$= \sum_k P\{\text{state after next is } j \mid \text{next state is } k\} P\{\text{next state is } k \mid \text{current state is } i\},$$

or

$$P\{\text{state after next is } j \mid \text{current state is } i\} = \sum_k a_{ik} a_{kj} = (A^2)_{ij}.$$

This can be iterated. In the above example,

$$\begin{pmatrix} 1 & 0 & 0 & 0 & 0 \\ 1-p & 0 & p & 0 & 0 \\ 0 & 1-p & 0 & p & 0 \\ 0 & 0 & 1-p & 0 & p \\ 0 & 0 & 0 & 0 & 1 \end{pmatrix}^3$$
$$= \begin{pmatrix} 1 & 0 & 0 & 0 & 0 \\ p(p-1)^2 - p + 1 & 0 & -2p^2(p-1) & 0 & p^3 \\ (p-1)^2 & 2p(p-1)^2 & 0 & -2p^2(p-1) & p^2 \\ -(p-1)^3 & 0 & 2p(p-1)^2 & 0 & p - p^2(p-1) \\ 0 & 0 & 0 & 0 & 1 \end{pmatrix}.$$

The matrix on the right-hand side offers the transition probabilities of three rounds into the future. For example, the entry

$$a_{25} = p^3$$

represents the probability that a system starting at state 2 will move from state 2 (two blocks behind) to state 5 (miner wins) in three rounds.

Some Markov chains (like the example above) have terminal states. As long as there is a chain of positive probabilities that lead to a terminal state for each state in the system, the system will arrive in a terminal state with probability 1 as the process is repeated indefinitely. In the above example, the game must end at some point. Iterating the transition matrix infinitely many times results in matrices which converge to a matrix which should have non-zero entries only in terminal columns.

Some Markov chains do not have terminal states (consider the flea in Example 8.1). These processes are analyzed using steady-state distributions: Suppose there is a vector $\vec{\mu} = \{\mu_1, \mu_2, \ldots, \mu_n\}$ with the property that

$$\vec{\mu} = \vec{\mu} A,$$

that is, $\vec{\mu}$ is a right eigenvector with eigenvalue 1. This is called a *steady-state distribution*. In the flea example, suppose instead that there are 30,000 fleas. If the vector $\vec{\mu}$ represents the number of fleas at each vertex and

$$\mu_1 = \mu_2 = \mu_3 = 10{,}000,$$

then the expected number of fleas after each flea hops once again is going to be again 10,000 on each node. Note also that if 30,000 fleas start on one node and randomly jump, the distribution at first will appear not random, but as the game is repeated, the distribution of fleas will move closer toward the steady state until only random noise is what separates the distribution from the steady state.

The flea triangle is an example of an ergodic system: Eventually we expect a flea to pass through every node, regardless of where the flea started. The mining game is not. A miner starting in terminal state will never play enough rounds of the game to arrive in a non-terminal state.

8.5 Selfish mining analyzed via a Markov process with cuts

The selfish mining as we have presented above has infinitely many states. Because selfish mining is a repeated process with every would-be terminal state reverting back to the starting state, it is ergodic. When an honest block is mined compelling the selfish miner to reveal all of their hidden blocks, the game state returns to the original state and selfish mining starts again. There are no terminal states. We are interested in computing the steady state. That is, if we start the game now, 10,000 blocks into the future, what is the probability that we will be at each state of the game? There will be a steady-state distribution describing these probabilities:

$$\begin{aligned}
\mu_0 &= P(\text{game is at state 0}),\\
\mu_{0'} &= P(\text{game tied at (1,1); selfish miners' block is not first-seen}),\\
\mu_1 &= P(\text{selfish miners have secret one-block lead}),\\
\mu_2 &= P(\text{selfish miners have secret two-block lead}),\\
&\text{etc.}
\end{aligned}$$

One can explicitly compute all of the transition probabilities. For example, the probability of moving from state (0,0) to (1,0) is p, the probability from moving from state (1,0) to (1,1) is $1-p$, etc.

At first one may be tempted to compute the eigenvector for the process. But this can be difficult. Instead, there is a trick that makes this easier, which is used in the computation [15]. Instead of computing the eigenvector, draw several *graph cuts* on the state diagram. A graph cut is simply a partition of the set of states into two bins, and represents a line that can be crossed one way or the other in each transition. By the definition of the steady state, the probability of crossing the cut in one direction must necessarily be equal to the probability of crossing in the other direction – this is the only way the distribution will remain steady. This means we compute the probability of both of these two events (crosses one way, crosses the other) and turn these into a useful equation.

In the selfish mining example, we draw a cut between a bin containing the two states $(0,0)$ and $(1,1)$ and a bin containing the rest of the possible states. Note that the only way to move away from the set of states $(0,0)$ and $(1,1)$ is to move from $(0,0)$ to $(1,0)$: The state $(1,1)$ moves to the state $(0,0)$. On the other hand, one can go to state $(2,0)$ from $(0,0)$ – this happens with probability $(1-p)$, and one can go from $(1,0)$ to $(1,1)$, which also happens with probability $(1-p)$. Since the probability of moving from $(0,0)$ to $(1,0)$ is p, we have the following equation:

$$p\mu_0 = (1-p)\mu_1 + (1-p)\mu_2. \tag{8.6}$$

Next we isolate the state $(1,1)$ as a bin of its own. With probability 1 this transitions to $(0,0)$. The only way to get to state $(1,1)$ is the event that happens with probability $(1-p)$ from state $(1,0)$. This gives us

$$\mu_{0'} = (1-p)\mu_1. \tag{8.7}$$

Next consider the cut involving states $(0,0)$, $(1,1)$, and $(1,0)$. To move from this, there is one way: Start at $(1,0)$ and move to $(2,0)$, which happens with probability p. To enter, however, this can only happen from state $(2,0)$. So the resulting equation is

$$p\mu_1 = (1-p)\mu_2. \tag{8.8}$$

In fact, for any $k \geq 2$ we also get

$$p\mu_k = (1-p)\mu_{k+1}. \tag{8.9}$$

Our next step is to combine the information contained in equations (8.6), (8.7), (8.8), and (8.9) to get information about the steady state.

One way to do this is to express everything in terms of μ_1. Directly substituting the left-hand side of (8.8) with the rightmost term in (8.6), we get

$$p\mu_0 = (1-p)\mu_1 + p\mu_1.$$

So we have

$$\begin{aligned} \mu_0 &= \frac{1}{p}\mu_1, \\ \mu_{0'} &= (1-p)\mu_1, \\ \mu_2 &= \frac{p}{(1-p)}\mu_1, \\ \mu_{1+k} &= \left(\frac{p}{(1-p)}\right)^k \mu_1. \end{aligned}$$

Now, we add all the probabilities up, knowing that they must sum to 1:

$$\mu_1\left[\frac{1}{p}+(1-p)+1+\sum_{k=1}^{\infty}\left(\frac{p}{(1-p)}\right)^k\right]=1.$$

This leads to

$$\mu_1\left[\frac{2p^3-4p^2+1}{p(1-2p)}\right]=1.$$

However, we do not actually need to compute μ_1. What we can do is compute the ratio with the top and bottom in terms of μ_1.

To compute the expected portion of blocks won by the selfish miners, we look at the states at which they may notch blocks. If they find the next block from state $(1,1)$ with probability p, they notch two blocks. From state $(2,0)$ with probability $(1-p)$ they fail to find the block but end up notching two blocks by publishing their secret chain. For states $(k,0), k>2$, with probability $(1-p)$ a single block is notched. So these add up as

$$\begin{aligned}&2p\mu_{0'}+2(1-p)\mu_2+\sum_{k=2}(1-p)\mu_{1+k}\\&=p\left(3-2p+\frac{1-p}{1-2p}\right)\mu_1\end{aligned}$$

after some computations. For the honest miners,

$$\begin{aligned}&(1-p)\mu_0+(1-p)\mu_{0'}\\&=(1-p)\left(\frac{1}{p}+2(1-p)\right)\mu_1.\end{aligned}$$

Dividing the expected number of blocks to the selfish miners by the sums of the expected blocks to both types of miners, we get

$$\frac{p(3-2p+\frac{1-p}{1-2p})}{(1-p)(\frac{1}{p}+2(1-p))+p(3-2p+\frac{1-p}{1-2p})},$$

which one can check is equal to $S(p)$ defined by (8.5).

8.6 Selfish mining with partial cooperation

In the above example it was assumed that the selfish miner is trying to break ties alone. However, it is very possible that they will be assisted by other miners who mine on top of the selfish chain at state $(1,1)$. When the selfish miner publishes an equalizing block, perhaps within seconds or milliseconds of the others, there is the possibility that honest miners may still mine on top of the selfish miners' block. One possibility is that these miners are indeed honest, but due to network latency reasons, may actually be shown the selfishly mined block first, in which case they will ignore the other block that arrives milliseconds later. This effect would be something that a selfish miner would be wise to attempt to maximize, by deploying listening and broadcasting nodes throughout the globe, ready to release the block the instant any nodes on the network hear of a competing block.

Another reason (discussed in §9.1.1) is that the miners may be mining strategically and amenable to fee-undercutting. For example, if a selfishly mined block leaves higher-fee transactions in the mempool, a strategic miner may choose to build on this block instead. Another possibility is that if the selfish miner is powerful enough (closer to 50 %) it may be more profitable for them to continue a race, even when they have fallen behind. This adds risk to anyone attempting to mine on a block opposing the selfish miners, and they may simply chose to mine on top of the selfish chain as this would be the safer of two options.

The parameter γ describes the ratio of miners not in the selfish pool that will mine the selfish chain when the equalizing block is posted. This gives three groups:

p, selfish miners;

$\gamma(1-p)$, miners who join selfish miners in a tie; and

$(1-\gamma)(1-p)$, miners who never join the selfish miners.

To compute the effectiveness, we repeat the computations from §8.2 noting that

$$\mathbb{E}[f_{\mathrm{SM}} \mid (1,1)] = 2p + \gamma(1-p),$$

leading to

$$\mathbb{E}[f_{\mathrm{SM}} \mid (0,0)] = p^2\mathbb{E}[f_{\mathrm{SM}} \mid (2,0)] + (2p + \gamma(1-p))p(1-p),$$

while

$$\mathbb{E}[f_O \mid (0,0)] = (1-p) + p(1-p)(2(1-p) - \gamma(1-p)).$$

Thus

$$\begin{aligned} S_\gamma(p) &:= \frac{p^2\frac{2-3p}{1-2p} + (2p+\gamma(1-p))p(1-p)}{p^2\frac{2-3p}{1-2p} + (2p+\gamma(1-p))p(1-p) + (1-p) + p(1-p)(2(1-p) - \gamma(1-p))} \\ &= \frac{4p^2 - 9p^3 + 4p^4 + \gamma(p - 4p^2 + 5p^3 - 2p^4)}{p^3 - 2p^2 - p + 1}. \end{aligned}$$

Exercise 8.3. Find γ such that the break-even p for selfish mining with γ is 0.25.

Note that there are variations of the selfish mining strategy that can be explored. "Stubborn mining" is a variation of selfish mining where the selfish miner may continue to mine from behind or use different block-withholding strategies when ahead.

8.7 Defense against selfish mining

It makes most sense for the selfish miner to restrict the pool participants to those that the organizer can trust. If the pool is unrestricted, it would be trivial for any pool participant to release to the public any block that the selfish pool finds.

Selfish mining, if it is successful, will be detectable. It is blatant. There is no way to hide it if it is successful: Stale blocks are quite rare in the absence of selfish mining, whereas selfish mining requires not only frequent stale blocks, but frequent two-block reorgs which appear suddenly. Everybody will see this. The question remains whether the "honest" majority can do anything about it.

The problem, as we will discuss in §10.2.2 and §13.2.2, is that coordinating a response requires some sort of authority or consensus outside of the protocol and could destroy features of the network. For example, if the network detected an attack, one proposal to frustrate the selfish miner would be to refuse to accept two-block reorgs, rejecting them on the spot. However, this would be tantamount to a rolling one-block checkpoint (see §13.3). Such a rule is not in the spirit of Bitcoin, and would leave the protocol vulnerable to chain splitting at very low cost. It is not clear that anything can be done to stop selfish mining without introducing centralization or other serious problems for the protocol.

9 Strategic mining

As we saw in the discussion of selfish mining, the rules defining Nakamoto consensus are not always incentive-compatible: Not publishing blocks is a rational profit-maximizing strategy that can lead rational miners to deviate from the standard protocol. More generally, there may exist quite a wide variety of deviant strategies. Recalling Definition 8.1, we provide a complementary definition.

Definition 9.1. We say a miner is practicing *strategic mining* if they are using any strategy other than default compliant in order to obtain a better outcome.

It is difficult to describe the complete set of strategies that miners can take. A strategy is any function from the set of all game states and beliefs to the set of all possible actions the player can take. For Bitcoin mining, the most general idea of the state includes information about the following:

- the current set of viable chains,
- beliefs about others players' strategies,
- the mempool visible to the miner,
- beliefs about mempools available to other miners,
- the hashpower cost function available to the miner,
- beliefs about hashpower/cost available to other miners,
- the time since the last block was mined on each viable chain,
- the time since difficult adjustment on each viable chain,
- beliefs about how the value of Bitcoin will respond to different game states,
- outside opportunities (for example competing blockchains) to devote hash to.

The miner's actions at any given time will involve the following:

- whether to hash or power-down machines,
- whether to hash Bitcoin blocks or use hashrate for other purposes,
- which chain to work on,
- which transactions to include in a block,
- whether to participate in a pool,
- when to publish a block if found.

As of 2022, the default strategy is default compliant, that is, it appears that most miners are using the protocol as intended. It is unclear given the current set of incentives that there is significant benefit to using other strategies. It is important to point out that the emergence of deviant strategies will create the opportunity for further deviant strategies to emerge. The goal of this chapter is to explore dynamic pathways in which deviant strategies may emerge and what would lead deviant strategies to become dominant.

https://doi.org/10.1515/9783110772845-009

9.1 Petty compliant strategy and fee-undercutting

The default strategy dictates that miners choose which block to mine upon based on the "first-seen" rule. If miners see a block appending the longest chain, they begin working to further append to that block, and do not switch unless they see a new longest chain. However, there is no way for the protocol to determine which block individual miners actually saw first. They can just as easily ignore a block if they have a good reason to mine on another block at the same height.

One reason that a miner may have a preference between two blocks is that if one block leaves up for grabs higher-fee transactions in the mempool, a miner mining on top of this block may earn more revenue by crafting a block appending to this block instead of the first-seen block. In this case, if a rational miner has a choice, they will opt for the chain that maximizes their fee revenue by mining on top of it.

Following [10], we define the *petty compliant strategy*.

Definition 9.2. A miner follows the petty compliant strategy if they follow the default compliant strategy, with one exception: If two chaintips exist at the same height, the miner will mine on the chaintip that allows them to include more fees in the next block.

9.1.1 Fee-undercutting

To model the profit-maximizing game for a strategic miner in the presence of petty compliant miners, we use the following notation. At any point in time, let:

$$
\begin{aligned}
f_{\text{tip}} &:= \text{fees that were harvested in the latest block,}\\
f_{\text{mem}} &:= \text{maximum amount of fees available in the mempool that can be fit into a block,}\\
p_{\text{PC}} &:= \text{hashrate of miners using the petty compliant strategy,}\\
p_{\text{DC}} &:= \text{hashrate of miners using the default compliant strategy,}\\
p_S &:= \text{hashrate of a strategic miner,}\\
\chi &:= \text{block subsidy.}
\end{aligned}
$$

Assume for this game that

$$p_{\text{PC}} + p_{\text{DC}} + p_S = 1$$

and that the strategic miners are operating as a single pool. This simplification makes this a single-player optimization game, which simplifies the presentation: We only need to determine the single pool's strategy.

If the latest block is published by a miner outside of the strategic pool and $f_{\text{tip}} > f_{\text{mem}}$, then there is an incentive for the strategic miners to mine a block at the same height of the latest block: By choosing to include f in fees in the block, with

$$f_{\text{mem}} < f < f_{\text{tip}},$$

they are claiming a block with higher fees. This gives more rewards for the strategic miners provided their block stays in the main chain. Now because

$$f < f_{\text{tip}},$$

all petty compliant miners will switch to mining this block, increasing the probability that the reorg is consummated. This is called *fee-undercutting* (it is sometimes also referred to as *fee-sniping*, although technically speaking, fee-sniping can happen without undercutting).

We can determine when fee-undercutting is a preferable strategy: With probability

$$p_{\text{PC}} + p_S$$

the strategic block will be appended, supplanting the previous tip, and strategic miners will reap an expected reward of

$$(p_{\text{PC}} + p_S)(\chi + f).$$

Had they mined on the tip, they would have received a reward of $\chi + f_{\text{mem}}$ with probability assumed to be 1. So the criterion for undercutting to be profitable is that

$$\chi + f_{\text{mem}} < (p_{\text{PC}} + p_S)(\chi + f_{\text{tip}}).$$

This can take a variety of expressions, for example,

$$\chi < \frac{(1 - p_{\text{DC}})f_{\text{tip}} - f_{\text{mem}}}{p_{\text{DC}}}. \tag{9.1}$$

So when the block subsidy is small relative to the fees and the hashrate of DC miners is small, such a situation becomes likely. On the other hand, if the block subsidy is large in relation to fees, undercutting is very unlikely to be a wise choice.

Note that if p_{DC} is very near zero, the question becomes only whether the available fees in the mempool are lower than the fees in the last mined block, in which case it will be preferable to use fee-undercutting. Intuitively this should be true for at least some positive time after each block is produced.

If 50 % of miners are using a default compliant strategy, we see that the condition becomes

$$\chi + 2f_{\text{mem}} < f_{\text{tip}}.$$

In particular, the remaining fees must exceed not only the block subsidy, but in addition to that, twice the fees available in the mempool. When the block reward is significant, this appears unlikely.

If 10 % of miners are using a default compliant strategy, the condition becomes

$$\chi < 9f_{\text{tip}} - 10f_{\text{mem}}.$$

Such an inequality becomes increasingly likely in the future when fees eclipse block reward as the main source of mining revenue.

9.1.2 Fee-sniping vs. fee-undercutting

Fee-sniping generally refers to mining blocks that do not extend the longest chain in order to collect a larger share of fees. For example, if someone posts a very large fee, one miner could mine that single transaction into a block. Even if this is not undercut, it still could be sniped: If the fee is large enough, other miners may be so inclined to also attempt to claim the fee. If they have enough hashrate so that their probability of finding the next block makes the attempt worth it, they can do this without relying on petty compliant miners for the help.

9.1.3 Emergence of petty compliant miners

Fee-undercutting is unlikely to be a profitable strategy until a considerable percentage of miners begin using a PC strategy. Without PC miners helping out, there is little hope that a block mined in hopes of undercutting will escape becoming stale. On the other hand, in the absence of undercutting, there is not a decisive incentive for miners to begin using the PC strategy. At the moment, the only time this would be worthwhile is during an incidental fork.

9.1.3.1 Adjusting mining to allow for chain splits

As of 2022, Bitcoin fees are small, but if there is variation in fees included in different blocks, there could be a small advantage to switching to the latest chain. Variation in fees between blocks mined at slightly different times does occur: These can be attributed to variations in available fees in the mempool and selection algorithms used by the miners when constructing a block.

To get an idea of the possible upside to PC mining, suppose that an incidental fork happens with probability ε and that the average absolute difference between the left-over fees in the two blocks created is δ in units of expected miner revenue per block (expected fees + block subsidy). Then by employing the petty compliant strategy, a miner will on average increase their revenue by a multiple of $\frac{\varepsilon\delta}{2}$. For example, if $\varepsilon = 0.0016$ and $\delta = 0.05$, the factor describing the increase in revenue will be roughly

$$0.0004.$$

This is quite small, and arguably not enough to influence miners' behavior.

All in all, if other miners are mining a default compliant strategy, there is a very small, but nonetheless positive benefit for miners to switch from a default compliant strategy to a petty compliant one. But in the near future, it is not very clear that there is serious incentive for miners to consider using a PC strategy. This poses a chicken-and-egg problem for strategic miners hoping to use an undercutting strategy.

We will revisit this in the setting of diminishing block reward in §11.4.

9.1.4 Selfish mining as incentive to use petty compliant mining

While naturally occurring forks are somewhat rare and are unlikely to present lucrative reasons to move towards petty compliant mining, selfish mining intentionally presents forks to the network. Further, if the selfish miner mined their block first, perhaps minutes ago, it is likely to contain lower fees. So not only does the event happen more frequently, one would also expect that switching to the selfishly mined chain would result in higher revenue for the petty compliant miners. This suggests selfish miners and petty compliant miners would enjoy an asymmetric but mildly symbiotic relationship. While the petty compliant miners are hurt overall by the selfish miners, they are hurt slightly less if they take advantage of fee opportunities, and would be in a better position than default compliant miners. If less than 33 % are mining selfishly and there are a few petty compliant miners, these petty compliant miners would gain from the arrangement.

This applies especially to smaller miners. A miner not large enough to selfishly mine is always better off with a petty compliant strategy. They will be unable to influence the behavior of the selfish miners on their own.

Example 9.1. Suppose a selfish miner with hashrate $p < 33\,\%$ begins selfish mining. This sets up a prisoners' dilemma among the remaining miners. By cooperating with default compliant miners, the remaining miners can force γ towards 0 (recall §8.6), which would lead to a loss of revenues by the selfish miner. Eventually the selfish miner would be choked out and give up the campaign. However, if a few miners defect and begin petty compliant mining, it will quickly become profitable for the selfish miner, and they will continue to take a disproportionately large share of the network.

A long-term strategy for a miner attempting to use selfish mining is to simply begin selfish mining, perhaps prematurely and at a loss. By presenting an opportunity for miners to switch to petty compliant, the quantity γ can be lifted to a value at which selfish mining becomes profitable.

Selfish miners may also engage in a slightly more aggressive form of undercutting. The selfish miner may send a transaction to themselves, with some small fee. They keep a conflicting transaction, sourcing from the same UTXO, sent also to themselves. They can then mine this transaction on their secret block and create a child transaction with a more significant fee. Because this enticing child transaction would be in direct conflict with a transaction mined by the honest miners when the next block is found and compatible only with the selfishly mined block, it would give added incentive for petty compliant miners to extend the selfish chain. This type of transaction may also be described as a bribe.

9.1.5 Measures against undercutting

Some measures against undercutting have been proposed. One involves miners prudently and defensively deciding how many fees to include (covered in more detail in §11.4). Another proposal is to make use of *lock time*, a transaction feature that prevents transactions from being spent before a certain block height. When a transaction is published, a lock time can be used to ensure that it is spent in the future, not in the past. This prevents a miner from including the transaction in a past block. This itself only provides a small motivation to the average user.

A miner who mines a block may include an *anyone-can-spend* transaction that is not spendable until the next block. (An anyone-can-spend transaction is an output that does not require the signature from a specific private key in order to move. Any miner who is paying attention would choose to move it into their own custody.) In this case, the transaction cannot be spent by a miner trying to undercut the block; it can only be spent by a miner mining a further block. The miner hopes that this induces all other miners to extend their chain rather than undercutting. The danger is that if someone undercuts the block, someone else may still go ahead and take advantage of the anyone-can-spend transaction. A miner considering this strategy should compare to a strategy of simply leaving some fees in the mempool – these fees will not come out of the miners' pocket in the case of a reorg.

It has been proposed to implement locks at the wallet level, so the users include by default a lock that prevents the transaction from being recorded at a height that was already mined. This would discourage aggressive fee-snipers who not only want to grab large fees from the previous block, but also include transactions that are appearing. This is a more aggressive form of fee-sniping that necessarily is not undercutting.

Another proposal (which has not been realized) is to allow coinbase transactions to mature before 100 blocks. Unlike other transactions that do not depend on the particular heaviest chain, the coinbase transactions depend on the particular chain. If coinbase transaction were to become immediately spendable, miners could include anyone-can-spend output from the coinbase transactions. These would be paid out only if the block remains on the heaviest chain.

9.2 Other deviant mining strategies

9.2.1 Accepting bribes

Petty compliant miners never attempt to create a split: Rather they are using a method to maximize their revenues when a tie already exists. But miners could take one step back. Do they have opportunities to increase their profit by accepting bribes in order to deliberately mine a weaker chain? The answer depends naturally on the hashrate of the miners, the distribution of other miners' strategies, and the rewards available.

9.2.1.1 Single-block bribery

Suppose a Bitcoin user sends a large transaction, and this transaction is mined. For whatever reason, perhaps the transaction was never supposed to be sent or sent to the wrong address, or perhaps the user is wishing to do something malicious, the user would like to rescind this transaction. So immediately, they send a conflicting transaction spending the same output, but including a much larger fee λ. Now miners who are thinking strategically will consider including this transaction in the block, which if mined immediately would be matching the height of the longest chain. If λ is significantly larger than the expected revenue from the next block, the rational miner should consider it, especially if the miner has large hashpower.

In particular, if

$$(\lambda + \chi)P(\text{this block will end up in the accepted chain}) > \chi + f_{\text{mem}}, \tag{9.2}$$

it is a good idea to create the chain split.

One of the most obvious reasons that the block has a high probability of being extended would be that the miner itself has a large portion of the hashrate. If the bribe is five times the expected rewards for mining and the miner has 25 % of the hashrate, immediately we see that (9.2) holds.

A healthy fraction of petty compliant miners will increase the probability that the block will be accepted, provided it is mined carefully. Certainly other strategic miners will covet the fee; however, it is not so clear that they can undercut it. If the fee is large enough, the miner could be wise to include only the single transaction. If only the single transaction is included, there is no way to take a slice of the lucrative block without removing the fee. In fact, by excluding transactions from the block, they are setting up a fee-undercutting situation, despite the fact they are taking in a larger chunk of revenue. Petty compliant miners will choose to mine their block.

We conclude that accepting a bribe λ becomes wise if

$$(p + p_{\text{PC}})(\lambda + \chi) > \chi + f_{\text{mem}}.$$

Example 9.2. If 50 % of the miners are petty compliant, any bribe λ that is double the expected block rewards (fees + subsidy) would cause strategic miners to attempt to mine the bribe.

9.2.1.2 Bribes incentivizing a deeper reorg

What about a bribe that attempts to reorg from two blocks back into the chain? It is not so clear that the most naive presentation of a bribe will be very effective. If a bribe is meant to induce a block being created that will not equalize the longest chain, the block that mines the bribe would fall short of even appearing on the radar of PC miners. The strategic miner who would accept a bribe must be assured that blocks are likely to appear on top of that block as well. This requires a bit more finesse or faith in other strategic miners; if a miner takes the large fee and leaves other transactions, there would be an opportunity for the other miners to undercut some of the fees, extending the rogue chain. There is also the possibility that the same miner who mines the first block will mine a second block that will be appealing to petty compliant miners.

The most obvious way is to create a lineage of transactions. If each transaction has a fee λ, the first miner could mine the parent transaction, the parent and the child, or the parent, child, and grandchild (more generally, using child transactions to encourage miners to include their parents in a block is called *child-pays-for-parent*). If undercutting miners are present, it may make sense to only include one transaction per block: attempting to include all three invites undercutting. So when the first block is mined, a bribe still remains available to mine the second block, and after this block is mined there is a still a bribe available to greet the miner who continues the process by mining the third.

Actually, there exists a number of technical tricks for the bribing party to structure a string of bribes. A simple way is to create a direct lineage of transactions, each with the large fee and each taking as input the transaction before it, where the output of each is locked for a single block (using the nLock feature). Thus no two transactions can be spent in a single block. Another option is to create a set of sibling transactions all from the same UTXO which determines the split. Each sibling transaction would be locked for one, two, three, etc., blocks, but could then be spent by its own child transaction with a large fee attached. Another brutal option is to create a lineage of transactions that exceed half of the blocksize. Any of these methods would guarantee that the string of carrots stretches far enough into the future to insure miners keeping working towards the reorg.

The success of a bribe working then depends on the percentage of miners who are ready and willing to accept bribes. Following this logic, it may be possible to execute a bribe reaching somewhat far back, provided there are sufficient strategic miners present.

Example 9.3. Alice would like to reverse a transaction mined three blocks ago. Alice posts an ordered sequence of five bribes each containing multiple λ of the expected block reward as fee. What does λ need to satisfy in order for strategic miners to accept the bribes?

We assume that the strategic miners study probability and use the maximizing strategy as a Schelling point or openly collaborate. (If some of the strategic miners do not participate it will hurt the rewards for the rest – this has a stag hunt dynamic. We assume the strategic miners are an aggressive breed: They all hunt stag and assume that others also hunt stag.) For strategic miners, the question is whether they will mine m blocks before the non-strategic miners mine $m - 3$ for $m = 4, 5$, or 6. For this example, we assume that there are no petty compliant miners, but strategic miners will behave as petty compliant miners in order to break a tie, as fees will be left in the mempool by nature of the bribing mechanism. In particular, the decisive sixth block would be incentivized by available fees, not by bribes.

We assume they go with a slightly simplified strategy: A strategic miner will continue to mine the bribe chain until the main chain has added three additional blocks, in which case they give up and stop. This strategy does not consider how long strategic miners could continue to race by using fee-undercutting, which may or may not be a profitable strategy.

The probability space is a Bernoulli space (§2.4) which we can partition into terminating events. Events are grouped into:

E_4 := strategic miners produce four blocks before DC produce one

E_5 := strategic miners produce five blocks before DC produce two, excluding E_4

E_6 := strategic miners produce six blocks before DC produce three, excluding E_4 and E_5.

These are all winning events for the strategists, while for $m = \{0, 1, 2, 3, 4, 5\}$,

$$F_m := \text{strategic miners have produced } m \text{ blocks, when DC produce three}$$

is the set of losing events.

While the mathematical tools for working with Bernoulli spaces are numerous and fascinating (binomial formulas, negative binomial formulas, Catalan numbers, Dyck words, among others), for many applications the most straightforward way is to simply count the paths through the space in an ad hoc way. The following picture is sometimes called a trellis diagram. We will go through this example carefully counting the paths. Refer to Figure 9.1 in the following discussion.

Here we define the state space to be

$$\{(i,j) \mid \text{DC chain has } i \text{ blocks, strategists have } j\}.$$

The probability $P(E_4)$ is the easiest to compute: It requires four consecutive strategic blocks, thus

$$P(E_4) = p_S^4.$$

Note that this the probability of crossing the path labeled E_5 will be the probability of one success when the fifth failure happens (success measured from the perspective of honest miners). However, we must also exclude the path that was already counted in E_4. The event we need to exclude only involves a single path, namely the path which crosses E_4 and then moves down before moving up to cross E_5. Using the negative binomial distribution formula in §5.3.1,

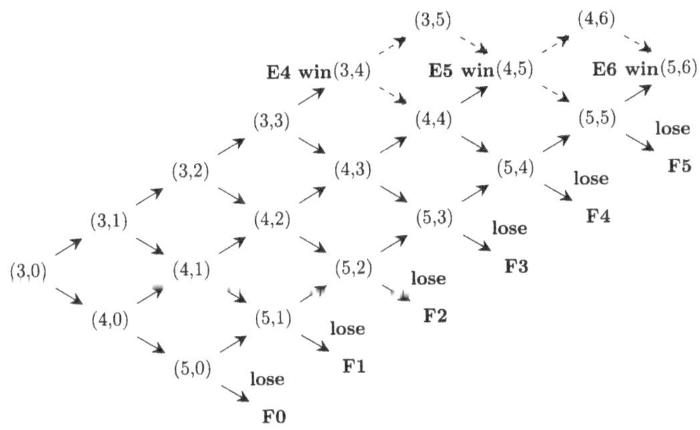

Figure 9.1: Trellis diagram.

$$
\begin{aligned}
P(E_5) &= \phi(1; 5, 1-p_S) - p_S^4(1-p_S)p_S \\
&= \binom{5}{4} p_S^5(1-p_S) - p_S^5(1-p_S) \\
&= 4p_S^5(1-p_S).
\end{aligned}
$$

Note that it may be just as easy to count the paths by hand. For E_6 we need to be slightly more careful in excluding paths. First we exclude the paths from state $(3,4)$ to state $(5,5)$: There are precisely three of these, and these are already included in the counting for E_4. The probability of a random path being one of the three leading through E_4 and then through E_6 is then

$$3p_S^5(1-p_S)^2 p_S.$$

Next we consider paths that go through E_5 (excluding those that have gone through E_4) and then through E_6. Once they have gone through E_5 there is a unique way to go through E_6, so to the total probability is

$$P(E_5)(1-p_S)p_S - 4p_S^6(1-p_S)^2.$$

Again applying the negative binomial formula, we get

$$
\begin{aligned}
P(E_6) &= \phi(2; 6, 1-p_S) - 7p_S^6(1-p_S)^2 \\
&= \binom{7}{5} p_S^6(1-p_S)^2 - 7p_S^6(1-p_S)^2 \\
&= 14p_S^6(1-p_S)^2.
\end{aligned}
$$

So the total probability of winning is

$$p_S^4 + 4p_S^5(1-p_S) + 14p_S^6(1-p_S)^2.$$

On the other hand, we can compute

$$P(F_m) = \phi(m; 3, p_S) - P\{\text{paths that went through } E_4,\ E_5,\ \text{or } E_6 \text{ and then } F_m\}.$$

Note from the trellis diagram that it is impossible for a path to cross a "lose" path to F_i and also to F_j for $i \neq j$. So

$$\begin{aligned} P(F_0) &= (1 - p_S)^3, \\ P(F_1) &= \binom{3}{2} p_S (1 - p_S)^3, \\ P(F_2) &= \binom{4}{2} p_S^2 (1 - p_S)^3, \\ P(F_3) &= \binom{5}{2} p_S^3 (1 - p_S)^3. \end{aligned}$$

However, the next computation is slightly more interesting because we must also exclude a path passing through E_4 and F_4: There is only one, so we have

$$P(F_4) = \left(\binom{6}{2} - 1\right) p_S^4 (1 - p_S)^3.$$

Also note that we have counted the number of paths that arrive at (5, 5) without running through E_4 or E_5 when computing $P(E_6)$: There are 14 such paths. So we conclude

$$P(F_5) = 14 p_S^5 (1 - p_S)^3.$$

Now we take the approach outlined in §2.4.2 to compute the expected difference in rewards for strategic miners if they pursue this strategy. We define

$$f := \text{rewards for strategic miners accepting bribe, subtracting rewards if not}$$

and would like to compute $\mathbb{E}[f]$.

In event E_4 the strategic miners win as many block rewards as they would have won had they mined honestly, but they gain collectively 4λ in bribes. Note that by quickly supplanting the three blocks, they are leaving a bribe available to everyone, which will be claimed by a strategic miner with probability p_S. Thus (denoting in units of expected block reward)

$$f(E_6) = 4\lambda + p_S \lambda.$$

In event E_5 the strategic miners win as many block rewards as they would have won had they mined honestly, but they gain collectively 5λ in bribes leaving no more bribes on the table. Event E_6 is the same. So

$$f(E_5) = f(E_6) = 5\lambda.$$

In each of the losing events F_m the miners lose m block rewards they would have won had they mined honestly. That is,

$$f(F_m) = -m.$$

Combining the probabilities,

$$\begin{aligned}
\mathbb{E}[f] &= p_S^4 4\lambda + 4p_S^6(1-p_S)^2 5\lambda + 14p_S^6(1-p_S)^2 5\lambda \\
&\quad - (3p_S(1-p_S)^3 + 12p_S^2(1-p_S)^3 + 30p_S^3(1-p_S)^3 + 56p_S^4(1-p_S)^3 + 70p_S^5(1-p_S)^3) \\
&= 2p_S^4(45p_S^4 - 90p_S^3 + 45p_S^2 + 2)\lambda \\
&\quad - p_S(1-p_S)^3(70p_S^4 + 56p_S^3 + 30p_S^2 + 12p_S + 3).
\end{aligned}$$

Inspecting this expression, we see that it becomes profitable to accept the bribe whenever

$$\lambda > \frac{1}{2}\left(\frac{1-p_S}{p_S}\right)^3 \frac{(70p_S^4 + 56p_S^3 + 30p_S^2 + 12p_S + 3)}{(45p_S^4 - 90p_S^3 + 45p_S^2 + 2)}.$$

Suppose that 50 % of miners are mining strategically. Then the bribe should be a multiple of at least

$$\frac{1}{2}\frac{(70(\frac{1}{2})^4 + 56(\frac{1}{2})^3 + 30(\frac{1}{2})^2 + 12(\frac{1}{2}) + 3)}{(45(\frac{1}{2})^4 - 90(\frac{1}{2})^3 + 45(\frac{1}{2})^2 + 2)} \approx 2.8961.$$

Supposing that 80 % of miners are mining strategically, then the bribe should be a multiple of at least

$$\frac{1}{2}\left(\frac{1-\frac{4}{5}}{\frac{4}{5}}\right)^3 \frac{(70(\frac{4}{5})^4 + 56(\frac{4}{5})^3 + 30(\frac{4}{5})^2 + 12(\frac{4}{5}) + 3)}{(45(\frac{4}{5})^4 - 90(\frac{4}{5})^3 + 45(\frac{4}{5})^2 + 2)} \approx 0.22095,$$

which is not incredibly large. If strategic miners expect that 80 % of miners are mining strategically, they will be very quick to accept bribes. In this case the briber will have to spend only 1.1 times the block reward (five bribes of 0.22 block rewards) to accomplish the reorg.

On the other hand, supposing that 20 % of miners are mining strategically, the bribe should be a multiple of at least

$$\frac{1}{2}\left(\frac{1-\frac{1}{5}}{\frac{1}{5}}\right)^3 \frac{(70(\frac{1}{5})^4 + 56(\frac{1}{5})^3 + 30(\frac{1}{5})^2 + 12(\frac{1}{5}) + 3)}{(45(\frac{1}{5})^4 - 90(\frac{1}{5})^3 + 45(\frac{1}{5})^2 + 2)} \approx 72.69.$$

These computations emphasize that strategic miners need to use the presence of other miners in their computations and take the most aggressive strategy as a Schelling point. This also highlights how consolidation of mining can tilt the playing field: A single entity controlling 25 % of the hashrate can have a large effect on whether strategic mining happens or not.

Note that we have made a number of simplifications discounting the incentives for strategic miners to accept the bribe. First, the strategic miners are not bound to continue pursuing this if the other miners take a decisive advantage after the first few blocks are mined (for example, they could abandon the pursuit at state $(5, 0)$, cutting losses). We also are not considering the motivations of individual miners within the pool of strategic miners to keep extending the contesting chain when the bribes run out. If the individual miners have notched bribes in the contesting chain, they do not necessarily need bribes in order to continue to work to have this chain accepted – in particular, F_5 may not be a giving up point for the strategic miners that have won bribes. This is particularly likely if the strategic miners consist of large players controlling a significant amount of hashrate. We also do not account for the common-sense event that the briber would be watching and continue to offer more bribes to break a tie if necessary.

9.2.1.3 Pay-to-win and AIM

Bribing does not need to be performed using only the Bitcoin blockchain. Because other smart contract platforms afford a wide range of computations, one could create contracts that pay bribes directly on other chains. These vectors can be called *pay-to-win* (P2W) [26]. Due to the flexibility of smart contract platforms, the creative attacker can use their imagination when constructing options for paying out bribes. The paper [27] defines *Algorithmic Incentive Manipulation* (AIM) as the use of “either credible threats, or conditioned rewards denominated in cryptocurrency units, to incentivize certain actions, within a targeted cryptocurrency system, to be taken by capable actors.”

The technical details of AIM are beyond our scope, but suffice to say that bribers may use ingenuity and smart contracts to structure commitments which can be rather expressive. Bribing commitments need not be as straightforward as Example 9.3, which uses Nakamoto consensus itself as the mechanism.

9.2.2 Block withholding

So far, we have largely steered away from the mechanics of pools, as these fall outside of the Bitcoin protocol. Miners join pools in order to pool the risk of not finding a block. Miner rewards are often distributed back to the miners according to partial proof of work. When a miner in a pool creates hashes, they should frequently produce hashes that are several orders of magnitude less than the blockchain difficulty. By communicating these to the pool, they “prove” to the pool that they are operating on blocks for the pool. Clearly the miner does not communicate trillions of hashes to the pool as this would be impractical.

This can present an opportunity for some deviant strategies. Suppose a large miner or group of miners is trying to maximize the proportion of the blocks which they mine.

It is possible for these miners to split their resources between a pool mining honestly and competing pools. For the competing pools, the miners could produce partial proof of work, but then decide to simply not communicate winning blocks when found, if there is some strategic reason for doing so. On an individual level, the miners would be reducing their share of the rewards of any pools they infiltrate, but this may still be advantageous to their own pool. Naturally this depends on the mechanics of the pool. If pools pay directly for partial proof of work instead of proportionally distributing the rewards, this strategy may be especially harmful to the compromised pools.

9.2.3 Protecting one's own interests

Once a race begins, a bribe is accepted, or for any other reason a split occurs, miners who always mine the longest first-seen chain may be putting themselves at a disadvantage. It is possible that blocks they have mined have been temporarily supplanted: If there is a tie, clearly, they would like to break it for the chain containing their own transactions. They may even attempt to perform undercutting in order to catch their chain up.

Example 9.4. Suppose that mining pool A mines a block, which is immediately countered by a single block coming from a selfish miner. Suppose the selfish miner then is able to append their block a fraction of a second before an honest miner finds an honest block that is broadcast throughout the network. There is again a tie, but most miners will have seen the block mined on the selfish chain first. The default compliant strategy dictates they mine the selfish chain. However, miners in pool A are going to prefer to ignore the first-seen rule: If they successfully mine the next block, they will be preserving the block they previously mined. It is clearly in their interest to do so.

9.2.4 Mine from behind

There are situations when a miner will continue to attempt to win a race, even while behind. If fees are unusually large, a miner may even attempt to snipe these for themselves. The computation would depend on the size of the pool, the size of the petty compliant miners, and the size of the fee.

Example 9.5. Suppose that a mining pool with 30 % of the hashrate records a block with a fee that is four times the expected block reward at the same time as a pool with 10 % of the hashrate, and then the latter pool is the first to find a block to extend their chain. Suppose that petty compliant miners outside of these two make up 20 % of the total hashrate. Should the mining pool with 30 % attempt to mine from behind?

As with many of these computations, if they do not win the next block, the computation is immaterial. So we assume they win the next block. If they have decided to mine from behind and use undercutting, they are setting up at least a 50 % chance (their own hashrate plus the petty compliant) that their block will be preserved after the tie is broken. If we assume that the smaller mining pool would not be challenging that, the expected value is clearly larger for this pool when mining from behind.

This is a simple example, but we can analyze this in much more depth. This is similar to the examples in §5; however, now we are allowing for other strategic and possibly petty compliant miners.

9.2.5 Extended example

Consider the following simple game: The Red pool has hashrate p, the Blue pool has hashrate q, and the remainder of the miners are default compliant. A transaction with a large fee of λ times the expected block reward is mined by one of the Red and Blue pools. When is the other pool wise to mine from behind?

To keep things simple, the following analysis will assume other fees are small relative to the block subsidy and we will exclude petty compliant miners. Note that if there are multiple strategic miners fighting over a given large fee, each would probably be careful to be excluding other fees that would put them at risk of undercutting, perhaps even mining empty blocks.

In this game, each pool has two most obvious strategies: Either give up when they fall behind or mine from behind until they fall two blocks behind. This gives us four possible combinations and we could build a payoff table, but to keep the discussion brief we discuss the outcome when both pools are mining from behind and leave the remaining cases as an exercise.

Suppose both pools are mining from behind. This can be analyzed using a Markov chain. Once the game has started, there are essentially six states to this game:

1. Red has won,
2. Blue has won,
3. Red is up by one,
4. Blue is up by one,
5. two chains are equal but Red was first-seen,
6. two chains are equal but Blue was first-seen.

The transition matrix will be

$$\begin{pmatrix} 1 & 0 & 0 & 0 & 0 & 0 \\ 0 & 1 & 0 & 0 & 0 & 0 \\ 1-q & 0 & 0 & 0 & q & 0 \\ 0 & 1-p & 0 & 0 & 0 & p \\ 0 & 0 & 1-q & q & 0 & 0 \\ 0 & 0 & p & 1-p & 0 & 0 \end{pmatrix}.$$

This is a big and slightly unwieldy matrix. Observe that squaring it yields

$$\begin{pmatrix} 1 & 0 & 0 & 0 & 0 & 0 \\ 0 & 1 & 0 & 0 & 0 & 0 \\ 1-q & 0 & -q(q-1) & q^2 & 0 & 0 \\ 0 & 1-p & p^2 & -p(p-1) & 0 & 0 \\ (q-1)^2 & -q(p-1) & 0 & 0 & -q(q-1) & pq \\ -p(q-1) & (p-1)^2 & 0 & 0 & pq & -p(p-1) \end{pmatrix}.$$

The final two pairs of states are "block-diagonalized." This is due to the fact that these represent "even" and "odd" states of the game. To deal with a slightly less messy situation, one could consider only the even states together with the terminal states and then replace a single round of the game with two. We get the matrix

$$\begin{pmatrix} 1 & 0 & 0 & 0 \\ 0 & 1 & 0 & 0 \\ (q-1)^2 & -q(p-1) & q(1-q) & pq \\ -p(q-1) & (p-1)^2 & pq & -p(p-1) \end{pmatrix}$$

describing the transition probabilities between states 1, 2, 5, and 6 after two blocks are mined.

Similarly, the matrix

$$S := \begin{pmatrix} 1 & 0 & 0 & 0 \\ 0 & 1 & 0 & 0 \\ 1-q & 0 & q(1-q) & q^2 \\ 0 & 1-p & p^2 & p(1-p) \end{pmatrix} \tag{9.3}$$

represents the transitions between terminal states and odd states 3 and 4 after two iterations.

We begin by computing the probability that Red wins, given either state 3 or state 4. Let

$$\begin{aligned} v_3 &:= P\{\text{Red wins} \mid \text{red leads by } 1\}, \\ v_4 &:= P\{\text{Red wins} \mid \text{blue leads by } 1\}. \end{aligned}$$

Extending to the terminal states, $v_1 = 1$ and $v_2 = 0$. Conditional expectations give us

$$\begin{aligned} v_3 &= (1-q)v_1 + q(1-q)v_3 + q^2 v_4, \\ v_4 &= (1-p)v_2 + p^2 v_3 + p(1-p)v_4. \end{aligned}$$

We can set this up as a matrix:

$$\begin{pmatrix} 1-q(1-q) & -q^2 \\ -p^2 & 1-p(1-p) \end{pmatrix} \begin{pmatrix} v_3 \\ v_4 \end{pmatrix} = \begin{pmatrix} 1-q \\ 0 \end{pmatrix}. \tag{9.4}$$

Now sticking with Red as our protagonist, we are also interested in how many blocks they may throw away while losing this game. Letting

$$f := \text{number of blocks lost by Red in losing effort,}$$

we would like to compute

$$e_m = \mathbb{E}[f \mid \text{game starts at state } m].$$

Looking at the matrix S in (9.3) and using (2.6), we obtain

$$e_3 = q(1-q)\Big(e_3 + \frac{p}{1-q}(1-v_3)\Big) + q^2 e_4.$$

The term

$$\frac{p}{1-q}(1-v_3)$$

is determined as follows. The two-step path from state 3 to state 3 must go via state 5 and then back to state 3. The first block is mined by Blue, and the second block is mined by someone who is not Blue. There is a $p/(1-q)$ probability this miner is Red. The factor $1-v_3$ is the probability that a system in state 3 will end up with Red losing.

Similarly,

$$e_4 = p^2(e_3 + 2(1-v_3)) + p(1-p)(e_4 + (1-v_3)).$$

In order to move from state 4 to state 3 necessarily two consecutive blocks must be mined by Red. These will be lost with probability $(1-v_3)$. Combining,

$$\begin{aligned} (1-q(1-q))e_3 - q^2 e_4 &= pq(1-v_3), \\ -p^2 e_3 + (1-p(1-p))e_4 &= 2p^2(1-v_3) + p(1-p)(1-v_4), \end{aligned}$$

or in matrix form

$$\begin{pmatrix} 1-q(1-q) & -q^2 \\ -p^2 & 1-p(1-p) \end{pmatrix} \begin{pmatrix} e_3 \\ e_4 \end{pmatrix} = \begin{pmatrix} pq(1-v_3) \\ 2p^2(1-v_3) + p(1-p)(1-v_4) \end{pmatrix}.$$

Example 9.6. In this Red/Blue example, specify that $p = 0.25$ and $q = 0.2$. Solving (9.4) we get

$$\begin{aligned} v_3 &= 0.95592, \\ v_4 &= 0.07353. \end{aligned}$$

We also get

$$e_3 = 0.01317,$$
$$e_4 = 0.22159.$$

Thus we can make a comparison. Starting from one block behind, the expected profit from mining from behind is

$$0.07353\lambda - 0.22159.$$

This becomes positive whenever

$$\lambda > 3.014.$$

A similar calculation from the perspective of Blue yields

$$v_3 = 0.92647,$$
$$v_4 = 0.04412,$$
$$e_3 = 0.01914,$$
$$e_4 = 0.18999.$$

So Blue will use this strategy from behind whenever

$$\lambda > 4.31.$$

This gives us an idea of whether Red or Blue is willing to play this game when falling one block behind. If the reward is sufficiently large, mining from behind will be a dominant strategy.

We observe that employing a mine-from-behind strategy will never advantage the competitor in the short run (in the long run, taking into consideration the difficulty adjustment, wasting blocks does have some small benefit to the competitor). At worst, the competitor will have blocks supplanted; at best there is no change in the short run.

Exercise 9.1.

(1) Prove that

$$\begin{pmatrix} I & 0 \\ B & C \end{pmatrix}^k = \begin{pmatrix} I & 0 \\ (I + C + \cdots + C^k)B & C^k \end{pmatrix}.$$

(2) Show that if

$$\lim_{k\to\infty} C^k = 0,$$

then

$$\lim_{k\to\infty}\begin{pmatrix} I & 0 \\ B & C \end{pmatrix}^k = \begin{pmatrix} I & 0 \\ (1-C)^{-1}B & 0 \end{pmatrix}.$$

Exercise 9.2. Suppose that in the examples in this section, a hashrate of r belongs to petty compliant miners. How does this change the transition matrix?

Exercise 9.3. Compute the payoffs for Red and Blue if only Red is mining from behind.

9.2.6 Deferring to the alpha dog

Consider the following situation. There is a single large miner and many smaller miners, a significant portion of whom are strategic. Suppose that a transaction with a moderately large fee is presented to the miners in the mempool. Obviously the large miner will try to maximize their fee revenue. In this section we argue that there are situations where other miners could be induced to exclude this transaction, if the value is within a certain range.

Suppose a miner with very small mining power takes this transaction and claims the fee. The large miner can either mine the tip or attempt to claim the fee for themselves by mining from behind. To understand their motivations, we work by backwards induction. Once we have determined the strategy of the large miner, we can determine the strategy of the smaller miners. If the larger miner will be attempting to supplant a smaller miner's block when they include the fee, the smaller miner must make a risk calculation to determine if the fee should be included.

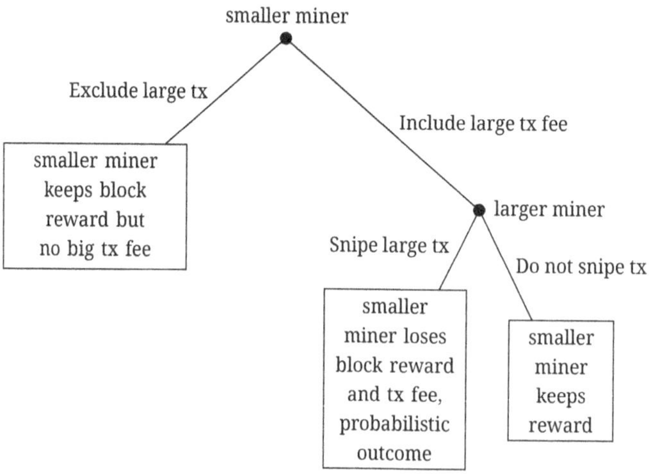

Suppose that the large miner is successful in also mining the large fee, splitting the chain into two chains of equal height. We claim that the strategic miners, with the exception of those who mined the other block, will switch to mining the tip favoring the larger miners. Why? Once the larger miner has signaled a willingness to mine from behind, there will be a non-trivial probability that they will produce two consecutive blocks, with or without the help of strategic miners. In comparison with opposing the smaller miner, the risk is quite small. We formulate this as follows.

Conclusion 9.1. If a split has occurred, one fork is favored by a larger pool, and that larger pool has signaled willingness to mine from behind, then strategic miners will attempt to append the chain favoring the larger pool, all else being equal.

This highlights another advantage afforded by centralization.

It is always possible that the larger miner greedily took many more fees, in which case they are naturally undercut by the first-mined block. But if the fees are comparable, rational miners will opt for the option that leaves them with the lowest risk of having their block supplanted. This fact influences the large miner's strategy, allowing for a more hawkish approach.

Let p be the hashrate of the large pool and let p_S be that of the strategic miners not in the pool and not in the pool that mined the first block. We assume that the small miner who took the first block has negligible hashrate.

We can model this like a Win–Lose–Pay game in §2.5. Suppose the larger miner equalizes, demonstrating a commitment to a mine-from-one-block-behind strategy. There are three possible outcomes after the subsequent two blocks have been mined:

1. A block from a non-strategic smaller miner is mined, followed by a block mined by any smaller miner. This happens with probability $(1 - p - p_S)(1 - p)$.
2. A block from a non-strategic smaller miner is mined, followed by a block mined by the large miner. This happens with probability $(1 - p - p_S)p$.
3. The larger miner or any strategic miner finds the next block, ending the game. This happens with probability $p + p_S$.

In terms of the Win–Lose–Pay game, (1) is a Lose, (2) is a Pay, and (3) is a Win. Letting

$$\begin{aligned} P &:= p + p_S, \\ Q &:= (1 - p - p_S)(1 - p), \\ R &:= (1 - p - p_S)p, \end{aligned}$$

the results from §2.5 give an expected loss of block rewards

$$E = \frac{RQ}{(P + Q)^2}.$$

Since the equalizing block is lost with probability $Q/(P+Q)$ we can determine that the expected number of blocks wasted in this effort is

$$\frac{Q}{(P+Q)^2}.$$

Thus the expected returns to the miners, given they have tied, will be

$$\frac{P(P+Q)\lambda - Q}{(P+Q)^2}.$$

Note that as the proportion of strategic miners to the rest of the pool grows, the expectation of winning becomes close to 1. For example, if a miner has 10 % of the hashrate, while 89 % of miners are strategic and small and 1 % is either protecting their own objective or not strategic, the pool is wise to pursue a mine-from-behind strategy to chase a fee that is a few percent above the usual block reward.

As a side note, it is interesting to consider the extreme case in which all miners are strategic and there is only one atomic pool. By "atomic" we mean to distinguish from "continuum" miners. Continuum miners are those whose operations are almost infinitesimal, say each miner produces a millionth of the hashrate, whereas atomic miners produce something non-negligible, say 2 %. We claim that mining from behind will always be a strategy for the atomic miner, even without variations in the block reward: As long as this pool is the only atomic pool, there is no risk to the miner in mining from behind. If they do not mine the block, there is no loss, but if the they do mine the block, their block will be the one on the longest chain with probability nearly 1. In the process, they will supplant other blocks and this will create the same effect as selfish mining: After the difficulty adjustment this pool will fare better. We leave the quantitative details to the reader.

Turning our attention back to the smaller miner, the smaller miner may practice strategic mining by intentionally leaving the higher fee in the mempool. This is a simple risk computation. In the notation above, the probability of the smaller miner keeping a fee of λ is $1-p$ (the probability that the larger pool fails to equalize the next block) plus $pQ/(P+Q)$ (the probability that the larger pool equalizes but then loses). The probability of losing the entire block to the larger miner is $pP/(P+Q)$. So if the fee satisfies

$$\left(1 - p + \frac{pQ}{P(P+Q)}\right)\lambda - \frac{pP}{(P+Q)} > 0,$$

the miner should include the fee. Also, if

$$P(P+Q)\lambda - Q < 0,$$

the larger miner will not contest it, so it is safe to the smaller miner. If

$$\frac{Q}{P(P+Q)} < \lambda < \frac{pP^2}{(1-p)P(P+Q)+pQ},$$

it will be wise for the smaller miner to leave the fee alone. It is easy to produce examples for p, p_S so that this range is non-empty.

Exercise 9.4. Find example values of p, p_S such that for $\lambda = 0.2$ it is wise for the larger miners to pursue the fee and it is prudent for smaller miners to leave the fee alone.

9.2.7 Responding to exogenous demand for hashrate

Entities who own ASICs may choose to produce hashes that are worthless to Bitcoin. Other chains with similar hashing algorithms exist, and machines can be temporarily re-purposed if the expected profits are better elsewhere. In general economic terms, following better returns on one's capital is a not a particularly "devious" thing to do. However, when miners turn their backs on Bitcoin for greener economic pastures, this unfaithfulness can lead to security issues, especially when the incentives are orchestrated by an attacker. So this falls squarely into the category of strategic mining. If someone is attempting to perform a double-spend or perform some other untoward action with the Bitcoin blockchain, they are interested in having less resistance from other miners.

Many ASICs can be deployed on other consensus algorithms that use the SHA256 hashing algorithms, for example Bitcoin Cash. If the Bitcoin price drops considerably relative to the difficulty, while another coin has increased in value, it may be profitable to switch to mining other chains.

So one strategy available to an attacker is to encourage miners to switch to other chains: This could be as simple as posting high-fee transactions in other blockchains consistently or buying up the other asset to increase the price if the market is thin enough and Bitcoin profit margins are low.

This may be preferable to directly bribing miners. Rather than bribing the miners to do something with the Bitcoin blockchain itself, they could offer a reward for the miner's work, inducing strategic miners to stop mining Bitcoin and produce other hashes. This is necessarily less efficient than a direct bribe, but miners may be more amenable to it. Besides not being directly implicated in the untoward action, miners can be paid regardless of whether the attack is successful, compared to on-chain bribery, which requires the reorg to be consummated in order for any rewards to be shared. Manipulating rewards on other blockchains is one way to do this, but one could imagine other ways (for example using algorithmic incentive manipulation) in which an attacker would directly pay for hashes that are completely worthless. We use the term *gelded mining* to describe proofs of work that are no more than a proof that hashrate was wasted.

Example 9.7. An attacker creates a smart contract (perhaps on another network like Ethereum) that rewards miners with a bounty for producing invalid blocks that are incompatible with the current chain: Fixing a height of N blocks, the contract will pay for any blocks mined that would make a valid $N + 1$ block if not for the fact that the actual chaintip was well beyond height $N + 100$. If the reward for producing these worthless blocks exceeds the rewards available for mining the chaintip, rational miners will devote their hashrate to this, freeing up the attacker to selfishly mine, double-spend, bribe, or do whatever else they want to do with the Bitcoin blockchain. This is also an example of P2W [26].

9.2.8 Harmonic mining/switch mining

As will be discussed in §12.3.1, there can be interesting dynamics when the profit margins are small. In this case a player with a large hashrate can exploit the difficulty adjustment in a much different way.

Suppose a difficulty period begins. A miner with hashrate p observes the hashrate and computes that they expect to spend $q < 1$ block rewards to mine every block reward (so the profit margins will be $1 - q$). Suppose instead of mining, they drop their hashrate completely, wait for the difficulty to adjust, and then begin mining at the lower difficulty. Is this a good strategy? We explore this here, making a simplifying assumption that fees do not play a big role in the block reward.

Observe that the hashrate will decrease by a fraction p. So the next difficulty adjustment is expected to happen in

$$T_1 = \frac{2{,}016}{1 - p} \times 10 \text{ minutes},$$

roughly (we are replacing the exact value with the expected value, which should be somewhat close after 2,016 trials; see Example 2.7). At this point the difficulty will decrease by a fraction p, and the miner rejoins. Because the difficulty is smaller, the cost of producing a block reward will now be

$$(1 - p)q,$$

so the profit margins are now

$$1 - (1 - p)q.$$

Mining at full capacity for this second period, the difficulty period will zip by, taking only (an expected)

$$T_2 = (1 - p)2{,}016 \times 10 \text{ minutes}.$$

So after

$$T_1 + T_2 = \left(\frac{1}{1-p} + (1-p)\right)2{,}016 \times 10 \text{ minutes},$$

the miner will have mined 2,016p block rewards, at cost 2,016$p(1-p)q$, so it will have made a profit of

$$2{,}016p(1-(1-p)q).$$

To determine if this was a good idea, we use the revenue ratio. Their average profit per 10 minutes over the two periods will be

$$\frac{p(1-(1-p)q)}{(\frac{1}{1-p} + (1-p))}.$$

We may compare this to their average profit per 10 minutes had they not rested; this would be

$$p(1-q).$$

Remark 9.1. These computations are not expectations; rather, they use the expectations to build a guess. The calculation above involves a dangerous linearization of non-linear calculations. Technically speaking, the expectation of non-linear functions is not the non-linear function of the expectation; in particular, for a non-constant positive random variable X, Jensen's inequality (2.10) tells us that

$$\mathbb{E}\left(\frac{1}{X}\right) > \frac{1}{\mathbb{E}(X)}.$$

A more thorough and exact analysis would require looking at the full probability distribution of T_1 and T_2 (see §2.7 and Example 2.7). We are content to work with the approximation in order to show heuristically that this sort of strategy is not out of the realm of possibility.

Example 9.8. Suppose that $p = 0.2$ and $q = 0.93$. Then the average profit for 10 minutes using a resting strategy is approximated by

$$\frac{0.2(1-(1-0.2)0.93)}{(\frac{1}{1-0.2} + (1-0.2))} \approx 0.02498.$$

Without using the resting strategy their average profit would be

$$0.2(1-0.93) = 0.014.$$

This suggests a benefit from resting.

One can compute from Example 2.7 that the standard deviation for the difficulty adjustment period is about 2.2 %. Thus random variation can play a non-trivial role when profit margins and a given miner's hashrate are all around this range.

We have discounted fees in this scenario, but this requires a simple modification, left as an exercise.

Exercise 9.5. Suppose that the average fee per block is given by f, where f is in units of the block subsidy. If we assume that the same amount of fees is collected over time, then the fees for the blocks mined when the miner is resting will be larger, while the fees in the following period will be smaller. Compute the appropriate factor and then a formula which determines if the resting strategy makes sense or not.

It is important to note that this strategy is not only theoretical: This has happened on the Bitcoin Cash blockchain, resulting in some wildly oscillating hashrates (see [54]). Bitcoin Cash is naturally more vulnerable because the hashrate is lower. When miners practice "switch mining," in which they alternate between mining Bitcoin and Bitcoin Cash, the smaller blockchain will feel more oscillation than Bitcoin during these events.

9.2.9 Merge-mining, layer 2, sidechains, and strategies playing out on other blockchains

Bitcoin can only process a few transactions per second. This is a result of the blocksize and a protocol that uses a difficulty adjustment calibrated to 10-minute intervals between blocks. This limits the scalability of Bitcoin as a layer 1 payment method (Visa, by comparison, can do 65,000 transactions per second). There is no possibility that Bitcoin could "compete" with a centralized payment network purely on layer 1 transactions; it clearly cannot. However, if layer 1 is used as a settlement layer, one can imagine possibilities for a very large number of Bitcoin-denominated transactions that take place off of layer 1 in real time. Layer 2 networks are being developed that could allow Bitcoin to scale to a larger payment capacity. Perhaps the most well known is the Lightning Network. The Lightning Network is a network based on a specific protocol that allows users to send each other bitcoins in time comparable to a Visa transaction, without requiring on-chain confirmation. This is done by anchoring nodes on the Lightning Network to specific UTXOs on the Bitcoin blockchain. This includes a mechanism to punish users who attempt to cheat the protocol and settle on layer 1 in dishonest ways. The details are quite interesting but a bit beyond the scope of this book. Lightning, if successful, may allow Bitcoin's use to scale (cf. [2, 1]).

Sidechains, like layer 2 protocols, are protocols which allow Bitcoin to be allocated to a sidechain purpose that can be manipulated by another protocol before being returned to the Bitcoin blockchain. Merge-mining (versions of which have been proposed for Bitcoin) allows confirmation for transactions from other chains to be included on the main Bitcoin blockchain. For a chain with less security, merge-mining represents an opportunity to etch transactions into a chain that has a much higher security budget. For Bitcoin, this could mean higher transaction demand and hence higher fees for miners.

How this will play into strategic mining is somewhat difficult to predict. Enmeshment with other chains' security protocols may create an aperture for more deviant strategies, because now other protocols, which may have different rules, may have different reasons to want to rewrite the chain. Bribes may be offered and executed in a different way, reorgs may happen for different reasons, and transactions may be reordered for strategic reasons that the Bitcoin blockchain cannot see. We suspect that under these situations, new incentives towards strategic mining will emerge. However, it is certainly not clear that this will be significant (this is a wide open question and topic of debate in 2022).

9.3 Will there be a transition from default compliant to a fully strategic regime?

As of 2022, most miners follow the default compliant strategy. This is a best strategy given beliefs about other miners following the same strategy and is also in line with the general ethos of Bitcoin. Default compliant makes the most sense as a Schelling point. As we have seen, as more miners become willing to deviate from a DC strategy in order to protect themselves or gain an advantage, more opportunities for strategic mining will arise.

Not only is the default compliant strategy the best strategy given that others are using the same strategy, the dominance of DC is reinforced by the fact that most participants are using the default reference implementation of Bitcoin software, which does not afford easy access to deviant mining strategies. The strategic miners' goal would be to nudge miners away from using the default implementation.

A landscape in which deviant strategies abound would be more likely to favor miners with a larger portion of the hashrate, for reasons outlined previously in this chapter. Such a miner may be interested in effectuating this. How would a miner with a somewhat larger amount of hashpower proceed if they were to try to create a landscape in which more deviant strategies become possible? The trick is to coax default compliant miners towards strategic behaviors by grooming them with occasional rewards. It would be necessary for miners to begin to craft their own software that implements deviant strategies, which may or may not be a trivial undertaking. A strategic miner with a strong interest in making this come about could publish their own open source strategic mining software in order to facilitate the process. As discussed above, the naturally occurring fee-undercutting in selfish mining may spur some miners to take a petty compliant strategy. A devious miner may begin by occasionally undercutting in hopes that miners are paying attention. Even less expensive, offering on-chain bribes has no penalty if the bribe is not accepted. A failed fee-undercutting attempt, on the other hand, does result in lost block rewards. Note that the use of partial proof of work can be useful in gauging the distribution of strategic miners. A briber can offer to pay bribes (perhaps with AIM) for partial proof of work. If there is enough interest, this would be "proven" in the partial proof of work. This would signal to other potential strategic miners that the bribed block is likely to be extended. In other words, partial proof of work can be used to determine quickly if a bribe is likely to succeed, encouraging strategic miners as they work to accept the bribe.

9.3.1 Block optimization and transaction selection

Another place where deviation from the reference implementation could seep into miners' strategies is in mempool transaction preference. Block optimization and transaction selection is an unsettled problem in Bitcoin mining. The most basic strategy is to skim the largest fee rate transactions and include these in the block. This works fine, most of the time. This may or may not involve rejecting transactions that are in conflict with previously seen transactions. Slightly relaxing this to improve revenue, a miner has the option to use replace-by-fee (RBF) and child-pays-for-parent; these are optional in the reference implementation.

RBF allows miners or nodes to choose to prioritize a transaction that offers larger fees to miners whenever the transaction is in conflict with other transactions sourcing from the same UTXO. RBF can be completely innocent, used to increase priority of a payment. Similarly, child-pays-for-parent transactions are an alternative to RBF. If a transaction is languishing in the mempool due to low fees, any recipient of the transaction can create a child transaction with larger fee rate. Note that as most transactions have multiple outputs, any of the outputs can create the additional transaction. A miner wishing to claim a fee from a transaction must also include any parents that are not yet in the blockchain.

Necessarily, this is something miners must compute and optimize in real time. There could be a number of choices between RBF transactions and child-pays-for-parent transactions, which can be as complicated as the sender would like. Note that both children and parents of a transaction need not be non-unique. A miner may choose to tweak their optimization software using a number of tools to perform the optimization. Deviant miners with a long-term goal of profiting from more deviant mining should create conditions in which small deviations are profitable. The easiest way to do that is to determine where the standard Bitcoin Core software would fail to optimize and deliberately create these situations. Currently, the optimization is somewhat simplistic and has room for improvement [14]. By deliberately creating such transactions, one could nudge miners into using software which could then be used for more deviant situations. This would begin as "wasting money" analogous to the strategy of leaving dollar bills on the ground at a park to encourage people to be on the lookout for dollar bills on the ground.

9.3.2 Overpaying pools

Typically, miners may pay a small premium (say 0.1 %) to a pool operator who distributes 99.9 % of the rewards to the pool back to the miners, whose share is apportioned by partial proof of work. Naturally, this fee will be competitive and may approach zero. However, it need not stop at zero. It is possible for a pool to give miners a small premium to mine with the pool. Rational miners will clearly join the mining pool that charges the lowest fee; this fee could be negative. It is possible that the pool operator is not acting maliciously in doing so, but miners should be suspicious. There is nothing inherently dishonest about it, and they may tell themselves that they will leave the pool the moment it appears that the pool is doing something untoward.

There may be good reasons to overpaying. By keeping a cohesive pool together, if the pool operators and participants are honest, this may help steady the network and ensure proper functioning. If the operator is making good profits or has a significant interest in keeping Bitcoin operating (perhaps because of their own holdings), overpaying may be a small price to pay. It is also possible, depending on how the pool operates, that the operator may have control over certain aspects and be able to prioritize certain transactions. For example, a financial institution that frequently uses the blockchain may organize their own pool in a way so that their transactions are given first priority; this can involve out-of-band payments to the pool operator. While today Bitcoin transactions remain straightforward, Ethereum transactions are more complicated and depend more delicately on how they are ordered, giving advantages to block validators with larger amounts of control; this excess is called *miner-extractable value* (MEV) and is a centralizing incentive in Ethereum (note that "MEV" also stands for *maximal extractable value*, which is essentially the same thing).

Overpaying may establish an interesting signal. Miners who join this pool are signaling that they may be interested in being paid a premium for potentially strategic (or maybe not) activities. At least, they are not running the opposite direction to avoid these activities. Miners in such a pool who are honest would have to monitor the pool's activity. This can be difficult to do 24 hours a day. Miners may be complicit in a 51 % attack or accept a bribe without knowing.

The pool operator may be paid by an outside organization who is interested in censoring certain transactions; this does not need to be made explicit. In fact a resourceful censor may make off-the-book payments to a number of pools for this very purpose. Individual miners may choose to accept the premium and not worry too much about how the pool operators are making ends meet.

9.3.3 Direct signaling and optimizer training

A deviant miner with large enough power or with censorial aspirations may just go ahead and attempt to censor or to mine from behind. This can impose some pain on DC miners. If a miner has 25 % of the hashrate, over 6 % of mine-from-behind attempts will be successful, which could represent a significant drop in revenue for a mining pool, potentially painful when profits margins are small. The natural response would be to alter their behavior in order to minimize these losses.

Simple actions include not mining transactions that may cause a large miner to supplant their block and choosing to mine the fork of a miner who they believe would mine from behind.

Eventually one would expect that rational miners would begin using sophisticated software to make decisions about what transactions to include or which chain to mine. Machine learning is a well-suited tool for this. If short reorgs are not infrequent, machine learning models would seek to determine which transactions or other events prompted the reorgs and craft their blocks wisely so as to minimize the possibility of producing blocks that will go stale. Deviant miners and censors are wise to encourage this.

For example, if a powerful miner repeatedly expresses censorship against a certain type of transaction (perhaps sourced from a common parent UTXO) by supplanting blocks that mine these transactions, machine learning could pick up this commonality linking these transactions. The software could do this in real time and instruct a miner to exclude these transactions, at least in the absence of sufficient fees to make up for the risk. This can all be done under the hood. Similarly, if a large miner is mining from behind to claim larger fees or undercutting, other miners would like to know in real time based on the fees available in the mempool if they are putting themselves at risk by including too many fees. These sorts of problems are by now easy to optimize with machine learning.

9.3.4 Purge pooling

In §13.2.6 and §13.8.2 we discuss a *purge attack* and a *purge pool exploit*. A *purge pool* is an alternative mempool full of transactions that are incompatible with transactions that have already been confirmed. These transactions would typically route back to the sender but include a very large fee. The idea is related to the *wide and shallow double-spend* exploit in which many transactions are double-spent in one reorg [29].

Example 9.9. Alice sends 2 bitcoins to an exchange, and the exchange credits her with the 2 bitcoins. Alice then creates a transaction that double-spends the 2 bitcoins, sending 1.5 bitcoin back to herself, leaving a significant fee of 0.5 bitcoin. Honest nodes then reject the transaction as it has come in far too late: The source UTXO is not unspent, so these would not be broadcast into the mempool. The second transaction costs Alice nothing to produce, and could be stored by a miner who would be interested in mining it if the original transaction is no longer on the longest chain.

Anybody can try a late double-spend; it is essentially a "Hail Mary" attempt that is very unlikely to work. However, it still could: If there were to be a deep reorg, the miners performing the reorg get to choose whatever transactions they would like. Miners would obviously prefer the transaction with a higher fee, if they are aware of it. The honest nodes will have rejected the transaction, but a loose network of devious miners could maintain a pool of such transactions ready for the event that someone is willing to perform a large reorg. Such transactions would be costless to generate for the spenders. They would benefit in the event of a large reorg by potentially getting some portion of their old UTXO funds back. Whether spenders find this attractive or not may depend on their belief about the market fragility hypothesis (see §10.1). If they believe that a deep reorg will not significantly impact the value of Bitcoin, they will be advantaged to recover a fraction of their previously spent Bitcoin during a reorg.

Note that if this practice becomes widespread, this alternative mempool of high-fee double-spends can amount to an extremely large reorg bounty which may become overwhelmingly attractive to a group of miners who would be willing to collude to collect the bounty (see §13.8.2). A miner who would be in position to take advantage of a purge pool exploit has an incentive to create, maintain, and advertise the purge pool, including soliciting double-spends.

9.3.5 Precedent: MEV in Ethereum

While Bitcoin does not see deviant strategies as of today, its closest competitor, Ethereum, does. Ethereum has a very robust fee market and is a well-used smart contract platform which affords complicated decentralized applications such as decentralized token exchanges. Smart contracts generally allow for very sophisticated approaches to reordering transactions, front-running transactions occurring in complicated contracts, and benefiting from short reorgs. MEV [11] is defined by the extra amount of profit that a strategic miner could make by using clever strategies. *Blockchain-extractable value* is defined by the extra amount of profit available to users. Sometimes this can mean arbitraging bribing opportunities in clever ways.

Example 9.10. For some decentralized exchanges, traders place buy orders by interacting with a smart contract via a transaction. This buy order is open until it is either filled by another user or canceled by another transaction initiated by the same user who generated the original buy order. Suppose a market shock happens and the market price crashes. The user sends a "cancel" order to the contract, in the form of the transaction, and this is included in the next block. However, another user, who owns the crashing token, may then send to the contract a sell order intended to fulfill the buy order which was just canceled. If this seller includes a very large fee, miners may be incentivized to ignore the block that was just mined which canceled the order. Instead the miners accept the bribe and process the sell order. Following this sell order, the seller who just unloaded a crashing asset would then be happy to provide another small bribe to incentivize miners to continue on the preferred chaintip.

Suffice to say, the attack surface is large and complex, especially with complicated smart contracts whose components may depend heavily on the order of transactions. While smart contracts are technically possible within the Bitcoin protocol, they are not involved in a significant portion of the transactions at this time. Ethereum was originally designed with smart contracts in mind, while Bitcoin was not. If smart contracts do begin to comprise a large share of transactions, the opportunity for MEV and BEV will expand with it, as will the likelihood of fully strategic miners.

Recently, researchers have detailed an Ethereum exploit called the riskless uncle maker (RUM) and presented evidence that the exploit has been taken advantage of by Ethereum miners [55]. The exploit, which takes advantage of Ethereum's proof of work, is similar to selfish mining but carries less risk to the miners. (Note: Ethereum successfully migrated from proof of work to proof of stake on September 15, 2022.)

10 What discourages strategic mining?

Strategic mining does not play an important role today. What keeps strategic mining from developing?

10.1 Market fragility hypothesis

As illustrated in Chapter 9, increased strategic mining may result in unpredictable dynamics, especially when the block subsidy is low. Bribes may become lucrative and effective.

In the early days of Bitcoin, it was commonly believed that 51 % attacks, or even the potential of 51 % attacks, would destroy the confidence in the network and doom the Bitcoin experiment to a failure. In 2014, the ghash.io pool temporarily gained over 51 % of the hashrate, which is said to have shaken the confidence in Bitcoin in ways that were felt by the market. The pool immediately downsized to avoid such an appearance (in context, this was in the midst of the unraveling of Mt. Gox following the end of the 2013 bull run, so it is not entirely clear the market was responding to mining pool activity).

One could conjecture that deviant behaviors such as bribery, censorship, 51 % attacks, selfish mining, deliberately mining from behind, or any other exploits would result in a shaken confidence in the protocol, which would have a negative effect on the asset price. Because most premiums paid for devious actions are denominated in Bitcoin and because many of the miners have capital investment with future returns denominated in Bitcoin, it is often assumed that miners will not take actions that will harm the confidence in Bitcoin. We formalize this.

Definition 10.1. The market fragility hypothesis is the belief that Bitcoin's price would suffer significantly when egregious instances of strategic mining (such as blatant 51 % attacks) emerge.

This will become important in later chapters: If taken literally it presents a clear attack vector. However, it may serve to create institutional norms that discourage deviant mining in the first place.

Miners who take the market fragility hypothesis seriously will tend to avoid situations that jeopardize future returns according to this theory [25]. Attacking the blockchain would damage their ability to profit from capital expenditures in mining equipment.

10.2 Repeated games and institutional norms

In absence of grooming defined in §9.3, default compliant mining serves as a good Schelling point. If the opportunities for strategic mining are few and far between, and

https://doi.org/10.1515/9783110772845-010

not especially lucrative, miners may have little incentive to risk long-term damage to the system while pursuing short-term rewards.

As miner profit margins decrease, it is feasible that some of these miners may be open to deviant opportunities. This may present a tragedy of the commons situation. If bribes are offered, the miners who take advantage of the bribes in the short term may benefit themselves a small amount that is nonetheless larger than their share of the damage they are causing to the entire system.

One way to counter this is as follows. Miners can agree, explicitly or as a Schelling point, to punish behaviors that would harm the system. This is itself a form of strategic mining. For example, if a malicious reorg is detected, miners can temporarily suspend Nakamoto consensus and choose to disregard the longest chain until it no longer becomes the longest chain. An attacker who bribes miners and temporarily marshals enough hashrate (perhaps with luck involved) to perform a 51 % attack would then find their resources wasted. The option to punish malicious actors by destroying their funds is often available in proof-of-stake protocols, and is called *slashing*. The Bitcoin protocol does not allow any methods for slashing.

While slashing is not an explicit option in Bitcoin, *tainting* could function as an effective deterrent. If someone obtains bitcoins via a double-spend, a bribe, or something else underhanded, this is something that can be observed and recorded. Such behavior is an indication that the holder of these UTXOs is not to be trusted. An exchange or financial institution can simply decide to not accept these UTXOs as Bitcoin. Nobody is forced to treat all Bitcoin the same. The decision of which UTXOs are acceptable is completely up to the individual institution. In an economic regime where most bitcoins trade hands between above-board individuals who do not transact in tainted bitcoin, tainted bitcoins could become essentially worthless. Thus any miners who break norms in order to give themselves an advantage may find themselves with block rewards that they cannot sell at market rates. There may be ideological opposition to tainting, but again, the decision of which UTXOs to accept is not something that can be forced on any user.

We distinguish tainting from other forms of punishment because it happens outside of Nakamoto consensus. Tainting may be considered a dangerous precedent, as it is a form of policing that is available to non-miners. If tainting proves an effective way to punish bad actors, outside entities such as regulators could attempt to orchestrate their own campaigns to taint what they consider malicious transactions. If given a choice, miners would prefer to self-police. The rest of this chapter will deal within mechanisms possible with Nakamoto consensus.

As blocks are found around 144 times per day, mining is a *repeated game*. Repeated games with many players have dynamics that are not completely captured in a round-by-round payoff table. Strategies that are not Nash equilibria in the short term may emerge as stable in the long run. Models for a repeated game are designed to examine what motivates long-term interactions. Such models consider the idea that a player will take into account long-term benefits of cooperation, which will typically be of greater value than defecting in the short term.

Repeated games are somewhat difficult to model mathematically: For finite but long horizon games, the results predicted often will not agree with what humans do in practice. For infinite horizon games, there are often infinitely many Nash equilibria for a given game, and describing these solution concepts may have little predictive power.

The prisoners' dilemma is the canonical example, and this has been studied both empirically and through simulations. Despite the fact that in the short term, both parties defecting is the dominant strategy and the unique Nash equilibrium, if the game is repeated players will tend to cooperate most of the time or use a *tit-for-tat* strategy. These can be studied using evolutionary game theory, which observes which sets of strategies tend to dominate if the game is modeled as many players playing the game repeatedly.

A tit-for-tat strategy is a strategy that rewards other players for cooperative behavior with cooperative behavior and punishes other players for defecting by defecting in the near future.

Finite horizon games are games in which the end point may be known. Game theory allows one to use backwards induction; however, sometimes the predictions following from a backwards induction argument can be counterintuitive or disagree with experimental results. For example, if a prisoners' dilemma game is to be repeated three times, the backwards induction argument is straightforward: In the last round, there is no benefit for cooperating; hence both players will defect. This is predictable. But because this is predictable, we also know that in round 2, players have nothing to gain by extending goodwill with cooperation – the other player will defect the next round regardless of the results of round 2. Hence both players might as well defect in round 2. But since round 2 is determined, players should also defect in round 1.

There is nothing special about a three-round game. For a 12,000-round game, the unique Nash equilibrium is the same, but in this case it is a bit counterintuitive: In a 12,000-round repeated prisoners' dilemma game, one would expect players to spend some time cooperating. Only as the last round comes into sight will one player attempt to strike first and doom the rest of the game to defecting.

For infinite horizon games, players have much to gain by cooperating and will often do so. Players can adopt more complex strategies, such as tit-for-tat strategy, in order to punish players that defect without causing a permanent state of defection. For example, one player may punish the other player who defected by defecting twice in a row, before returning to cooperate. Simulations and studies of evolutionary game theory suggest that in games like the repeated prisoners' dilemma, versions of the tit-for-tat strategy may be the most adaptive in the long run. Strategies that involve punishment do tend to enforce "social norms" which tend to optimize socialized benefits.

Does this insight apply to Bitcoin mining? Most of the time Bitcoin participants can mutually benefit from using a default compliant strategy. Bitcoin miners have much to gain well into the foreseeable future by mining honestly, and no strong incentives pull them towards deviant mining, so it is reasonable to suspect honest mining should persist for a while. However, when strategic mining arises, do we expect it to be punished?

10.2.1 Mining that can be punished

Collectively, honest miners can temporarily ignore the heaviest-chain rule to punish what are believed to be clear examples of dishonest behavior. For example, suppose a selfish miner posts two blocks consecutively seconds after a block is mined. If the other miners believe this falls into a pattern of selfish mining, they may simply ignore these blocks and continue to build on the other block.

If a miner accepts a large bribe in order to participate in a reorg, a majority of miners may choose to ignore that reorg. Eventually, the chain preferred by the majority will win out. The briber may avoid spending the funds in the bribe; however, the miners who accepted the bribe will have wasted their energy, discouraging further mining of bribes.

Fee-undercutting can be punished simply by being ignored. Those who encourage fee-undercutting by taking the bait can also be punished: Punitive miners can ignore the blocks mined on top of fee-undercutting blocks.

10.2.2 Problems with punishment

Punishment requires breaking from Nakamoto consensus protocols in a coordinated way. This will only be effective if it is unambiguously clear what should be done. If some honest miners are not involved in the punishment, they will be fighting against the punitive mining. Some miners may feel that this is a violation of the protocol and not participate in the punishment, some may not be interested in upgrading software or manually managing their miners to administer the punishment, and some may simply not be paying attention and continue mining in default mode.

Without centralized coordination, it may be difficult to determine who the selfish miner is. If two blocks are published within seconds, one cannot be certain which one is the "honest" one. In fact, this relativism is a fundamental problem that Nakamoto consensus considers in its design: The block that is built on the next will be declared the correct block. If a selfish miner simply waits, say, 20 seconds before publishing the next block, this may be enough to cast enough doubt on whether they are selfishly mining.

Punishing a UTXO implicated in a bribe by refusing to mine further transactions is stronger than tainting and essentially amounts to a miner-enforced blacklist – something generally frowned upon for ideological as well as logistical reasons: Who would maintain "the" blacklist? Vigilante justice is in general a bad idea; miners who are too eager to punish deviant miners can create their own problems within the ecosystem. This can be exacerbated by attackers who intentionally troll the punishers in delicate ways. An attacker can publish two blocks simultaneously; overzealous attackers could pit themselves against honest miners or one another in an extended hash war that damages network security and functionality. These division lines can be easy to identify and exploit, for fun and for profit.

For this reason, while tit-for-tat strategies that involve punishment seem to appear in evolutionary game theory situations, it is far from obvious that such strategies will emerge in mining games.

It does seem likely that in certain egregious cases, nodes (including mining nodes) may choose to simply invalidate certain blocks or reorgs that seem particularly untoward. This can be thought of more as a "defensive" strategy and less of a punitive strategy. This will be discussed more in §13.1.

Note that for double-spends, the party who was the victim can launch a counter-attack [36]. This is based less on community ideals about punishing malign actors, and more on a pure economic model: An exchange who is the victim of a double-spend may then engage a costly war of attrition against the spender. Since the war of attrition is costly and results in at least one loser, this alone may serve to discourage others from attempting a double-spend against this exchange.

Exercise 10.1. Using "the market" and "the miners" as two cohesive agents in a two-player game, use a game tree and backwards induction to explain the market fragility hypothesis.

11 Declining block subsidy

Bitcoin's block subsidy adds bitcoins to the total supply in an exponentially decreasing rate. The last halving took place in May 2020. Before May 2020, miners received a block subsidy of 12.5 bitcoins per block mined. In 2022, miners receive a subsidy of 6.25 bitcoins per block mined. Sometime in early 2024 this will be cut in half again, to 3.125.

Some heuristic math to keep track of this follows: The original block subsidy was 50 bitcoins. Once this period was over, 50 % of the total 21 million supply had been mined, and there was 50 % left. In the next time period the block subsidy was 25 bitcoins; this accounted for 25 % of the 21 million, and when it was over, there was 25 % left. From July 2016, when the subsidy dropped to 12.5 bitcoins per block, to May 2020, 12.5 % of the final supply of bitcoin was mined. In early 2024, only 6.25 % of the total supply will be left, and half of this (3.125 %) will be mined in a roughly four-year period from 2024 to 2028. This means the total annual increase in Bitcoin supply over this period will be about 0.78125 % per year, or around 164,000 bitcoins per year. This process continues in an obvious way. If we consider the current halving period (2022) to be the fourth halving cycle, then in the *n*th halving cycle we have:

Subsidy (in bitcoins)	$100 \times \frac{1}{2^n}$
Percentage of total mined in the cycle	$100 \times \frac{1}{2^n}$ %
Percentage of total left after the cycle	$100 \times \frac{1}{2^n}$ %
Rough yearly inflation in terms of total supply	$100 \times \frac{1}{2^{n+2}}$ %
Rough yearly inflation (in bitcoins)	$21 \times 10^6 \times \frac{1}{2^{n+2}} = 5.25 \times 10^6 \times \frac{1}{2^n}$

For the upcoming decades this gives us:

Years (approximate)	Subsidy	Yearly subsidy	% left at the start of the cycle
2024–2028	3.125	164,060	6.25
2028–2032	1.5625	82,031	3.125
2032–2036	0.78125	41,016	1.5625
2036–2040	0.39063	20,508	0.78125
2040–2044	0.19532	10,254	0.39063
2044–2048	0.09766	5,127	0.19532

So even though the final block subsidy will not be issued until around 2140 (or perhaps a bit earlier, due to some blue shift), the block subsidy declines quite rapidly: A week of the block subsidy in 2022 will amount to more bitcoins than an entire year's block subsidy in 2044.

https://doi.org/10.1515/9783110772845-011

This poses a very significant problem for security if the fee rates do not increase dramatically. *This is not a concern that will show up next century*; this will become a concern within two decades.

11.1 Security budget

The *block reward* is the sum of the block subsidy and the fees collected. This is not precisely predictable, due to variance in the fee market. This is usually denoted in units of bitcoins per block. The *security budget* is also some value per unit time, indicating how much miners are getting paid to mine blocks (while it is not technically a "budget," this is the commonly accepted term). The security budget can be given in dollars per year, in bitcoins per year, or even as a percentage of the total market cap. For example, in 2022, this rate is roughly

$$1.75\,\%\text{ market caps/year.}$$

When expressed in terms of market caps per unit time, this becomes the *security factor*.

Note that the transaction rate given by transactions over time allows us to compute the average security per transaction. Some estimates of the capacity are around 200,000,000 possible transactions per year. In 2021 there were nearly 100,000,000 actual transactions. We can compute the average security per transaction:

$$\frac{0.0175\text{ market caps/year}}{100{,}000{,}000\text{ transactions/year}} = 0.0175 \times \frac{21}{100}\text{ bitcoin.}$$

So the average security cost per transaction is roughly 0.003675 bitcoin, which is 367,500 sats. This means that roughly speaking, in September 2022, the security budget per transaction was around \$75 per transaction. Necessarily, in order to maintain that same security budget in the absence of the subsidy, one would expect to pay \$75 in fees.

Transactions themselves are definitely not a perfect metric as discussed in §1.2.9: "Transactions" is by no means equivalent to "payments" [4]. A single transaction can have hundreds of outputs [4]. This happens often when an exchange is processing withdrawals; the single transaction may represent many payments. Also, each transaction may contain scripting for layer 2 protocols; for example, the transaction may form a channel for the Lightning Network. In this case it is impossible for an external observer to determine how many payments have been facilitated. Nonetheless, in order to keep things simple, we count transactions as UTXOs, which is a metric that does not require conjecture.

The blockspace available in each block is a fixed constant. This constant, 1 MB, was put in place by Satoshi in 2010 and has not been changed. This constant appears very much to be fixed for the foreseeable future after a long and contentious debate over this parameter [5].

The available blockspace is going to be around 144 MB/day, or 52.5 GB/year. In practice, determination of how much space is being used for a given type of transaction becomes a bit technical for this discussion. A technology called segregated witness ("SegWit") was adopted in 2017 and allows for roughly twice the transaction data. The term *vByte* (for "virtual byte") is used by those studying the market for blockspace. A vByte is a weight unit that allows one to compare the fee rates of different transactions in the mempool. Each block has a capacity of 4 million vBytes per block. It is common to calculate transaction fees in terms of sats/vByte. In 2022, the market price for space on the next block ranges around 2–5 sats/vByte. By including the block subsidy, one can also compute the security per vByte, which in 2022 is about 160 sats/vByte. Clearly if the same security per vByte is going to be maintained into the future even while the block subsidy declines, the fee market will have to compensate by increasing by a factor of 20–50. Note that measuring security in terms of sats is orthogonal to Bitcoin's market price in terms of other assets such as the US dollar.

In 2021, about 21,500 bitcoins were collected as fees by miners, representing slightly over 0.1 % of the total supply and slightly more than 6 % of the total security budget in the same year.

The security budget measures the raw amount needed to attack, but the security factor is a better gauge of the size of the incentives to attack: Certainly a $50 trillion bitcoin would be more likely to be attacked than a $50 billion bitcoin if the cost to attack was the same.

It is fairly simple to turn estimates of future security budgets into estimates of transaction costs in the absence of the subsidy, if we make assumptions about the number of transactions per block. Approximately 52,600 blocks are produced per year. For example, if the annual security is $100 billion per year, the security budget must be roughly $1.9 million per block, or a little under $1,000 for a typical transaction if we assume around 2,000 transactions per block. This can also be expressed in cost per vByte.

That problems will arise if a fee market does not develop is clear. In efficient economic conditions, the cost to deploy the hashrate needed to control the market should be determined by the security budget: If the security budget is $100,000 per day, in theory, an attacker could overwhelm the security with a 51 % attack for slightly more than $100,000 per day, which is clearly unacceptable for a payment network that aspires to be a mainstay reliable global currency. A low-fee situation would likely mean that many ASICs which are unprofitable to run would be available for low prices; these could be deployed in such an attack. Thus it is crucial for the security of Bitcoin that fees increase in the coming decades.

Note that the demand for blockspace and the demand for the underlying asset are different markets. They are largely but not totally independent. Increased demand for Bitcoin as a store of value does not necessarily translate to increased demand for blockspace in the next 10 minutes. Conversely, if Bitcoin is frequently used as a cross-border medium of exchange for counterparties that immediately convert back to their local currency, there may be demand for blockspace but only fleeting demand for the asset itself.

There are many factors that would influence this. An efficient layer 2 system such as Lightning Network may increase demand for blockspace, although this remains to be seen. This would make sense in light of the Jevons paradox: Even though the Lightning Network would be a low-fee competitor for on-chain transactions, an efficient layer 2 may increase the demand by making the blockspace itself much more useful. It is worth pointing out that the blockspace is indeed limited: Even if we assume that the total capacity of 200,000,000 transactions per year is spent entirely for creating lightning channels, it would take over 40 years to create a channel for everyone on the planet who is alive today. In practice, widespread use of Bitcoin as a payment method and unit of account would mean broad use of layer 2 and custodial methods.

Even though transactors may be willing to pay higher costs, they typically will not do this voluntarily; they will submit transactions that best optimize their objectives. The security budget of Bitcoin is a not a concern that individual transactors take into consideration. In a low-subsidy regime, holders who never transact are essentially security free-riders. Because their supply is no longer being diluted, they are no longer paying for security; this is entirely left to transactors.

11.2 The mining gap

Transactions are created and communicated to the mempool. Each transaction has a sats/vByte fee associated to it. A simple algorithm miners use to create blocks is to simply order all transactions according to the sats/vByte fee and then include as many transactions as they can in a single 1-MB block. Each block created in this way will have fees exceeding a minimal fee rate, which could vary from block to block. Presumably all transactions available in the mempool with fee rate above this value will be included in the block, while those below the minimal fee rate will remain in the mempool after the block is mined.

When the block is mined, the transactions with the largest fee rates are skimmed off of the top of the mempool, and lower-fee rate transactions remain. This drop is significant if demand is low, but less significant if demand is tight. When analyzing the market for fees, the shape of the demand curve is extremely important. Transactors that are less eager can submit transactions that can be below the median minimal fee rate. By the random chance naturally occurring in block creation, it is possible that blocks with lower minimal fee thresholds will be found in a short period of time, in which the lower-fee rate transactions will still make it. A transactor who wants to have their transaction included in the next block should use a fee rate somewhat above the median minimal fee rate. In the case that more than 10 minutes elapse, the fee rates market can become more competitive.

In practice, one can look at recent blocks (say the last 1,000) and easily compute the distribution of minimal fee rates. A transaction exceeding the maximum minimal fee rate is nearly guaranteed to be included, while a transaction below the minimum minimal fee rate is extremely unlikely to be included in the near future.

While the subsidy exists, the reward to miners does not change drastically the instant that blocks are mined. In 2022, if a block containing 0.15 bitcoin in fees is mined, the total reward is 6.4 bitcoins. The instant this is mined, 0.15 bitcoin worth of fees are removed from the mempool, and some other value will replace them. Even if this fee reward falls by a factor of 2/3, the miner still has an expected block reward of at least 6.3 bitcoins. It is very unlikely that the expected cost of producing a block is between 6.3 and 6.4 bitcoins, so the miner's behavior should not change: They continue to mine whatever is in the mempool.

When the subsidy goes away this calculus changes. The bulk of the miners' reward is determined by the fees. As time changes, the value of the fees in the top 1 MB of transactions increases. The rewards for mining then can become very much time-dependent: If there is a significant drop of fees available, a miner may determine that mining bitcoin has a negative expected reward. In theory such a miner could then opt to power off some or all machines during the first few minutes after each block. Considering not only electricity costs but also cooling requirements necessary to run ASICs, this may be a workable strategy for many miners.

The result of this, as conjectured in the paper [10], is a *mining gap*. There will be a time between each block in which few miners are actually mining. As more transactions are posted, mining will intensify as the expected block reward increases. Note that transactors using RBF can always submit their transaction with low fees and walk these up as competition for the next block increases (RBF does not work backwards; if a high transaction fee was submitted but not accepted by the miner, a miner in a later block should use the largest-fee transaction available).

The size of the mining gap is important. It is also very difficult to precisely model. The paper [10] appears to assume an essentially unlimited blocksize. If blocksize is larger than the mempool, most transactions with a positive fee will be included and the mempool will be cleared out with each block. The result will be a sawtooth-shaped reward function for miners. This would potentially create a significant mining gap. In reality, with transactions showing up in the mempool with different urgency and hence different fees, a robust but variable mempool is likely to develop. This is something which can be simulated given some idea of how fees show up on the mempool. It can also be observed in real time.

In 2022, there is no mining gap as the mining gap refers to an amount of time where it is not profitable for miners to mine. However, one can compute variance in fees. There is a gap in the available fees. One can observe this by noting that fees included in a block often fall below 0.01 bitcoin and often range above 0.2 bitcoin. If the block reward were made up mostly of fees instead of subsidy, this would likely create a mining gap, as block rewards would frequently jump by a factor of 3 or more as time goes on.

Exercise 11.1. Suppose that the blocksize is unlimited and fees available in the mempool grow at a constant rate of λ block subsidies per 10 minutes. Find the variance of the block reward.

11.3 Problems with a mining gap

The mining gap presents a serious security issue. For example, if the mining gap induces miners to not mine 50 % of the time, an attacker with 34 % of the hashrate could execute a 51 % attack by mining full time. Miners who wish to resist this sort of an attack will have to mine blocks with lower fee rewards, likely at negative profit margins. Some miners may choose not to make this sacrifice, creating potential for a free-rider problem or volunteer's dilemma. Effectively, the *hash deployment ratio* (see §12.1) may be a time-dependent variable with considerable variation within 20-minute intervals.

A miner executing a 51 % attack could do this over a long period of time, especially if operating stealthily. Consider the example of a miner with 34 % of the hashrate in a regime where economics dictate that most miners reduce their hashrate by 50 %. Suppose that difficulty has been calibrated so that when all miners are mining 50 % of the time, a block will be produced on average every 10 minutes. Suppose that 34 % of miners mine selfishly, but are mining at full deployment. The 34 % mining full time will be producing slightly more than the other 66 %; they can foist their long reorgs onto the chain frequently, effectively supplanting all competing blocks. In this case they are receiving 100 % of the miner rewards over that period, instead of 34 %. However, they doubled their electricity costs. Nearly tripling rewards while doubling electricity is still a fantastic trade-off

How selfish mining behaves for those without capacity to do a full takeover is difficult to model. On the one hand, a selfish miner who chooses to mine when others are not can occasionally develop a long lead; however, the fees they would be claiming would not be of the same proportions as those for "classical" selfish mining. The selfish mining computation leading to (8.5) describes the proportion of blocks the selfish miners keep after the difficulty adjustment; however, the fees harvested by a selfish miner who is mining the gap will not be in direct proportion to the number of blocks mined. On the other hand, the lack of fees included in previously mined blocks has an important side effect: If other miners are petty compliant, they will almost always prefer the block that was mined with low fees over the block that was found later. This means that a selfish miner incurs very little risk in not broadcasting their block. Since selfishly mined blocks will tend to have lower fees by virtue of the fact they were mined earlier, they benefit from petty compliant miners.

Short of selfish mining, miners with cheaper operating costs may simply choose to run continuously. While they may lose some money, other miners would lose considerably more and may be forced out of business. A selfish miner may use this as a tactic to increase market share by hurting their competition. This type of *pushout attack* would allow a miner with low operating costs to increase their market share more quickly.

In the high-block subsidy regime, selfish miners must wait until the difficulty adjusts to begin to see their reward: On average, 2,016 block rewards are distributed every two weeks – when the selfish mining starts this will slow down, but it will readjust after the difficulty adjustment, at which point the selfish miner enjoys a larger share of these 2,016 block rewards over the next two weeks. In the low-block subsidy regime, however, the selfish miner does not need to wait two weeks. If the majority of the rewards paid out are fees, the higher fees are likely to be paid out anyways, even when transactions slow down. So selfish miners in a low- or zero-subsidy regime are less reliant on the difficulty adjustment.

11.4 Undercutting

As we see from (9.1), as long as the block reward is large compared to the fee revenue, considerations involving fee-undercutting are unlikely to arise. However, as the block reward decreases, it becomes much more difficult to model. Simulations are often used to develop theories about what will happen in this case (perhaps unsurprisingly, different simulations result in different conclusions; see [10, 18, 19]).

Available fees become a major consideration in the absence of a block subsidy. The amount of available fees is a time-dependent random variable which may come with quite a bit of variance. A very significant factor in how mining behaves in this regime is the shape of the demand curve for transactions.

Miners typically will rank transactions by sats/vByte. The goal for miners is to cram the most fees into a block. If the demand is somewhat consistent and high, this is the most favorable for stability of the network. For example, if low-priority transactions were priced at 150 sats/vByte and high-priority transactions at 165 sats/vByte, it may make sense that the blocks that are mined after a longer gap may contain more high-priority fees, while the ones mined shortly after the previous block may contain more low-priority fees. Because of the Poisson nature of blocks appearing, low-priority transactions should eventually be mined, thus it is reasonable that some transactions would make smaller bids.

In this case, the difference between the fees harvested in the previous block and the fees available for the next block might not be that large; perhaps there is a 10 % difference. In this situation we can examine the condition

$$p_{DC}\chi < (1 - p_{DC})f_{tip} - f_{mem} \tag{11.1}$$

and see that unless nearly all miners are using a PC strategy, fee-undercutting (for a typical, non-selfish miner) may not be a profitable strategy.

On the other hand, it is possible that blockspace is simply not in high demand. In this case the miner of a block may claim a large portion of the fees available. The ratio of f_{tip} to f_{mem} would then be somewhat large. In the extreme case, when demand for blockspace is small (or equivalently, blocks have unlimited size), miners would claim all available fees. If the block subsidy is non-existent, fee-undercutting would then be a profitable strategy. This was modeled in [10]. In this case, strategic miners have to balance between collecting fees and not being so greedy as to entice other strategic miners to steal what they just stole. The model [10] predicts this will lead to a considerable backlog. However, this is (at least from our perspective in 2022) an extreme situation where the mempool is cleared each block and the block reward is literally zero.

Studying a less extreme model [19] suggests that with some amount of default compliant miners and a positive block reward, strategic miners can choose the next f_{tip} to be large enough so that the undercutting condition (11.1) is not satisfied. Define

$$f_{\text{mem}\,+} := \text{max fees in the mempool that fit in a block.}$$

We define

$$f_c := \frac{p_{\text{DC}}\chi + f_{\text{mem}\,+}}{(1 - p_{\text{DC}})}. \tag{11.2}$$

By including up to f_c fees in a block, the block should remain unmolested by undercutters. The backlog only occurs if

$$f_c < f_{\text{mem}}.$$

This may or may not backlog transactions, as miners can include lower-priority transactions and leave enough high-fee transactions available. However, this behavior would present a slight wrinkle in the fee market: There may be conditions in which a transaction with a lower fee rate will be included before one with a higher fee rate: Publishing a transaction with a large fee could actually result in the transaction being avoided by miners who do not want to be undercut. This slight wrinkle in incentive-compatibility is probably not a major problem. If miners did believe that including lower-fee rate transactions was hurting the mining revenue, they could simply go back to ordering transactions by fee rate and including transactions until the critical value f_c is reached. This would slightly decrease the throughput, without giving transactors an incentive to avoid high fees.

Building a model for the fee market is extremely difficult and depends on many parameters; we will not attempt this here. However, it does seem reasonable that miners would use the strategy of including transactions ordered by fee rate so that (11.2) is satisfied. This shields the block from undercutting while also not encouraging transactors to lowball fee rates.

11.4.1 Private transaction brokering

So far in this discussion, we have assumed that the market action determining the fee rates and which transactions appear in the next block is visibly happening in the mempool. However, there is nothing prohibiting the emergence of private or external transaction brokering or even blockspace futures markets. An entity wishing to pay to have their transaction mined can pay miners in a private transaction (perhaps in other currencies) to include the transaction. This will make the fee unsnipable, which transactors may prefer. Large miners could sell blockspace in advance: This would allow the transactors to hedge against transaction fee spikes while the miner would hedge against a drop in demand.

While there are no provisions in the Bitcoin protocol that afford this, nothing that happens outside of the protocol is a violation of protocol. We suggest that there is incentive for such a market to form.

In some conjectured growth scenarios in which fees are sufficient to maintain the security budget, many Bitcoin transactions are between large institutions, settling fairly large amounts of money. If layer 2 protocols are in use for smaller, more immediate transactions, the layer 1 transactions may not need to settle with the same urgency.

Miners who solicit transactions directly do not incur the risk of fee-undercutting, so they can offer slightly below average mempool rates: This slight premium would be worth the insurance and would attract transactors who do not want to roll the dice competing with the mempool, which will exhibit variation due to naturally occurring variations in transaction demand and the randomness involved in finding a block.

Miners would further be incentivized to enter into these relationships if there is a significant mining gap. Rather than resting for a few minutes until the mempool offers sufficient fees to power up machines, the miner could wait until they have enough privately brokered fees and mine continuously until they find a block. If a miner has 6 % of the hashrate, they will in normal circumstances produce a block roughly every three hours. But in the presence of a mining gap, they would expect to find a block in a shorter time period; they would begin running continuously from the moment they have gathered enough transactions.

Miners who are interested in taking up a large share of the market could also use private arrangements to drive transaction volume from other miners. By offering to process fees at a fee rate somewhat lower than the going rate for transactions clearing the mempool, they will attract customers. If this happens, the blockspace market may behave somewhat like the Bertrand model described in §7.2. The miner with the lowest costs will be incentivized to offer to process transactions at slightly higher costs. Other miners who cannot offer to mine blocks at this fee rate without taking a loss may be forced out. As with selfish mining, this sort of undercutting could be a tactic used by miners with cost advantages as part of a pushout attack.

Exercise 11.2. Describe how a market situation in which transactions are privately brokered could lead to an example of the Edgeworth paradox (§7.2.2).

11.5 Summary

In a low block-subsidy regime, even if fees stay well away from zero, the picture is somewhat unpredictable and difficult to model. Researchers use simulations to make predictions (cf. [10, 18, 19]); however, every model naturally makes a number of assumptions and choices for many parameters. Selfish mining becomes somewhat complicated to analyze. The dynamics of profit margins and costs for each separate mining firm also become important, as larger miners with cost advantages can push out smaller miners.

The best-case scenario would be that the fee market is strong, consistent, and not too heavily sloped. This would lead to a smaller mining gap and fewer opportunities for undercutting. This in turn may slow or limit the spread of strategic mining.

Historically, fee rates have fallen during periods when speculative interest is waning. If this persists, bear markets will offer weak security periods for attackers. To remain secure, Bitcoin must develop a robust and consistent demand for not only the asset Bitcoin itself, but also for the blockspace allotted by miners.

12 The Flattening ASIC cost curve

In recent history, the production of ASICs has lagged behind the price of Bitcoin. Producers who have secured space in production foundries have been able to sell their equipment for a healthy profit. One would expect market forces to begin to compress profit margins of ASIC producers and users.

12.1 Hash deployment ratio

If Bitcoin becomes a more established and dependable source of revenue for miners, mining hardware becomes a reliably profitable enterprise. Economic forces will bring more producers into ASIC production while the price continues to rise geometrically. Eventually the price growth must slow, and hardware supply will catch up. As more hardware enters the market, the value of the current hardware decreases as margins compress.

At this point, electricity costs will dominate miner concerns. Mining equipment will derive positive value in the short term if it can be used to mine profitably, and in the longer term as a form of an out-of-the-money call option or as back-up security. At any given point in time, some amount of mining technology exists which is otherwise operational but has obsolesced due to competition. Taking the ratio of the active hashrate to the total possible hashrate, we get the *hash deployment ratio* (HDR):

$$\text{HDR} := \frac{\text{current hashrate}}{\text{hashrate if all ASICs were running}}.$$

This represents an important security variable. The complementary ratio, that is, the percentage of hashrate not being used, will be called the *dormant hashrate*. If the HDR is near 100 %, then any attempt to divert active hashrate towards an attack would be visible in its absence. A stealth 51 % attack would necessarily be accompanied by a 51 % drop in visible hashrate. However, if this ratio is less than 50 %, it would be theoretically possible to acquire and deploy sufficient hashrate to accomplish a 51 % attack without causing a change in the observed hashrate.

One measure of the cost to perform a 51 % attack is simply

$$\text{Cost to attack} := \begin{array}{c}\text{cost of purchasing 51\% of the total hashrate}\\ +\\ \text{cost of operating hardware for the duration of the attack.}\end{array}$$

If the equipment is unlikely to turn a profit based on electricity prices and the Bitcoin price, the market price for the machine itself may be quite low. An old ASIC may be quite inefficient, say if it returns 1 bitcoin for every 3 bitcoins of electricity put in (this depends, naturally, on the price of electricity available to the particular miner). But this is an expense over time. It would only take around an hour to execute a six-block reorg

https://doi.org/10.1515/9783110772845-012

with 50 % of the hashrate at 50 % HDR. In a single day, one would expect to be able to perform a series of very damaging reorgs that cut much deeper. If the goal is to sabotage the network, such a goal may be accomplished in a short period of time. If the security budget for Bitcoin is $20 billion/year, the security budget for four hours is around $9 million. If the attack is coordinated among confederates around the world, one can simply use whatever energy is available. This may be costly, but only for the small amount of time that it is deployed.

If an oversupply of ASICs is significant and persistent, it may make sense for an entrepreneur to attempt to acquire as many as possible. If the resulting reserve of hash force comprises a game-changing amount, this could be rented to attackers or exploiters, or even deployed against attackers for a fee. This would put the owner in an enviable king-making position. Suffice to say, an oversupply of ASICs leading to a low HDR represents a vulnerability.

12.2 Purge pools

Recall the purge pool discussed in §9.3.4. Suppose such a pool is created and populated with would-be double-spends. As more blocks are mined on the main chain, more transactions will be considered "confirmed" and then the original UTXO sources from these transactions are free to be nominated for the purge pool. The freshest alternative transactions may be only seven to 10 blocks old, and some may be much older. In a healthy mining scenario with HDR near 100 %, deeper reorgs will be considered off the table. It would take some serious effort to perform a 20-block reorg, as this would require convincing a significant portion of the active hashrate to join the attack. Directly bribing miners for a very deep reorg could be quite expensive. For a purge pool transaction to be viable as part of a purge exploit, it should be in a certain range of depth, no more than a few blocks deep. When HDR is high, the window for transactions to be vulnerable to reorg and also deep enough to be confirmed is quite small.

This changes drastically if the HDR drops below 50 %. The aggregate fees offered in a purge pool may be quite significant: The senders of the double-spend UTXO are getting "free money" if the purge event happens, so they may be willing to offer a large fraction of the transaction value as the fee; for example, if they use "split the proceeds" as a heuristic, whoever accomplishes the purge exploit would be taking home half of the spoils. It is very feasible that there could be hundreds of bitcoins offered as a purge reward for transactions that are 10–15 blocks deep. In this case, the reward for mining a 20-block reorg could be orders of magnitude larger than the reward for honest mining. This may move many ASICs from the "non-profitable" category to the "incredibly profitable" category. If someone is able to marshal enough dormant hashrate to generate a larger hashrate than the active hashrate for an extended period of time, they are in a position to be handsomely rewarded for writing a reorg.

In the low-HDR regime, secrecy is more important. Honest miners could have their own supply of dormant machines. If they see an attack coming, there is motivation to add to their chain to protect their own rewards. So a purge writer may want to keep the reorg secret. In fact they may want to create a chain that leads by quite a bit before foisting it onto the network, as this may discourage the honest miners from tapping into reserves in order to sprint out against the attacker.

Also crucial in a low-HDR regime is the following. The rewards in the purge pool can grow quite large as deeper reorgs become viable. Instead of focusing on reorgs only 7–15 blocks deep, a miner who can marshall enough dormant hashrate may consider much deeper transactions, say 100 blocks deep. This massively extends the rewards and the incentive to perform the attack.

Exercise 12.1. Using a block explorer, determine how much Bitcoin was transferred in the last five blocks. Assume that 0.1 % of this is nominated to the purge pool and that the purge fees are half the value of these transactions. Determine this reward and compare it to the total block rewards over the last 15 blocks.

12.2.1 Protection against a purge attack when HDR is low

Typically, those on the receiving end of Bitcoin transaction may wait for a confirmation time of six to 10 blocks before considering the transaction final. These numbers are meaningless when more than 50 % of the network is available to attackers (this is discussed further in §17.1.1). While the recipient may always extend the confirmation time, it remains unclear when it becomes "safe." One may argue that there is a large value N such that a reorg beyond that depth will either destroy Bitcoin or be rejected by the nodes; for example, were a 1,500-block reorg presented, a mild constitutional crisis would most certainly ensue. The economic majority of nodes and miners could easily decide to reject the reorg (see §13.8).

Alternatively, honest miners could voluntarily, or by loose agreement, agree to hold at least 50 % of the total available hashrate. If at any time there is an attack, they could (again by loose agreement) agree to continue to write the original chain, but this time with full deployment of their resources, which would be less than profitable block for block. This may require some altruism, presenting a volunteer's dilemma if some miners are counting on other honest miners to strengthen the main chain. It would also require that this course of action was so blatantly obvious to all honest miners.

Honest miners could also monitor the purge pool and add hashrate if the purge pool becomes gravid with rewards. This may also require some altruism on the part of the honest miners, who would be mining blocks with inefficient equipment.

12.3 The dynamical system determined by the difficulty adjustment

Suppose that $F : \mathbb{R} \to \mathbb{R}$ is a continuous function. One may consider what happens with the following process: Start with a value x_0, then let $x_1 := F(x_0)$, then let $x_2 := F(x_1)$, and so forth. In some cases the process will converge to a *fixed point*, that is, a value x such that $F(x) = x$. In other cases the process will converge to periodic orbits (say $F(x) = y$ and $F(y) = x$ but $x \neq y$) and for other functions F the values will diverge chaotically alternating towards positive and negative infinity.

Example 12.1. Consider $F(x) := x^3$. This has three fixed points: $x = 0$, $x = -1$, and $x = 1$. Note that for initial values x_0 with $|x_0| < 1$ the process converges to 0, while starting with $x_0 > 1$, the process diverges to infinity.

Example 12.2. Consider $F(x) := 2 + \frac{8}{x}$, with domain $x > 0$. This has one positive fixed point, obtained by solving

$$2 + \frac{8}{x} = x.$$

If we take any arbitrary positive starting point, iteration converges very quickly to $x = 4$.

This example illustrates the behavior of a dynamical system converging to a unique fixed point as asserted by the following theorem.

Theorem 12.1. *Suppose that* $F : [a, b] \to [a, b]$ *is a differentiable function. If* $|F'| < 1$, *then* F *has a unique fixed point.*

In Example 12.2, note that the conditions were not strictly true when we started:

$$F'(x) = \frac{-8}{x^2}$$

becomes small enough in magnitude, provided that $x > \sqrt{8}$. One can check that, for example, the interval $[3, 5]$ satisfies $F([3, 5]) \subset [3, 5]$ and $|F'(x)| < 1$. Thus anything in this basin of attraction will converge to the fixed point $x = 4$. Anything outside of the basin of attraction will eventually end up in this basin of attraction (while not true in general, one can check this is true for this example).

Consider the function F that takes a current hashrate, computes the difficulty adjustment, and then determines the new hashrate based on economic considerations. This is a function from the space of hashrates to the space of hashrates. The goal of this section is to present this function as a dynamical system and determine criteria under which it would be stable.

To present this discussion in the clearest way, we begin with a discussion of units and their role. In principle, the *choice* of units should not play a role. The function F defined by redetermining the hashrate has as domain the space of hashrates and as range the space of hashrates. The “space of hashrates” does not come with a particular unit, just like the “space of lengths” does not require we use feet, inches, or meters. However, when we make a choice, say, exahash per second (EH/s), we are prescribing descriptive coordinates to the (1D) space of hashrates.

Units of value must be used in the economic calculations: These include the US dollar, Bitcoin, and block rewards. These may vary in their conversion rates over time, but at a fixed point in time, the conversion rate can be considered to be roughly constant. For this discussion the most convenient unit will be block rewards (BR).

We can pick any large measure of hashes, say EH. When discussing hashrate, we need a time unit. The most convenient unit for us to use will be 600 seconds, which we call a block period (BP).

Suppose that the total network hashrate $H_0 = h_0$ EH/BP is roughly constant for a two-week period (note we are distinguishing between the hashrate H_0 and the real number h_0 describing the hashrate in terms of a unit).

Consider the quantity

$$A := \frac{1}{H_0} \times \frac{\mathrm{BR}}{\mathrm{BP}}.$$

In the ensuing adjustment period, the expected rewards for producing any value H_1 of hashrate will be precisely

$$H_1 \times A,$$

or

$$\frac{H_1}{H_0} \times \frac{\mathrm{BR}}{\mathrm{BP}}.$$

Note that this holds regardless of what units are used for computing the hashrate.

If we consider the hashrate of all possible producers of hash, we can consider a function $\mathrm{mc}(H)$ which describes the marginal cost of producing more hashrate based on the current amount of hashrate being produced, see Figure 12.1. Assuming that cheaper hashrate is produced before more expensive hashrate, moving to the right on this curve describes the cost of adding to the total hashrate. The cost is in units of value/time, as this is typically not a one-time cost, but an operating cost per unit time. There is a maximum hashrate that could be produced; if everyone on the earth did everything they could to produce hashrate tomorrow, it would result in some finite number. If we fix a unit describing hashrate (say EH/BP) for the x-axis and a unit describing value per time (say BR/BP) for the y-axis, there should be some value $h = h_{\max}$ where we see a vertical asymptote.

Given some value per hashrate (on the y-axis), there will be a unique point at which the marginal cost of producing hashrate is equal to the marginal value of producing more hashrate; this determines the total hashrate.

For a fixed adjustment period, the network hashrate should be determined by the hashrate H_1 such that

$$\mathrm{mc}(H_1) = \frac{1}{H_0} \times \frac{\mathrm{BR}}{\mathrm{BP}}, \tag{12.1}$$

where H_0 is the hashrate over the last adjustment period.

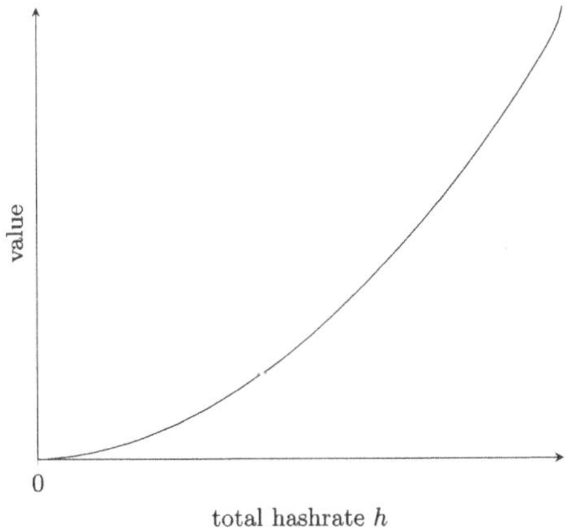

Figure 12.1: Marginal cost of hash.

Fixing units for H as EH/BP we can express the value function

$$v(h) := \frac{1}{h\text{EH/BP}} \times \frac{\text{BR}}{\text{BP}}$$

on the same plot; this function v determines a value of producing hash during the next period v, based on the amount produced in the current period.

In Figure 12.2, the point of intersection represents an equilibrium. At the h value above which intersection happens, the marginal cost of hashrate is equal to the value of the marginal hashrate in the upcoming difficulty period. So if the current hashrate is h, then the hashrate in the next difficulty period will also be h.

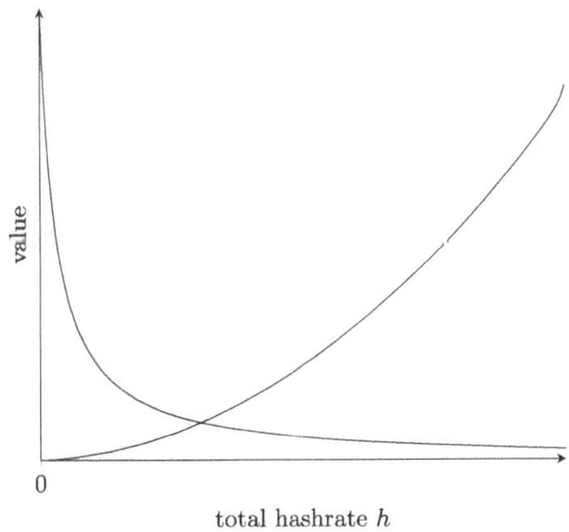

Figure 12.2: Value of hash meets marginal cost of hash.

We can describe a dynamic system as follows:

$$F(H) := \text{solution } H_1 \text{ to (12.1) with } H_0 = H.$$

Given any units (say EH/BP) we describe how this F is computed using the graph: Given any value h_0, plug this value h_0 into the function v to get a value $v(h_0)$ which determines the value that hashrate will be worth after the adjustment. From the point $(h_0, v(h_0))$ move horizontally to where the horizontal line intersects the curve $(h_1, \mathrm{mc}(h_1))$. That is, solve h_1 for

$$\mathrm{mc}(h_1) = v(h_0).$$

Then define $F(h_0) = h_1$.

Note that we can iterate this process. If we then move vertically along the line $h = h_1$, we intersect the curve at $(h_1, v(h_1))$. Move horizontally to discover where $\mathrm{mc}(h_2) = v(h_1)$. The result is $F(h_1) = h_2$. Tracing the picture should yield a "spiraling" pattern: The points $h_1, h_2, \ldots$ should typically alternate on opposite sides of the equilibrium value h_*.

If the dynamic system is stable, there will be some h_* such that

$$\lim_{i \to \infty} F(h_i) = h_*.$$

Example 12.3. Suppose that

$$\begin{aligned} \mathrm{mc}(h) &:= e^h, \\ v(h) &:= \frac{10}{h}. \end{aligned}$$

To find the dynamical system this describes, we need to solve

$$e^{h_1} = \frac{10}{h_0},$$

which gives

$$\begin{aligned} h_1 &= \ln\left(\frac{10}{h_0}\right) \\ &= \ln 10 - \ln(h_0). \end{aligned}$$

In other words,

$$F(h) := \ln 10 - \ln(h).$$

Now this has a solution, approximately

$$h = 1.745.$$

Note what happens if we start with $h_0 = 1$:

$$\begin{aligned} h_1 &= F(h_0) = \ln 10 \approx 2.3026, \\ h_2 &= F(h_1) \approx \ln 10 - \ln(2.3026) \approx 1.46852, \\ h_3 &\approx 1.9183, \\ h_4 &\approx 1.6511, \\ h_5 &\approx 1.8011\ldots \end{aligned}$$

Iterating, we get a sequence: ... 1.7142, 1.7636, 1.7352, 1.7515, 1.7421, 1.7475, 1.7444, etc. This sequence is converging to the fixed point $h \approx 1.745$.

Note that for

$$F(h) = \ln 10 - \ln(h)$$

the derivative is

$$\frac{dF}{dh} = -\frac{1}{h},$$

which satisfies

$$\left|\frac{dF}{dh}\right| < 1$$

for any $h > 1$. So once a value $F(h)$ lands in the region $[1, \infty)$ we can be assured that the system will converge to the fixed point.

The following example shows what can happen if $|F'(h)| > 1$.

Example 12.4. Suppose instead that

$$\mathrm{mc}(h) = 5 + \frac{1}{10}h, \tag{12.2}$$

$$v(h) = \frac{10}{h}. \tag{12.3}$$

To find the next difficulty, we solve

$$5 + \frac{1}{10}h_1 = \frac{10}{h_0}$$

to get

$$F(h) = \frac{100}{h} - 50.$$

Now

$$F'(h) = \frac{-100}{h_0^2}.$$

This will satisfy

$$|F'(h)| > 1$$

whenever

$$h < 10.$$

There are two solutions to the equation

$$h = \frac{100}{h} - 50.$$

One is negative ($h = -51.926$) and out of bounds as far as we are concerned, while the other is $h = 1.9258$.

But near $h = 1.9258$, the dynamical system is unstable, because

$$|F'(1.9258)| = 26.964.$$

We can see what happens if we start with a slight perturbation of h: If $h_0 = 1.925$, then

$$\begin{aligned} h_1 &= 1.9481, \\ h_3 &= 1.3321, \\ h_4 &= 25.069, \\ h_5 &= -46.011. \end{aligned}$$

At this point the values no longer make sense for a difficulty adjustment problem. So if equations (12.2) and (12.3) correspond to a difficulty adjustment, the adjustment will lead to extreme instability in the hashrate.

One may conjecture that the shallow slope of the function $\mathrm{mc}(h) = 5 + \frac{1}{10}h$ leads to instability, while the steeper slope of $\mathrm{mc}(h) = e^h$ lends itself to a more stable system. This intuition is correct: Shallower slopes of mc lead to larger values of $F'(h)$.

We can make this precise: Pick units for F and h. Recall that $F(H)$ is determined implicitly by

$$\mathrm{mc}(F(H)) = \frac{1}{H} \times \frac{\mathrm{BR}}{\mathrm{BP}}.$$

We can differentiate this implicitly with respect to hash

$$\frac{d}{dH}\,\mathrm{mc}(F(H)) = \frac{d}{dH}\left(\frac{1}{H} \times \frac{\mathrm{BR}}{\mathrm{BP}}\right)$$

using the chain rule

$$\frac{d}{dF}\operatorname{mc}(F(H))\frac{dF}{dH} = -\frac{1}{H^2} \times \frac{\mathrm{BR}}{\mathrm{BP}}.$$

(Observe here that we are performing this differentiation "coordinate-free." You can verify this by picking any units for hashrate and computing a derivative with respect to that particular hashrate; see Exercise 12.3.)

We conclude that

$$\frac{dF}{dH}(H) = -\frac{1}{\frac{d}{dF}\operatorname{mc}(F(H))}\frac{1}{H^2} \times \frac{\mathrm{BR}}{\mathrm{BP}}.$$

This quantity, which is in units of hashrate per hashrate, is a scalar, and this number does not depend on the choice of units. In particular, if we agree to express mc in terms of BR/BP, the number

$$\Gamma = -\frac{1}{\frac{d}{dF}\operatorname{mc}(F(H))}\frac{1}{H^2} \tag{12.4}$$

determines the stability or instability of the dynamic system determined by the difficulty adjustment. If this number is larger than 1 in magnitude near a fixed point, that fixed point will be unstable.

Exercise 12.2. Describe units for h, $\operatorname{mc}(h)$, and $v(h)$ and explain why setting $v(h) = \operatorname{mc}(h)$ makes sense in terms of units.

Exercise 12.3. Suppose Jane runs 1,500 m. The time it takes her is given by

$$\text{Time(rate)} = \frac{1{,}500\ \mathrm{m}}{\text{rate}}.$$

Using m/s as a unit, express the function Time as

$$\text{Time}\left(r\,\frac{\mathrm{m}}{\mathrm{s}}\right) = \frac{1{,}500\ \mathrm{m}}{r\,\frac{\mathrm{m}}{\mathrm{s}}}$$

Compute

$$\frac{d}{d(r\,\frac{\mathrm{m}}{\mathrm{s}})}\left[\frac{1{,}500\ \mathrm{m}}{r\,\frac{\mathrm{m}}{\mathrm{s}}}\right]$$

directly using the limit definition of derivative. Then suppose you have some other unit, with coordinate ρ. Compute

$$\frac{d}{d\rho}\text{Time}(\rho)$$

by converting ρ to r and using the previous expression, given in invariant form. Conclude that

$$\frac{d}{dR}\,\mathrm{Time}(R) = \frac{-1{,}500\ \mathrm{m}}{R^2}$$

regardless of the choice of unit.

12.3.1 Phase transition to Galloping Gertie

Units will not affect the stability or instability of the system: All we need to compute stability is the number given in (12.4). As long as the slope of $\mathrm{mc}(h)$ remains steep enough near the equilibrium, we expect the system to remain stable. If the system is stable, perturbations in the value of Bitcoin and the cost of hashrate or naturally occurring variation in difficulty, which happen frequently, will not upset the dynamics.

However, it could be that over time, as more ASICs are produced, the costs of adding more hashrate could grow less steeply. It is a bit difficult to develop a precise model for this, so we are only going to offer a certain family of curves with a given shape: some positive minimum cost ε to produce hashrate and then hashrate production cost growing quadratically. This family of curves affords enough flexibility to demonstrate a phase transition.

Consider a two-parameter family of mc curves given in some coordinates h by

$$\mathrm{mc}(h) := (\varepsilon + \lambda h^2)\frac{\mathrm{BR}}{\mathrm{BP}}$$

and assume units are normalized so that

$$v(h) = \frac{1}{h}\frac{\mathrm{BR}}{\mathrm{BP}}.$$

This determines a function

$$F(h) = \left[\frac{1}{\lambda}\left(\frac{1}{h} - \varepsilon\right)\right]^{1/2}.$$

Solving for the fixed points

$$\varepsilon + \lambda h^2 = \frac{1}{h}$$

involves solving a third-order polynomial and is probably best handled by a computer. We can compute the derivative of F using implicit differentiation to get a tidier formula:

$$\frac{d}{dh}\left[\varepsilon + \lambda F^2 = \frac{1}{h}\right],$$

which yields

$$\frac{d}{dh}F(h) = \frac{-1}{2\lambda F(h)h^2} = \frac{-1}{2[\lambda(\frac{1}{h} - \varepsilon)]^{1/2}h^2}.$$

To illustrate a phase transition in the dynamics, fix $\varepsilon = 0.3$. If we take $\lambda = 1$, we can iterate the function

$$F(h) = \left[\left(\frac{1}{h} - 0.3\right)\right]^{1/2}$$

starting with a wide range of h_0 values, and we will see that this process quickly converges to a value around $h_* = 0.900366$. Testing numerically, we can check that the basin of attraction includes the interval $(0.1, 3.0)$. Starting with values outside this interval, iterating the function will result in h values spiraling out of the domain of F.

At the fixed point, we can compute the derivative:

$$F'(0.900366) \approx -0.68503,$$

which is less than 1 in magnitude, as it should be for a stable equilibrium.

Now we can decrease the parameter λ; this corresponds to the marginal cost growing less steeply. At the value $\lambda = 0.3$ we observe that there is a fixed point, with value

$$h \approx 1.272556,$$

but the basin of convergence has shrunk to approximately

$$(0.335, 2.99).$$

We can also check the derivative of F at the fixed point:

$$F'(1.272556) = -0.80876.$$

Now try $\lambda = 0.11$. The basin has shrunk further. Iterating from $h_0 = 1.44$, the system does not converge. Similarly for $h_0 = 1.90$. For values within the basin, one observes that it converges to $h_* \approx 1.65903$ and that

$$F'(h_*) = -0.99544.$$

Now we try $\lambda = 0.105$. We can solve the equation

$$0.3 + 0.11h^2 = \frac{1}{h}$$

numerically; we get $h_* = 1.659$, but

$$F'(h_*) = -1.0189.$$

This is not a huge jump from $\lambda = 0.11$, but it means that the system is now unstable. There is no basin of attraction, which is easy to check. Plugging in the value 1.659 (which is a truncated version of the actual solution) and iterating, one observes that the process will result in undefined values within a few hundred iterations.

In real terms, this means that if the mc curve becomes too flat near where the equilibrium happens, the difficulty adjustment could behave chaotically: jumping from a low difficulty (which entices many miners) to a large difficulty (which discourages miners). This could easily cause a situation in which the value of producing hash drops below the marginal value of producing hash for all but a few miners. This results in periods of bad security (very low HDP) and long block times.

The field of discrete dynamical systems is a very rich and interesting area of mathematics. We only offer a taste here, illustrating how small changes in parameters of the system can result in drastically different long-term behavior. For a nice introduction, see the book [13].

Exercise 12.4. Prove Theorem 12.1.

12.4 Financialization of hashrate markets

Corporations who mine Bitcoin, especially on low profit margins, have an incentive to hedge risks. A mining corporation faces costs in the form of energy and capital costs for ASICs and recoups revenue in terms of Bitcoin. The revenue depends on the fee market, the market price of Bitcoin, and the hashrate of other miners. Each one of these is a variable that could make the difference between a corporation making huge profits and suffering large losses. One would expect that futures markets for all three of these variables will develop. These could help stabilize profit margins for miners, but could also play into the difficulty adjustment in interesting and unpredictable ways.

There are a number of ways that futures contracts can be structured. The corporation could buy contracts that pay out if energy prices spike. In this case, if the energy is too expensive to create hash profitably, they could sit out the market until energy prices are more favorable. They can also buy hedges against total hashrate: If the hashrate spikes in the future, they could structure a contract so that they are paid when the hashrate is high. This allows them to not hash and cover their capital expense losses. To hedge against price collapse, miners could buy put options for Bitcoin.

It is unclear how future markets will exactly affect the shape of the hashrate curves, but it is clear that financial markets should be taken into consideration. While most markets have natural occurring competition, the automatic adjustment at discrete intervals creates a unique type of dynamical system.

It would appear that sophisticated financial instruments such as futures contracts would allow corporations to exist on compressed margins and at the same time put them in situations where the total hashrate is extremely sensitive to perturbations in energy costs, new hashrate, fee market demand, or changes in demand for the asset. This remains an important area to watch in the coming decades.

13 Attacks

In this chapter we discuss attacks on Bitcoin. To clarify what is meant by “attack on Bitcoin” we use the three pillars identified in a 2013 paper [31, §3] as the “triad necessary for success of the Bitcoin economy.”

This triad consists of:

1. *Consensus about rules*: Everyone must agree on what rules are being used in Bitcoin.
2. *Consensus about state*: Everyone must agree on the ledger (up to perhaps the latest block or two).
3. *Consensus that bitcoins are valuable*: Everyone should agree that the bitcoins assigned to owners of keys in the ledger are something valuable.

These three pillars are useful in describing and organizing attacks. Attacks can chip away at any or all of three of these pillars. Often, an attack on one can cause weakness in others, which may set off a vicious cycle. Attempts to shore up one may cause weakness in another. We emphasize that successful attacks are attacks on the consensus around the rules, state, or value, as opposed to attacks on the rules, state, or value themselves. Bitcoin lives and dies by consensus.

We define an *attack on Bitcoin* to be any actions intended to cause confusion about the rules or the state or to cause the value of bitcoins to be fundamentally diminished. A successful attack will be an attack that accomplishes any of these objectives.

On the other hand, we define an *exploit* as a possible or successful attempt to work within the agreed-upon rules and consensus mechanisms of Bitcoin to arrogate to oneself an unfair share of either Bitcoin or something else of value. For example, selfish mining, if intended only to distort the apportioning of mining rewards, is fundamentally not an attack but an exploit. However, selfish mining becomes an attack if the goal is to sow unease in the value of Bitcoin. A double-spend, if intended to defraud, is an exploit, while a double-spend, if intended to shake confidence in the certainty of the ledger, is an attack. These need not be exclusive. A clever double-spender may short the market while simultaneously attempting to steal back spent coins.

Note that we are departing from standard terminology. What we are calling “exploits” are often called “attacks” (although not attacks *on Bitcoin*). The protocol itself is amoral; declaring that an act is dishonest or against the good spirit of Bitcoin requires some amount of universal clarity of judgment. Exploits may or may not pose a risk to the system itself, so they do not necessarily constitute an attack on Bitcoin.

13.1 Why have there been no attacks yet?

Before beginning a more detailed discussion of attacks, we offer three categories of reasons why serious attacks on the Bitcoin are yet to be witnessed.

https://doi.org/10.1515/9783110772845-013

Attacks are prohibitively expensive and/or logistically impossible. Simply put, the security budget (around $10–15 billion per year) is too high for any one party to attempt to overcome. Beyond the nominal amount, the logistical issues, in particular, the cost of getting one's hands physically on ASIC machines necessary to perform such an attack, present a much larger challenge than funding an attack budget.

Miners or other members of the ecosystem will never participate because it will destroy the ecosystem. This argument is essentially the *market fragility hypothesis*: *Any successful attack or untoward behavior by miners will result in a crisis of confidence in the network, leading to plummeting market prices of the Bitcoin asset.* Although we propose this as a hypothesis, it is certainly not a binary proposition. The market response to various attacks and exploits will depend on the particular attack or exploit.

Nodes will manually override any obvious attacks, rendering the attacker's efforts worthless. At the end of the day, Bitcoin is a consensus, which may rely on Nakamoto consensus, but is not inexorably bound to Nakamoto consensus. Nakamoto consensus need not be a self-destruction pact. If users observe what appears to be an egregious abuse of the protocol's hard-coded rules, the community can decide to simply ignore the bad actors and right the network as a response to an attack. We call this the *soft-forkability hypothesis.*

These statements involve some hand-waving. One of our goals in this chapter is to nail down what each of these statements can and should mean. The fact that nobody knows exactly how Bitcoin would respond provides strategic ambiguity. Much like a mixed strategy, the lack of clarity as to what a response would be removes the ability of attackers to count on given incentives: The game tree has many undetermined leaves. While Bitcoin does not do this intentionally, strategic ambiguity does provide some protection against an attack.

Remark 13.1. This is not meant to be an exhaustive presentation of everything that can go wrong. We are not dealing with technical and physical attacks: eclipse attacks, denial of service, physical sabotage, hacking/malware attacks, back-doors, five-dollar-wrench attacks, developer malfeasance. Nor are we diving into the details about how the peer-to-peer network actually operates. Nor do we discuss collisions, cryptography, or quantum computing. These require discussions focused more on technical issues and less on game theory.

13.2 Nation-state/authoritarian attacks

The most straightforward attack to discuss is the nation-state attack. This attack would be waged by a nation-state who perceives Bitcoin as some sort of threat. The nation-state makes the calculation that spending resources attacking Bitcoin will pay off by mitigating whatever damage Bitcoin is causing to the nation-state (or more precisely, the damage Bitcoin may cause to the individuals who exercise control over the nation-state).

A nation with the objective of obliterating Bitcoin can choose from one of several (non-exclusive, non-exhaustive) paths forward:

- acquiring mining equipment, possibly at high cost, to achieve a majority of the hashrate,
- coercing miners within their jurisdiction to participate in an attack,
- establishing incentive structures (bribes) for miners beyond their jurisdiction to participate in an attack,
- reducing demand by changing laws to make holding or using Bitcoin undesirable (for example, throttling on-ramps),
- leaning on allies to change laws to make holding or using Bitcoin undesirable,
- loudly and openly declaring their intentions in an attempt to scare investors and miners away from Bitcoin.

Acquiring the equipment to produce a supermajority of the hashrate is ideal and would likely result in a fatal blow. This would allow the attacker to permanently throw the state of the ledger into confusion. This in turn would devalue Bitcoin, by making it essentially unusable.

13.2.1 The empty-block attack

The empty-block attack is the most straightforward to describe. An entity with over 51 % of the hashrate simply begins mining and includes no transactions. They ignore all other mined blocks. This is a very special form of a direct frontal attack (see Chapter 14). The longest chain will always be the one controlled by the nation-state, and this will not contain recent transactions.

The failure of transactions to be added to the ledger violates the principle of *liveness*. Liveness as defined in [49] is the property that "New blocks will continue to be added and valid transactions with appropriate fees will be included in the blockchain within a reasonable amount of time."

A payment system without liveness is largely worthless as a payment system, so one would imagine that the value of Bitcoin would deteriorate, as there would be no consensus that the network is valuable as a medium of exchange.

13.2.2 Obvious criticism of the empty-block attack

Executed literally, an empty-block attack would most definitely fail. This is a situation in which the soft-forkability hypothesis should apply. Bitcoin node operators and miners who are not participating in the attack could simply ignore the empty blocks and only mine blocks containing transactions. This could be accomplished via a straightforward software patch that invalidates empty blocks, or node operators could invalidate the empty blocks manually (note that enforcing an additional rule requiring blocks not be empty results in a soft fork). One would not expect a nation-state with the ability to perform an empty-block attack to make it so simple to thwart.

In an empty-block attack, there is not confusion about the ledger; it is empty past a certain point. This would likely cause the value to plummet. This is remedied by ignoring the empty blocks. However, slight variations would make the ledger uncertain: Instead of mining empty blocks, the nation-state could mine blocks either containing very few actual transactions or mostly junk transactions. The mined blocks could each contain four transactions from an honest mempool and 1,500 transactions that move around a few sats controlled by the nation-state. In this case, there would be no obvious consensus for the honest nodes. One may attempt to claim that a block containing only four honest transactions should be invalid, but what about eight, or 30? Honest miners would be forced to constantly decide which blocks are valid and which blocks are not. While the community may develop guidelines, these guidelines will always lead to ambiguities that would fracture the network. If such guidelines are easy to create, they would already be in place. For example, a guideline that each block must be full is easily overcome; the attacker can fill it will meaningless transactions. If an attacker creates a series of blocks and observes which ones are rejected and which ones are accepted by the network, it becomes a fairly simply machine learning problem to attempt to generate blocks that are accepted with probability close to 50 %; this would constitute an effective *edge-case attack*. Edge-case attacks are attacks that cause confusion about the ledger, assuming that users agree on the rules. A well-landed edge-case attack will result in confusion about both the ledger and rules as the community scrambles to define new rules.

There is only one clear solution to this, and this is an anathema to Bitcoin: Some sort of leader or panel of leaders must be chosen to determine which blocks are valid and which are not. This would require constant vigilance and would often halt the network as incoming blocks are analyzed. One version of this is to whitelist miners – if there is a select group of miners who all acknowledge each other as honest, the blockchain can continue without fear of external hashes destroying the consensus. However, this would destroy the permissionless nature of Bitcoin mining, and may cause some loss of value.

If the whitelisted miners reside outside the jurisdiction of the nation-state, there may be nothing the nation-state can do. However, a whitelist of miners is a weak point; any attacker can attempt to compel or denial-of-service attack these miners individually. Sufficiently motivated attackers could even attack physically, if they know where to attack.

Conclusion 13.1. In the face of a persistent majority attack on the ledger, the Bitcoin network must submit to some sort of governing body in order to maintain consensus on the state of the ledger.

13.2.3 Layer 2s under an empty-block attack

When the blockchain is well functioning, layer 2 protocols such as the Lightning Network typically do not require the layer 1 chain to be fully live. Lightning transactions are peer-to-peer and refer to layer 1 transactions that have been previously recorded. However, in order to function fully and discourage cheating, the Lightning Network uses an enforcement mechanism that must be executed on layer 1.

In theory, if nobody on the Lightning Network were to cheat their peers, the network could remain a functioning closed system with no on- or off-ramps. It is unclear how valuable this would be. One would expect that any other layer 2 protocols that require the ability to plug back into the main chain would have similar problems. If Bitcoin's layer 1 appears to be out of commission for the foreseeable future, one would expect that confidence in the Lightning Network would also be eroded; the trustless nature requires an enforcement mechanism that is only available on layer 1.

13.2.4 The seesaw attack

A *seesaw attack* is described as a method to attack the consensus about the ledger. The attacker would create one or more chain splits and then consistently attempt to extend the shorter of chains, extending these into the lead before pulling back and mining a different chain.

Like a modified empty-block attack, it would be very difficult for the community to agree on guidelines that could not be abused. For example, if the rule were to "reject any block that overcomes a two-block deficit," the miners could attempt to maintain chains that are only one block behind or publish blocks in a coordinated manner so that the network is split based on their observational point within the network (this rule amounts to a checkpoint; see §13.3).

Repeated reorgs would throw the ledger into confusion and make the network unsuitable for payments, in particular if each reorg includes different and incompatible sets of UTXOs. Again, as Conclusion 13.1 states, the most likely way out of this is to choose leaders to navigate through this attack.

13.2.5 Bribe-flooding attack

Miners are motivated by the security budget. An attacker attempting to destroy certainty about the ledger can try to induce miners to constantly create reorgs. By bribing miners to constantly reorg the chain, the attacker could blur the Schelling point provided by Nakamoto consensus for an extended period of time, which may lead to dysphoria. This requires no ASICs, only money, and could be effective when making payments on other blockchains. This would also create a prisoners' dilemma dynamic among miners: If the blockchain is being attacked, one can either participate by accepting the bribes or miss out by fighting the bribes. If the network survives the attack, it is still better to have accepted the bribes. If the network collapses, it is better to have accepted the bribes. But as long as no single miner believes their decision to accept bribes or not determines the fate of Bitcoin, they will be motivated to accept the bribe (this prisoners' dilemma probably should not apply to public-facing corporations in the industry who need to maintain goodwill). In theory, this requires only a multiple of the security budget and can be enacted in a number of ways. For example, if an honest block is mined, the attacker could offer to pay double the subsidy for an equalizing block and then double the subsidy for the block that consummates the reorg. Miners may attempt to come up with clever ways to milk the attacker without damaging the system, but without forming some centralized planning committee to do this, the tragedy of the commons dynamic will remain in play.

13.2.6 Empty-block purge attack

An attacker may intermittently use deep reorgs with empty blocks to throw confusion onto the ledger. For example, suppose an attacker secretly mines 31 empty or garbage blocks and then publishes these to supplant 30 blocks in the longest chain. According to Nakamoto consensus, the chain with 31 empty blocks is the correct chain, essentially eliminating confirmations for several hours worth of transactions. Any transactions that have been excluded by the attacker are kicked back to the mempool. These can be double-spent by anyone who is paying attention. If you sent your Bitcoin to an exchange and this was credited after 12 blocks by the exchange, you can now resubmit a transaction to the mempool, sourcing from the same UTXO, which directs the output back to yourself. Anyone in a similar situation could do this.

This is called a *purge attack* [24] and can serve as both an attack and an exploit. If many Bitcoin users attempt to take advantage of this, there will be massive confusion about the ledger. The attacker is encouraging others to use a double-spend exploit. The more users that take advantage of purge opportunity, the more damage it will do to the credibility of Bitcoin as a payment network.

13.3 Checkpointing against a persistent 51 % attack

Checkpointing is a security measure that allows the community to all agree on some point in the past where the ledger is final. As it is, there is no finality in the Bitcoin protocol; an attacker who developed a powerful enough hashrate could rewrite much of the chain, and according to the rules, this would be correct. Other blockchains have add rolling checkpoints as a security measure to prevent deep reorgs. Bitcoin Cash uses a rolling 10-block checkpoint.

Checkpoints can be implemented in a number of ways. A simple way is to have each node keep track of which recent blocks they have seen and commit to reject reorgs that suddenly appear and supplant more than m blocks. This is simple but has the problem that it is subjective: Each node on the network has a different viewpoint and some may reject a reorg if they detect it seconds after another node who saw the reorg in time. An attacker with some precision could use this to split the network.

There are other ways: A central authority could simply arbitrate when a block has become m blocks deep. This does confer authority and a small amount of trust to a third party. There are some extremely unlikely, but possible scenarios in which such a position could be abused. In this case central authorities are also possible targets of denial-of-service attacks.

The question of whether checkpointing is a good security measure or a bad one comes down to beliefs about the likelihood and payoffs of different types of attacks. A well-resourced attacker can attack in many ways: One is to destroy confidence in the currency by creating a very deep reorg. Another very damaging attack is an edge-case attack based on enforcement of the checkpoint as an additional rule.

If an edge-case attack against checkpointing is successful, some external consensus mechanism would be required to manually reunite the network, lest the network remain forever fractured. A persistent attacker can repeatedly launch edge-case attacks until the network is sufficiently splintered. This brings us back to Conclusion 13.1.

Would a checkpoint be effective against attacks that are not persistent? Reorging the chain with the goal of fracturing the network requires a bit of precision; however, it may be more damaging than a one-off deep reorg. For example, suppose a 10-block checkpoint is in effect, and an attacker with slightly less than 50 % of the hashrate expends greatly to extend a stealth chain from height n to height $n + 11$, while the public chain is currently at chain $n + 10$. The attacker has two options – either publish immediately and hope that the 11-block reorg will cause ample damage to the network, or listen closely until block $n + 11$ appears on the main chain, at which point they publish their own $n + 11$ as soon as possible. If this is well timed, their block will reach some nodes before the main chain does. These nodes will be at odds with the network until they are manually rectified. Is this worse than simply creating an 11-block reorg?

An attacker with less than 50 % may be occasionally be able to perform long reorgs which would maximize the damage, but these attacks are not guaranteed to land on a regular basis. Still, a series of nine-block reorgs will be very damaging, especially if these purge transactions from the chain.

On the other hand, a rolling checkpoint does protect against certain exploits. With a 10-block checkpoint, a transaction with 11-block confirmation can be considered final. Even if through some effort and expense an attacker is able to overcome the odds and create what would be a 12-block reorg, the reorg would be rejected. Thus checkpointing does disincentivize double-spend exploits. With Bitcoin, while six is a commonly accepted number of confirmations after which a transaction can be considered "confirmed," this is not a magic number; there is no such number which guarantees finality under the strict rule of the protocol.

While checkpointing is unlikely to be effective against persistent majority attacks, it may be useful in an attack scenario that is less persistent; such scenarios are discussed in §13.7.3. If the expectation of damage caused by a low-probability attack is larger than the expectation of damage caused by an edge-case attack or ceding power to some authority in a rare near edge-case attack, rolling checkpoints make sense. Checkpointing could encourage attacks by making a damaging attack slightly more attainable, or could discourage attacks by making the most damaging type of attacks impossible.

One concern with checkpoints that is often mentioned is that it could cause confusion for nodes who are coming back online. For example, if someone shuts down their computer for 8 hours, they may be 50 blocks behind when they connect to other nodes – this would require special handling. In normal circumstances, the danger of confusion among nodes who are coming back online can be easily minimized. Excluding rare and unexpected events, this is a small technical issue. However, a full accounting must take into consideration unexpected events. In cataclysmic events that cause disruption or temporary fragmentation to the internet, the pure proof of work does provide a Schelling point. Of course, one would assume that a checkpointing system could be dropped in such conditions; the simplest protocol is always there to fall back on.

Philosophically, many Bitcoiners, in particular those who value the elegance of Nakamoto consensus, are opposed to checkpoints. Consider the simplicity of the heaviest-chain rule: A Bitcoin user may bury themselves in a cave for 18 years and return to the surface. When presented with possible chains, they have only two tasks, each of which can be done without any other information: (1) verify that each chain is valid from the genesis block and (2) compare the chainwork on each chain. They do not have to have known anything about what the consensus was at various points in history.

This property of Bitcoin is called *chain selection objectivity*. Bitcoin users always have a universal measuring stick which will determine which blockchain should be the correct one. This contrasts with *weak subjectivity*, a characteristic of consensus protocols that require cues from the social network to determine the correct chain. Proof-of-stake systems generally require weak subjectivity to maintain continuity: Network participants use information about recent states of the ledger in order to maintain a Schelling point. Proof of stake, which does not involve a costly mechanism such as proof of work in the chain selection process, suffers from a theoretical *long-range attack*, in which a chain that complies with the rules but may have deviated from the accepted chain months or years in the past suddenly appears and claims to be the correct one. Without something difficult to forge like proof of work, there is no way for someone with only knowledge of the protocol to step out of a time machine and determine which of two chains is correct. A long-range attack, while theoretical, is extremely unlikely as it requires most of the network to develop sudden amnesia over the recent ledger states.

13.4 Buyout attacks

A buyout attack (cf. [7]) is a simple but direct approach available to a wealthy attacker. The attacker offers a premium for all the existing capacity of ASICs. If they can achieve a majority via directly bidding on the existing equipment, they can use this to brute-force attack the network using their choice of attack designed to sow confusion on the ledger. If the attacker finds that the miners are unwilling to sell, even at high prices, the attacker can offer a premium for any new hardware as it becomes available.

With this sort of attack, there will likely be a point at which capitulation happens and the market collapses. If the miners who are unwilling to sell stick together, they can force the bidder to pay high prices, and even at these high prices, the buyer may only acquire a fraction of the capacity. This sets up a prisoners' dilemma for any miner being offered a premium for their machines. Clearly, if they all sell, not only will the price crash, but the attack will also be successful. In that case it would be sound advice to the miner to "cash out while you can." An individual miner with a small percentage who believes that selling will not lead to the tipping point may be willing to part with their equipment, slightly jeopardizing the network but benefiting the miner handsomely.

Once the attacker obtains 51 % of the hardware, the value of the remaining hardware should plummet to near zero. The race-to-the-door created by an attacker who has made a commitment to a buyout attack ends with a door slamming shut, which increases the urgency of the race to this door. The key is that the attacker must appear committed. Note that this is distinct from the "hostile takeovers" that occur in the business world: A shareholder who does not give up their shares still retains shares that have value, so the potential to set off a vicious cycle is not present to nearly the same degree.

In fact, the attacker need only acquire 33 % to begin to cause the hardware value to cave – at this point the attacker can use selfish mining to diminish the value of the remaining hardware, adding further urgency to the race to the door. Even without announcing the attack, an attacker can build up to it by mining to full capacity, even with inefficient equipment and at a loss, in order to bring down the rewards for competing miners. This is one form of a *pushout attack*, where a miner aggressively increases hashrate in order to make mining unprofitable for competitors. If a mining gap develops, pushouts will become easier as the attacker can force other miners to run machines at full capacity in order to obtain blocks.

The value of the hardware depends on the future expectation of profits. We define the *total security stock value* to be the expected revenues of all ASICs in existence minus the expected operating costs to run these. Note that this will be bounded by the lifespan of the equipment and the total revenue available in the blockspace market (total subsidy + fee) and will be diminished by electrical costs required to attain any blocks. In particular, as the block subsidy decreases, if the fees do not increase, the total market value of the hardware decreases. This is made worse if the energy costs rise as competition among miners becomes more stiff, leading to margin compression.

For a very rough estimation of the cost, suppose we say that the average lifespan of equipment in use today is N years (the actual figure varies, depending on who you ask within the industry). Roughly speaking, the total of all mining equipment today will return N years of revenue. The revenue per year is simply the security budget. From this, subtract the costs of mining for electricity, cooling, personnel, etc. If we assume that profit from this machinery attenuates linearly over the next N years, we get the following total expected profit from the current equipment:

$$\int_0^N \Pi_0(1 - t/N)dt = \frac{N}{2}\Pi_0.$$

Note that this scales linearly in the expected lifespan and also in the profit margin. Here Π_0 is the expected annual profit.

Example 13.1. If the security budget is \$20 billion, the profit margins are 20 %, and the equipment is expected to last 3 years, the value of the equipment will be only around

$$\frac{0.2 \times \$20 \text{ billion} \times 3}{2} = \$6 \text{ billion}.$$

So the market value of the hardware required to start selfish mining would be \$2 billion and it would be \$3 billion for a 51 % attack.

13.5 Nuclear option: new proof-of-work algorithm

If a nation-state or other powerful attacker has amassed an arsenal of ASICs in order to launch a persistent majority attack, there may be few options left for the network to defend itself. One proposal is to change the proof-of-work algorithm so that the ASICs become either useless or less useful. This may be something simple like adding one more hash function to the proof of work. If ASICs are not hardwired to this, they will become worthless.

There are many ways to do this, and some do not do away with ASICs completely. One proposal is to compute two proofs of work, one with SHA256 and another with a different hashing function, either in parallel or sequentially, and then take geometric means of the two proofs of work to get a difficulty target. For example, if the ASIC produces a proof of work with difficulty of order 10^{19}, while the additional algorithm produces a proof of work with difficulty on the order of 10^9, the geometric mean would be of order 10^{14}. Thus, shortly after the switch, the ASICs would still be valuable for the computation, but more advantage would be gained by improving the efficiency of the machines working on the second algorithm, diminishing any advantage conferred by a party who had accumulated ASICs. Another proposal is to require blocks to alternate between proof-of-work algorithms.

Attempting to dive into details, one finds very quickly that any specific proposal will be fraught with unknowns. If the hashes are computed in parallel, a new game theory regime emerges (no longer based on the same Poisson dynamics) that requires its own analysis. On the other hand it is a technical issue how to make sequential hashes possible without obsoleting the ASICs.

How this would play out is uncertain: In a short period of time, community leaders would need to emerge and present a new algorithm. This would then need to be agreed upon and then implemented and shipped out to all honest nodes and miners.

Suffice to say, this would certainly be a consummate mess. Miners and corporations that mine would lose substantial revenue and business, and it is unclear who they would be replaced by. New ASICs may be developed in short order, and powerful programmable hardware such as GPUs may surge in demand. This may work if the hardware required by mining is widely distributed and Bitcoin is very popular or necessary for people around the world. One could imagine an evil villain attempting to destroy a global economy, prompting citizens of all nations to band together by using their CPUs and GPUs to mine Bitcoin while the network recovers from the attack. This is somewhat far-fetched; centralization seems a much more likely outcome if Bitcoin users require that the network keep functioning.

The dominating force in hashpower would likely shift to a new locus. It is possible that some chipmakers would have developed versatile equipment that could be flash-programmed with some ASIC capabilities in anticipation of such an event. The community would have to trust that this would be appropriately distributed or that the parties who had the largest control would behave benevolently. Similarly, the community would be thrust into a position where they would have to trust leaders who are suggesting various protocol changes; how can we be sure they are not motivated by some hidden incentive?

Exercise 13.1. Show that (even without changing the proof-of-work algorithm) requiring two proofs of work in parallel to mine a block would result in a distribution of blocks that is not proportional to the hashrate.

Exercise 13.2. Suppose that there are two proof-of-work hashing algorithms and that a miner has portions p and q of the respective hashrates. If the "difficulty" is computed by the geometric mean of the two difficulties, is it possible to determine the probability that the miner obtains the next block based on p and q?

13.6 Less-than-brute-force attacks

13.6.1 Nudging towards a strategic regime

Brute-force attacks such as empty-block and seesaw attacks are designed to throw the chain into confusion in the short term and demoralize the users. A more patient attacker with sufficient financial resources (but perhaps not the ability to acquire 51 % of the hashrate) may try a more subversive attack: Erode the confidence in the network by encouraging other miners to adopt strategic mining.

A nation-state can either start with a purchase of their own mining equipment to command a small portion of the hashrate, or they may bribe or commandeer existing equipment. By aggressively fee-undercutting, they may encourage miners to switch to a petty compliant strategy. In the short term, this would be likely to be a waste of money. However, it would present a repeated opportunity for other miners who are interested to earn extra funds, eliminating the chicken-and-the-egg obstacle in the way of active fee-undercutting.

Whether other miners are willing to respond to these invitations would depend on their beliefs about market fragility with regard to undercutting and whether they see their participation as potentially damaging. If miners do not feel continual undercutting will harm the network, they may take advantage of the opportunities. If enough miners adopt the petty compliant strategy, more fee-undercutting is sure to follow. Miners who think that undercutting is bad but that it will eventually take hold are put in a prisoners' dilemma situation: either take advantage of the opportunity or miss out.

Eventually this will lead to a more strategic mining regime. Once miners are behaving fully strategically, the attacker may begin to present bribes. At first these may be ignored, but the miners controlled by the attacker may accept them. This only encourages more bribing and more participating in reorgs. Once miners signal willingness to participate in short reorgs, the attacker would hope this can be stretched to longer and more lucrative reorgs. Opening up the window of possibilities to include bribing is an attack itself, called a *slippery-slope attack* (see §13.7.1). The principle in play in a slippery-slope attack is that once a few miners have signaled a willingness to accept bribes and mine strategically, miners begin to fear this behavior is inevitable and change their behavior accordingly. Deterioration of norms would lead to more chaos and a crisis of confidence. Longer reorgs, if performed correctly, can be nurtured to become purge attacks, which would be devastating if performed on a regular basis.

13.6.2 FUD and social attacks

The third pillar of Bitcoin's consensus is the social consensus that Bitcoin is valuable. While the network is still nascent and is not fully enmeshed with the economy, parties interested in attacking the third pillar of Bitcoin may engage in "fear, uncertainty, and doubt" (FUD) attacks. This can be as simple as the threat of regulation or deliberate spreading of information (true or otherwise) that may give potential users reasons to pause before becoming involved. Social shaming of individuals or corporations who use the network, citing political or environmental reasons, also may be an effective tool. Humans have long organized themselves by ostracizing those who engage in unacceptable behaviors.

As long as Bitcoin is a speculative commodity, these attacks may be impactful. Bitcoin is a consensus mechanism, and if enough people do not agree that it is a good or valuable consensus, it can be abandoned. "Layer 0 is people" is a simple but accurate proverb describing how blockchains ultimately must function. People are always vulnerable to emotions and groupthink; such an attack vector cannot be ignored.

While security fundamentals such as encryption are well understood, the future of the economic security underpinning Bitcoin is not. Misbehavior by miners can be blown out of proportion by FUD attackers. As we discuss in §13.7.1, an attacker with minority hashrate can work with media to spin their own or others' misbehavior into a much darker and more foreboding picture. The ultimate FUD attack would be to convince the general public that security is fundamentally and fatally doomed.

As consensus around Bitcoin is controlled by people who often communicate on social networks, it bears mentioning that centralized control over media represents an attack vector. Most information presented online is curated algorithmically to the particular viewer at the device. This is especially true of social media, where viewers who believe they are getting a "slice of life" are in reality being pushed a particular feed designed to capture attention and prime them for further engagement with the app, website, or sponsors. This can be manipulated in a number of ways. It should be a fairly trivial application of machine learning to dial up or dial down content for users who can be identified as vulnerable to certain sentiments.

With very little creativity, one could imagine a motivated administrator or CEO of a social media giant using their power to act for or against a cryptocurrency. An algorithm could "divide and conquer" market bulls, while amplifying bearish voices, or vice versa. Even more subversively (yet trivial by modern artificial intelligence standards) an algorithm could push memes or images regarding cryptocurrency directly to users who will find them personally distasteful and create a visceral reaction against the coin. Someone who associates Bitcoin with the type of person they find despicable becomes more guarded and less likely to embrace Bitcoin. Unattractive social media profiles could find themselves unwittingly as the payload of a social attack against a cryptocurrency.

13.6.3 Regulations

Regulations present an attack on the third pillar if the net effect is to reduce the overall demand. The direction of the effect is not universal, as the net effect may be positive if the regulations confer legitimacy. Regulations could reduce demand by restricting the set of users or by adding friction to on-ramps. Carefully chosen taxes can also punish holders of the currency in a way that could force selling or discourage ownership in general. Restrictions on how businesses deal with currencies could make using Bitcoin in a business transaction setting prohibitively difficult, quashing serious hopes of adoption.

The claim that "Bitcoin cannot be banned" was true in 2011, but less so today. Since publicly traded corporations have invested billions of dollars in ASICs, a ban today by the right nation-states would destroy all security assurances.

Open-ended Question 13.1. How would a broad unrealized capital gains tax affect the adoption of Bitcoin?

13.7 Goldfinger attacks

While nation-states may have incentive to destroy Bitcoin because it poses a threat, the so-called *Goldfinger attack* is purely profit-driven. This attack is defined by the motivation to attack, not the method – the attacker gets to choose their method. Anyone who has a way in which to profit from a collapse in either Bitcoin's price or network operations has an incentive to attempt a Goldfinger attack.

The most direct way to profit from a Bitcoin collapse would be to short-sell Bitcoin. Short-selling, which is common for stocks, is the act of borrowing an asset with promise to return the asset at a future time. The short-seller then sells the asset. If the price drops, the short-seller can then repurchase the asset and return it, pocketing the difference. Methods to do this for traditional equities are well established.

The market for Bitcoin derivatives is still fairly nascent and short-selling opportunities are not as well developed. There are futures ETFs and other funds which can be sold short. Some exchanges offer opportunities to short Bitcoin directly. It is not clear what the maximum amount of short interest an investor could obtain today.

The number of proxies are growing rapidly as of 2022. Corporations which hold Bitcoin on their balance sheet would experience a loss if Bitcoin were to collapse. So would corporations whose entire business model is based on Bitcoin: This includes exchanges and mining farms, an increasing number of which are publicly traded.

Investors may also buy put options. A put option gives the holder a right to sell at a predetermined price – this may be less risky than a short. For example, instead of short-selling, an attack may pay for the option to sell 1,000 bitcoins at 95 % of today's price. If the price drops by 10 %, the attacker can purchase Bitcoin at 90 % of today's price and immediately sell these for profit.

The Goldfinger attacker faces a series of computations. First, for each possible attack vector, they must assess whether the market fragility hypothesis holds. This typically involves a probability distribution. For example, the attacker may conjecture that with 50 % probability the market price will drop by 30 %, etc., and use this as a basis to compute the expected earnings of an attack. Then they must asses the probability of an attack being successful and then compute the cost of the attack. Reality dictates they must also consider if they can *get away with it*, on a social, technical, and legal level.

Example 13.2. Suppose that in 2037, an attacker conjectures that performing a 12-block reorg will cause a drop in price of 10 %. The attacker estimates that they can assume a short interest of 0.8 % of the total market capitalization of Bitcoin. If the total market capitalization is $20 trillion, they can access $160 billion in short interest in Bitcoin. A successful short would then net the attacker $16 billion.

Note that in a purely "economic incentives" regime this attack may become likely as the price increases, accessible short interest increases, and the security factor dwindles. The block subsidy in 2037 at this price would be around $400,000 per block in US dollar terms. A sum of $16 billion would represent around 80 % of the annual security budget. On the other hand, a 12-block reorg could be accomplished within a few hours; a time period in which millions of dollars are spent on the block subsidy. The attacker is faced with a difficult but high-reward logistical problem: How to invest possibly billions of dollars to overcome millions of dollars of security?

Enacting a stealth 12-block reorg seems incredibly difficult. One would have to develop close to a majority of the hashrate while appearing to mining honestly or back-channel to major mining operations in secret near the time of the attack to avoid signaling the attack well ahead of time. This becomes much simpler if there is hashrate for rent or if miners are willing to switch chains if there is a price spike on competing blockchains. If there is dormant hashrate for sale, the Goldfinger may wish to seek this out.

Directly bribing miners may be the best option if this is available. In a purely economic regime, miners should balance their future profits against the rewards offered to them in a bribe. A bribe may be an easy sell when profits are small (ignoring social pressures to do the right thing, of course).

The Goldfinger attacker also has to contend with the possibility of a soft fork that corrects the reorg and must compute whether such a soft fork would harm or strengthen the market price. If the soft fork reifies some form of leadership or hierarchy structure within the Bitcoin community, it could lead to the end of the "leaderless" narrative and a further drop in price. On the other hand, there is no guarantee that the "leaderless" narrative actually holds up the price. A successful soft fork could prove that Bitcoin is agile, resilient, and anti-fragile in the face of attack, leading to a surge in confidence and pain to the short-seller.

Clearly this works best in a strategic regime. The best-case scenario for the attacker is that most miners are responsive to bribes or willing to curtail or redirect their mining. In particular, if miners are prepared to respond immediately to high payouts for more lucrative mining, an attack would require a much smaller percentage of hashrate to accomplish, especially for a short period of time such as 4 hours.

13.7.1 Slippery-slope attacks

Because some elements of a Goldfinger or nation-state attack coming in to place will presage the attack, it may be difficult to predict the exact moment when the market will respond fearfully. There may be a less expensive strategy: If the market believes they are moving along a slippery slope towards a full attack, some traders will want to front-run the event. If an attacker can bribe one miner, they perhaps can bribe 5 % of the network. If they can bribe 5 %, why not 30 % or 55 %? If they can bribe miners to execute a two-block reorg, why not three, four, or 10? The realization of what may have previously been an unlikely scenario will increase the perceived likelihood of other unlikely scenarios.

We will call this the *slippery-slope attack*. A slippery-slope attack is a minor deviation in protocols by miners, intended to suggest that further deviation is not only possible but likely. As attacks begin to move along a slippery slope towards full strategic attack, market confidence evaporates, making such an attack more likely.

The movement by the market as a response to a small but ominous change in miner behavior could be a manifestation of market fragility. The movement could be small and short-lived as market participants may want to "send a message" that the market fragility is real. Short-term traders would also be happy to seize an opportunity to profit from volatility in the asset price. Needless to say, it is very difficult to predict from first principles how the market would respond to strategic overtures by miners.

A clever Goldfinger attacker thus need not launch an attack on the scale of the security budget: They can instead bootstrap the attack by merely convincing the market that such an attack is possible, at which point the market will price in the vulnerability. Ultimately this is due to market perceptions and beliefs about security, and is extremely difficult to predict. A well-positioned Goldfinger may have insights and possible influence regarding the market's reaction and would be well-advised to exploit this.

13.7.2 Selfish mining as a slippery-slope attack

Selfish mining may function as a slippery-slope attack. If a miner or pool with 30 % hashrate begins to selfish mine, this could be a watershed moment. If the economic majority supporting the Bitcoin network does not respond negatively to selfish mining, there is nothing discouraging the selfish miner from continuing to selfish mine, building out more hashpower in order to increase the returns. Clearly, there is no a priori magic number beyond 33 % at which selfish mining has gone "too far."

The exploit that starts with selfish mining and snowballs to a full takeover will be discussed further in Chapter 14, in which the market fragility hypothesis is assumed to be weak. However, if market fragility is a strong force underpinning the security of the network, at some point between 30 % and 49 % the market would experience a significant drawback, providing a lucrative opportunity for a short-seller or put-holder.

For example, a corporation or small cartel controlling 35 % of the hashrate may purchase thousands of Bitcoin put options, begin selfish mining, exercise the put options, and then stop selfish mining, declaring themselves chastised by a negative market reaction. This need not be a corporation. Modulo some logistical details, a pool could form with the explicit goal to perform selfish mining to increase revenues. The unspoken goal of the pool could be to profit from negative reactions to the pool itself forming.

13.7.3 Low-probability attacks

We briefly return to the setting of Chapter 5, where we considered the probability of completing a double-spend exploit. We focus on the probability of landing the attack and the time it will take, not so much the expected winnings. These considerations are important to either Goldfinger or nation-state attackers.

Suppose an attacker has one objective: to successfully create a double-spend by foisting a stealth seven-block reorg into place, supplanting six blocks in the process. Setting aside other possible financial objectives, assume the goal is simply to prove this is possible. The attacker starts by computing two transactions, each to addresses controlled by himself. These might as well be large sums, not so much for profit, but in order to maximize the theatrics of the reorg. The attacker works on mining a block containing one of these transactions. As soon as this transaction is mined, the attacker publishes the contradictory transaction and continues mining their chain in secret. If the attacker is able to mine six blocks before the rest of the miners mine six blocks, the attacker wins; they can reveal their secret seven-block chain when the main chain has built six blocks. Assume that attacker uses a simple strategy which requires they give up when they have fallen six blocks behind.

Recalling the negative binomial distribution,

$$\phi(i; 6, p) = P\,(\text{attacker has mined } i \text{ when honest mine their sixth block}),$$

we note that

$$\sum_{i \geq 6} \phi(i; 6, p)$$

is the probability that the attacker will have already found at least seven blocks (including the original) when the rest of the network has mined six blocks. This satisfies

$$\sum_{i \geq 6} \phi(i; 6, p) = 1 - \sum_{i=0}^{i=5} \phi(i; 6, p).$$

Now to include the other probabilities, in the event that the attacker has found only i additional blocks, which happens with probability $\phi(i; 6, p)$, the attack must overcome a deficit of $6 - (i + 1)$ blocks to win. This probability of winning a race when starting from i blocks behind, giving up at 6 blocks behind, is given by

$$w(i) = \frac{\left(\frac{p}{1-p}\right)^{i+1} - \left(\frac{p}{1-p}\right)^{7}}{1 - \left(\frac{p}{1-p}\right)^{7}}.$$

The total probability of winning is given by

$$1 + \sum_{i=0}^{i=5} \phi(i; 6, p) \left(w(5 - i) - 1\right),$$

where

$$\phi(i; r, p) = \binom{i + r - 1}{r - 1} p^i (1 - p)^r.$$

We see that while a miner with 10 % of hashrate has an unlikely 0.059 % chance of completing this attack after mining their first secret block, an attacker with 20 % has a 2.3 % chance. An attacker with 30 % has a 15.4 % chance of success with this attack, which is by no means unlikely. A miner who controls 30 % of the hashrate will obtain dozens of blocks per day. If they make m attempts, the probability that all will have failed is

$$(1 - 0.154)^m.$$

So even after 10 attempts, there is a greater than 80 % chance that they will have succeeded at least once. This math tells us that a miner deploying 20 % of the active hashrate who launches an attack twice every day has a slightly larger than 50 % chance of having landed the attack after 15 days.

We are not computing the expected value that takes into consideration the loss. A more precise expression for the expected value of such an attack would be given by substituting a reasonable value one can expect to win via a short for the double-spend reward λ into formulas found in Chapter 5.

A miner with 20 % hashrate must expect that they will be losing quite a few blocks in futile efforts, but even over 15 days, the total revenue distributed will be 1/24 of the security budget, of which the miner can only expect to claim 20 %. So if the security budget is \$30 billion per year, a miner with 20 % of the hashrate expects to claim only \$250 million over a 15-day period. If the same miner can net tens of billions by exercising put options or short-selling, the destructive route may become attractive. The miner must weigh the damage in future revenue against the windfall today.

Clearly, beliefs about market fragility and soft-forkability hypotheses play a huge role in a miner's computation. A miner who sacrifices millions in revenue in failed attempts may be very disappointed to discover that the market is unfazed by the reorg or collectively decides to reject it without much controversy.

Exercise 13.3. Suppose a miner assumes that the market will crash if selfish mining is detected. Compute the expected number of blocks supplanted in reorgs by a selfish miner with 33 % of the hashrate over a 3-hour time period. What is the probability that a miner with only 20 % achieves the same number of uncle blocks?

13.8 Soft-forkability: can the community suspend Nakamoto consensus?

Whether or not the community could or would suspend Nakamoto consensus remains an extremely important open question today. Historically, there is precedent on the Ethereum blockchain: The famous DAO hack of 2016 left the community with almost no other options but to break the protocol and manually set things "right." Ethereum was nascent at the time and was clearly not a leaderless organization. Nonetheless it was a controversial move.

Bitcoin has had software trouble: In 2010 miners were able to "create" 184 billion bitcoins due to a software bug. Clearly, this was not an issue with the underlying protocol; it was a problem with how the software detected improper money creation by miners. The blockchain was manually fixed and the bug was patched. But Nakamoto consensus was not suspended; one can argue that the longest *valid* chain is and was always the correct one. In the 2016–2017 Blocksize wars, a user-activated soft fork (UASF) was threatened, but not enacted. Nodes on the network have the ability to use the command `invalidateblock`, which marks a given block as invalid. This command could be invoked manually or according to some script that identifies malicious blocks. This has the potential to throw the network into disarray. The underlying Schelling point is that all blocks can be checked solely on their proof of work and their adherence to the protocol rule. Once this Schelling point is abandoned, if the discountenanced block is part of the longest chain, the network will remain fragmented until either an alternative chain grows to supplant the invalidated chain or the remainder of the miners all agree to invalidate the block. Many Bitcoin purists are very reluctant to use `invalidateblock`.

In fact, before 2017, this was widely considered out-of-bounds. The prevailing opinion seems to have been that if there were to be some majority exploit (say a double-spend), the best option was just to leave it as is: Merchants and exchanges should develop their own risk mitigation strategies when dealing with large transactions. Looking back at the opinions of early Bitcoiners, for example on `bitcointalk.org`, the dogma seems to have been that in the case of a hostile majority attack, the best (only) option is simply to abandon Bitcoin: Bitcoin, according to the whitepaper, was designed to work with an honest majority of hashrate – attempting to operate Bitcoin against a hostile majority is like spooning water upstream.

However, while it is difficult to accurately poll the average Bitcoin user to gauge their opinions, since 2017 a narrative has emerged that nodes or users could actually reject a malicious attack on the network. This is not just a story that users tell themselves; it is a story that if widely believed may become true. This new security layer embraces the idea that Bitcoin has evolved from a small-scale experiment to something that can and should be protected using other aspects of social consensus.

The mechanism is imprecise; however, some simple thought experiments suggest this is true: If a scientist is able to program a quantum computer, at great expense, to start from the genesis block and mine thousands of blocks per second at today's difficulty levels, the chain they produce would quickly become the valid chain according to rules of Nakamoto consensus. Certainly those with any interest in preserving Bitcoin would simply invalidate this chain, and Bitcoin would continue.

One can propose less damaging (and hence more controversial) scenarios: What if a one-off double-spend reorg of 15 blocks took place? One can only speculate; however, the important thing for the network is that a decision would likely become clear in a short amount of time and the network would continue in agreement with that decision. If there is no will to overturn the reorg, Bitcoin will continue as is. This may embolden other attackers to attempt similar exploits. If there appears to be a will to manually overturn the reorg, nodes can try to quickly communicate with other nodes that they are invalidating the malicious blocks. If more than half of the mining nodes participate in reorging the reorg, they will eventually prevail, the malicious chain will cease to be the longest, and Nakamoto consensus will once again be the rule. Certainly, miners who feel they will be on the wrong side of the outcome are wise to side with whichever side they think is more likely to win.

For many miners this means a calculation about the value of future paths of Bitcoin. Which precedent is worse: the precedent of a 15-block reorg going unpunished or a precedent for nodes manually overriding the protocol due to a collective opinion that the current state of the ledger was accomplished by untoward maneuvering? There is hazard for both. By allowing the exploit to remain on the chain, the exploiter keeps their proceeds. This encourages more exploits. There is now reason to believe that brazen exploits will be honored. On the other hand, once the community demonstrates a willingness to suspend Nakamoto consensus, this suggests a potential slippery slope: Could the network be maliciously split by a slightly more controversial exploit or at the polite solicitation of a government? There is also the question of moral hazard. Will participants be willing to undertake more risky applications of the network if they believe any major attacks can be overturned, as happened with Ethereum's DAO hack (which has drawn analogies to government responses to the global financial crisis of 2007–2009)?

Prior to any precedent being set, the mere threat of a community reorg serves as an important disincentive for many exploits and attacks. Even if a miner has rejected the market fragility hypothesis with high confidence, it is difficult to reject the soft-forkability hypothesis with the same confidence; the strategic ambiguity provided by this fact alone may be enough to dissuade certain exploits.

Of course we are speculating, but if exploits or attacks that require intervention are relatively rare, they could be governed by a sort of "common law" that exists outside of the protocol. Clearly, this is insufficient in the case of persistent attacks, but if infrequently tested it could form a second protective reinforcement edifice which operates outside the protocol. A malicious attacker aspiring to disrupt consensus about the ledger may have to try many times at great expense to successfully execute an edge-case attack.

There are dangers that were obvious to Satoshi: Any choice about when or how to reorg is necessarily a political one. Bitcoin is largely motivated by the desire to get away from political wrangling. Any discussion about whether or not a problem warrants a community reorg is likely to take place on social media, and many participants will likely form opinions based on the opinions of more prominent members of the community. These could be CEOs of major corporations, major influencers, whales, or even elected or unelected officials. It does not take an imaginative leap to see how a pattern of reorgs could lead to the types of politicking we see outside of Bitcoin.

This becomes worse if proposed solutions have the potential to favor players who may have acted imprudently. Indeed, Bitcoin's genesis block, created by Satoshi, has a message: `The Times 03/Jan/2009 Chancellor on brink of second bailout for banks.` Bailouts are the ultimate anathema to the Bitcoin ethos. Many would consider a Bitcoin amenable to bailouts a complete apostasy.

The moral hazard does not require an untamed imagination. The latest upgrade to Bitcoin's core software, called Taproot, allows for slightly more functionality, making it slightly easier to create smart contracts. Clever financial engineers can develop complicated tools like those being used in other parts of decentralized finance. Corporations following the path towards maximal short-term profit can build layers upon layers of intricate DeFi, and even algorithmic stablecoins, assuming that certain black swan events (for example, a six-block reorg) would never happen. History has shown us that caution is thrown to the wind when there is good money to be made in financial engineering. One can imagine DeFi profiteers developing financial instruments much more delicate and dangerous than collateralized debt obligations or other derivatives that played a role in the 2008 global financial crisis. For this reason, among others, those with influence over the debates surrounding the addition of new features to the Bitcoin protocol tend to err on the side of extreme caution.

Bitcoin is not one fixed thing, however. What Satoshi wanted Bitcoin to be and what the early adopters wanted it to be may be different from what it will become. What it will become is determined by its trajectory and the desires of the "economic majority" of participants in the future. The fact that soft forks may deeply offend Satoshi does not preclude Bitcoin from following its own path.

Logistically, community reorgs may be difficult. If a six-block reorg is considered disruptive, it would stand to reason that the community would want to have the problem resolved within an hour. Clearly, the miners cannot wait and have this discussed at great length, lest the ledger be subject to even more violent deep reorgs. This requires vigilance by a majority of miners and node operators around the world or some system allowing deference to some form of leadership.

13.8.1 Additional security layers

In its purest form, Nakamoto consensus declares the chain with the most chainwork to be the valid chain by default and leaves miners free to operate within this specification. There is nothing forbidding extra-protocolar coordination by miners or nodes to correct misbehavior. Such a prohibition is ultimately impossible to enforce and therefore hopeless. Nakamoto consensus is not based on the absence of secrets or coalitions, but rather on the visibility of proofs of work. Miners or business leaders who are interested in creating a judicial layer to dissuade attacks and exploits do not have to ask permission to do so.

Noting that the community may be forced to make decisions in short time during an attack, there is potential value in the formation of an organization who would aim to shore up security. Certainly some Bitcoin advocates find this notion unpalatable, but if Bitcoin is to expand and gain more broad acceptance by financial institutions, an additional layer might be necessary to give added assurance that low-probability attacks are unlikely to materialize. If a significant portion (again the "economic majority") prefers the clarity provided by additional layers of governance, it may become integrated into the general ecosystem. The participants in the Bitcoin network who are opposed to this can stay or they can leave; telling corporations to simply get their own coin is fruitless. Bitcoin should remain, in principle, for anybody, even those who choose to trust others or defer to leaders.

Example 13.3. Suppose there exists a loose coalition of mining and financial corporations who all have a common interest in keeping the network running smoothly. They realize the possible chaos that would ensue if a reorg is performed and understand how difficult it would be to override a malicious reorg if the network does not act in a timely and unified way to correct this. They form a 24-hour rolling panel of nine members who are close to their nodes and able to make judgements within minutes as to whether a manual reorg should be performed or not. If five of them agree, the group signals a reorg is necessary, and all miners and nodes in the coalition defer automatically by default. Members of the coalition would install a simple patch that watches for this signal.

This would not be forced consensus or take away freedom of the mining operations to decide which chain to build on; it is not a contract or anything binding. Rather it acknowledges the fact that it may be difficult for all operators to assess and come to agreement on whether a particular reorg is damaging enough to warrant intervention, at any hour of the day or night. If they can defer this decision to a trusted party, consensus will happen much more smoothly.

This panel would be formed with the idea that it would be rarely or never used. If a majority of the mining defers to this panel, a malicious attacker would be discouraged on several fronts. The attacker would be unable to perform an edge-case attack – because a decision would be made quickly about whether a reorg belongs or not, the edge-case attack has little hope to disrupt the network. Nor would the attacker be able to perform egregious exploits: Obvious double-spends would be rejected by the panel with high probability. Similarly, a pointless but deep reorg with the only purpose of destroying the network or activating a purge attack would be likely to fail. The sacrifice is that the community must place a small amount of trust in the panel, trust that only rarely will be leaned on.

Some will immediately complain this arrangement gives the panel too much control. However, the panel can just as easily be overridden. In fact, overriding the panel if the panel rules in a bizarre way or appears compromised would require the same effort that it would take the network to override a malicious reorg without the panel. Further, the same "skin-in-the-game" argument that is often applied to miners in general holds: The panel is chosen by parties who are motivated to see smooth operation going forward. If the panel fails its duties, the default is to revert back to the absence of the panel. Political differences may emerge but these could be dealt with by a panel of nine in a short time. Some members of the panel may have a very light-handed policy, others may not, but if a healthy fraction of miners defers to the majority decision, the network should have a decision and continue without disruption. The goal of consensus is consensus, not prosecuting morality.

A potential problem that such a panel would pose would be the anonymity and physical security of the members. A well-motivated attacker can perform a denial-of-service attack on the panel members or do much worse. If the panel is the only safeguard against a malicious and lucrative attack, the attackers may go to great length to render the panel incapable. This is a concern that would need to be addressed. However, this is not insurmountable; important people protect themselves in the real world every day.

The primary function of such a panel would be to diminish the reward that an attacker can expect to easily garner via an obvious reorg, thus discouraging low-probability attacks. Nakamoto consensus does not prohibit such a panel, and miners do not have to signal their deference, so it would be unclear until tested how much sway the panel actually holds. Drastically decreasing the rewards available to an attacker significantly boosts security.

This does lead to an uncomfortable question: If we can trust a panel to correct reorgs, why do we spend billions of dollars on proof of work anyways? Will this not just devolve into proof of stake? This is a valid question. To answer quantitatively, Bitcoin miners produce 144 blocks a day, i. e., 52,560 blocks per year, and these are secured on the tougher, more expensive security level. Very few, if any, breach this "hard" security layer and make it to the "soft" layer. If the soft layer were placed outside and were to face attacks and politically fraught decisions 144 times per day, cracks would form and it would quickly become untenable. But this is not the case; the soft layer would be rarely tested. Any decision made by such a panel would be closely scrutinized and would be more likely to be "right," or at least, not so obviously corruptible.

Open-ended Question 13.2. What would be the potential downside of such a panel?

13.8.2 Example: purge exploit and the buyer beware response

We revisit the idea of a purge pool as an exploit that the community may decide to ignore. This exploit would represent a collusive effort by deviant miners and spenders of

Bitcoin. If market fragility is not a big factor, perhaps because Bitcoin has become too large to fail, and at the same time the network is reluctant to soft-fork away any egregious exploits, the incentives could align. It becomes especially easy if a miner is able to acquire dormant inefficient hashrate that could be deployed only for the attack.

Example 13.4. Suppose that 20 years into the future, Bitcoin is held by many institutions and used by many as a medium of exchange as well. A major move away from Bitcoin would cause significant pain, so for the purpose of this example Bitcoin is "too big to fail." Exploits have happened from time to time, but the community has ossified around the Schelling point that intervention is not a viable option: Exploits or attacks are left in place, and Nakamoto consensus is never suspended.

Due to some market factors, suppose the revenue for miners has decreased significantly, many machines are offline, and the HDR drops to 20 %. A miner begins to purchase dormant machines, acquiring over a third of the dormant hashrate. Some transactors publish "Hail Mary" double-spend transactions, which are stored by the miner. When enough of these transactions accumulate within a relatively short window, say 12 hours, the miner fires up all the machines and performs a long reorg, collecting the large fees while returning much of the funds to spenders via their double-spend transactions.

At this point, the community consensus against intervention may be put to the test. There may be outcry from those who whose funds were "stolen" by this exploit. However, this group may consist of parties less likely to speak out. Claiming to have funds stolen is an admission that you have dealt with shady counterparties. Law-abiding institutions with above-board KYC relationships with their counterparties would not be affected by such an exploit.

The community need not intervene. Instead the community can re-emphasize that with the laissez-faire nature of Bitcoin, a "buyer beware" approach is the best remedy. The coins which have been double-spent, while still Bitcoin, will not be accepted so readily in the future, at least by merchants paying attention: It will be public knowledge that these UTXOs are owned by people who are willing to cheat – anyone who accepts these coins in return for anything of value must take precautions to insure they do not get cheated. This could have the effect that only the purloined coins have lost value, having associated themselves with a risk of future reorgs. This example demonstrates how *tainting* can serve as a reasonable non-binding community response against certain exploits, obviating other types of interventions and serving as a deterrent for double-spends.

13.8.3 The government to the rescue

In §13.8.1 it was assumed that the formation of the back-up security layer was somewhat organic, but this need not be the case. If Bitcoin were to become a critical piece of infrastructure for one or multiple nations, these nations would be wise to have a plan of action ready to go in case of an attack. If there is no clear alternative plan of action and the government presents a clear plan that works, an economically significant portion of the network could follow. This could creep towards overt centralization or a switch to a permissioned network, but businesses and individuals who are relying on Bitcoin transactions happening may be willing to play along.

13.8.4 Economic majority vs. mining majority

The term "economic majority" does not refer to a specific group of Bitcoin users in a universal way; rather, it makes more sense in relation to a given proposition that participants of the network could be debating. An *economic majority* is a portion of the users of a network that is significant enough economically to wield veto power over the network's continuance or adoption of any protocol rule. The mechanism for wielding this power, and hence the term, is necessarily somewhat vague and difficult to define (also, the "majority" may in fact be a misnomer; "supermajority" or "predominance" may be more appropriate).

In practice, if some rule is proposed and an economic majority finds this objectionable, the rule should not come into effect. If it were to come into effect, the economic majority would opt for a fork of the network that does not adopt the rule. The chain preferred by the economic majority would have more activity, more transactions, and more demand for the asset, leading to higher prices of the assets and more robust market for blockspace. Miners would opt for the stronger chain, which produces more revenue and is more secure. The weaker, less desirable chain would likely fall off in popularity, becoming less valuable and less secure.

This can also happen the other way: In 2017 an economically significant portion of users lobbied for a protocol upgrade that included the technical upgrade SegWit. When some miners and other players were hesitant, the users threatened to perform a soft fork. The users ultimately prevailed without a direct showdown; the threat of creating the fork was sufficient. Many overly simplistic interpretations of the blocksize wars of 2016–2017 have been offered. It certainly is not clear whether we saw universal principles playing out or individual actors using specific tactics in a narrow game. Negotiation involves many factors: what each party has to lose, what they have to gain, and what they believe their odds are in a showdown. If miners have little to lose by accepting the terms of a UASF, they are less willing to take the risk of bringing this to a full-fledged chain war.

In 2017, a revolt by a majority of miners against an economic majority of users could have destroyed the network. Market fragility was still a strong cohesive force: Certainly miners understood the risks of a nuclear war and knew the importance of preserving Bitcoin's credibility. Thus the miners were wise to follow the lead of what was perceived as the economic majority. One interpretation is that in 2017, miners bore the most risk, so they had less control. The balance of risk is something to watch in the future, especially if Bitcoin users and their nodes begin to develop a reliance on Bitcoin.

In a future regime where market fragility is not as strong, it remains unclear and completely situational as to how a conflict would play out.

Example 13.5. Suppose that a reorg of 10 blocks happens. A would-be economic majority, perhaps via a panel or another collective decision-making body, decides this reorg is "bad for Bitcoin" and should be overruled. So the nodes in this majority perform a soft fork, using `invalidateblock` to invalidate the blocks on the reorg chain. Miners, however, bristle. They feel that allowing a panel to assert themselves as authorities on the correct chain undermines the whole point of Nakamoto consensus and band together to resist: Instead of respecting the invalidation of the nodes on the network, the miners continue to mine the longest chain. A subset of miners may see an opportunity and attempt to mine the weaker chain, but miners from the mining majority work together to attack the chain using empty-block, seesaw, or edge-case attacks. If the miners persist, the users who want Bitcoin to continue to be unified must revalidate the blocks. The chain preferred by the miners will continue to function, so it will remain viable for anyone who is not too upset by the possibility of reorgs. This outcome is not certain: The miners may just as easily decide "not the hill to die on" and capitulate to preserve network unity. The strength of their resolve comes into play. This is why "economic majority" can be a very vague notion to define.

13.8.5 A tetralemma

We conclude this section by observing the following tetralemma.

Conclusion 13.2 (Bitcoin security tetralemma). In as much as an attacker can marshal sufficient hashrate, one of the following must be true:

1. Bitcoin is exploitable.
2. Bitcoin's value can be destroyed by a malicious actor.
3. Nakamoto consensus can be changed or suspended.
4. Some transactions are taintable.

Proof. First of all, let us suppose that (2), (3), and (4) all fail. If (3) fails, Bitcoin's rules are etched in stone, and nobody can suspend Nakamoto consensus. Miners and nodes respect the heaviest chain as the valid chain in all situations. If (2) simultaneously fails, then no amount of selfish mining, double-spending, fee-sniping, malicious reorgs, or other shenanigans will harm the value of Bitcoin. In this case, not only is there nothing preventing selfish mining, but the market does not respond negatively to it. An attacker with 35 % may begin selfish mining without fear of their blocks being considered invalid or losing value of bitcoins mined. Similarly, an attacker who can string together enough blocks to perform a reorg resulting in a double-spend can enjoy the fruits of their double-spend without worrying about a loss of faith in the market. By (4) any proceeds are untainted and can be spent across the network. Necessarily, we conclude (1) must be true, and the attacker can exploit Bitcoin for fun and for profit. □

14 Direct frontal takeovers

Since the early days of Bitcoin, participants in the security discourse have espoused various expressions of the market fragility hypothesis. The market fragility hypothesis rules out certain direct assaults on Bitcoin's decentralized security model – no one would try to take control of something if that thing were to become worthless. However, since 2020, as the price has risen and more above-board institutions have placed their bets on the currency, it becomes important to reckon with the idea that the market fragility hypothesis may not hold to the extent that it did in early years, when Bitcoin was a fledgling experiment.

One particular exploit, which most certainly would have been a resounding death knell in 2014, becomes available if the market fragility hypothesis fails. The term *arrogating exploit* refers to the situation where any party, which could be a single entity or a cartel, arrogates to themselves full control of Bitcoin's future transactions by acquiring 51 % of the hashrate and ignoring all blocks mined outside of the cartel, for purposes of controlling rather than destroying the network.

The motivation is simple. Once a cartel has arrogated full control of the mining, they can receive 100 % of the mining rewards for slightly more than 50 % of the costs. Further, as the competition begins to abate, the cartel can ramp down their mining expenditures and the difficulty at the same time, as long as they reserve enough equipment to counter any attempts to regain control by other miners. They then have a monopolistic pricing control on Bitcoin transactions, at very low cost, and have discretion to pick and choose who is allowed to make transactions on the network, which is a very enviable and powerful position.

The execution can take one of several simple forms. The most straightforward is as follows: A corporation continues to expand their hashpower by purchasing equipment directly or by merging with other mining operations. Once they near 33 % they begin selfish mining. As the size of their hashpower snowballs toward 51 %, they bring in more corporations as partners as necessary or simply purchase equipment from corporations who see the writing on the wall and bow out.

Another simple execution of the attack is for a miner controlling some percentage of the hashrate to communicate to other parties the intention to perform the attack and offer other miners a share if they join. Once the cartel reaches 51 % participation, the attack begins.

It appears too simple. But what stops it?

Such an exploit would represent the ignoble demise of the Bitcoin that Satoshi invented: It would be both effectively centralized and subject to censorship. So Bitcoin purists would declare that any power or wealth obtained through this is ultimately worthless. But this assumes market fragility. One has to consider that there are parties who may be using Bitcoin for reasons that are not tied to decentralization or its censorship resistance properties. If corporations, individuals, central banks, and the like are using Bitcoin as store of value, an inflation hedge akin to gold, there is no reason why

https://doi.org/10.1515/9783110772845-014

cartelization of the mining process would devalue this, unless they believe others will find Bitcoin less valuable in the future. The store-of-value narrative surrounding Bitcoin requires only a belief that an economically significant portion of investors around the world will continue to share a common interest in a fixed supply currency.

If the demand for Bitcoin is dominated by such institutions, there is good reason to believe that the market fragility hypothesis does not hold to the degree assumed by early Bitcoin advocates. Thus any corporation with enough resources would be motivated to take on such an exploit. If the exploit appears likely to pay a windfall, it would most certainly be financed by profiteers.

We are leaving legal issues aside: Would the US Department of Justice (or some other governing body) attempt to break up such a cartel by prosecuting anti-trust laws? Relying on the US Department of Justice for preservation of their network would be unwelcome for Bitcoin purists: This would imply some sort of capture. However, even government enforced decentralization may still be preferable to the alternative.

Assuming that price would not plummet and that government officials would not intervene, it would be up to a economically significant fraction of the network to organize against it. However, this would be somewhat difficult. The longest chain according to Nakamoto consensus will with high probability belong to the cartel, so the resistance may attempt to use `invalidateblock` on blocks in the cartel's chain. This would lead to a fork; there would be a subset of nodes that accept the cartel's chain as valid and a set of nodes that do not accept the chain as valid. The cartel then has two options: either let the fork be, expecting it to diminish, or demolish it by attacking it with empty-block, seesaw, or edge-case attacks designed to throw the ledger into confusion. If the cartel would like to maintain full control over Bitcoin, attacking the weaker chain is an easy and obvious choice. Eventually the resistance would fade away to attrition.

This assumes that an economic majority would not be horrified by the development. Indeed, this is something that the opportunist should gauge ahead of time and is baked into the market fragility hypothesis. Even if the economic majority objects to cartelization, unless they can safely exit their positions without suffering significant losses, they may be trapped. Lack of a readily available and viable alternative would pin institutions to Bitcoin, especially if their profit model relies on Bitcoin functioning day-to-day.

14.1 Institutional response

Many profit-seeking entities would be not be affected by Bitcoin mining transitioning to a single miner who would be as reliable as an unlimited number of Bitcoin miners. A single miner would likely keep Bitcoin running smoothly, without undercutting or reorgs. A single Bitcoin miner would have power to root out bad actors from the network, which may reduce ransomware and other criminal activity, making the network more pristine and above-board. Further, a single Bitcoin miner would be able to reduce the hashrate and with it the energy consumption, which would be important if Bitcoin is facing social or political pressure to reduce the environmental footprint. So a centralized miner may actually be preferred by some institutions.

On the other hand, this would put a much larger amount of control into the hands of the cartel. A large institution that formerly relied on fee rates determined by the mempool could be forced to renegotiate mining fees. The monopolist can use differential pricing, effectively charging each institution according to what they are willing to pay. This could be very costly, and the institutions would be loath to accept this.

The crucial condition here is that there exist no viable alternatives for the economic majority of Bitcoin users. If holders are unwilling to let go of Bitcoin because it has a history of serving as a value-preserving Schelling point, it will be very difficult to agree on a new Schelling point that has the same guarantees. It the institutions are using Bitcoin because of its unique network effect and the infrastructure that is built up around it, it may be impossible to switch to a substitute, at least in the short term.

So even though the monopolizing cartel may have an advantaged position in fee negotiation in the short term, they may choose not to exercise it: If the monopolist pushes their advantage too far, alternative networks who use other consensus mechanisms may develop to take Bitcoin's place.

A more general version of the market fragility hypothesis must hold, almost tautologically: If a cartel or single miner makes Bitcoin unusable, the value must plummet. Certainly the profit-seeking cartel does not want this and will strive to preserve most of the functional properties of the network that were present before takeover.

Open-ended Question 14.1. Is Bitcoin with an explicit cartel more or less secure?

14.2 Creeping corporate takeover

There could come a time when Bitcoin must respond to a threat. This threat could be political or social: It could stem from sources opposed to Bitcoin mining based on environmental concerns or on Bitcoin's role in facilitating ransomware or terrorism. Political authorities could threaten to subject Bitcoin to regulations that would damage the price, or Bitcoin could simply be discountenanced by polite society, leading to a secular decline in the demand over time. At this stage, a loose organization of some sort could assert some form of leadership. A "green energy memorandum" could be signed by members of the coalition, pledging to use only renewable or a bounded amount of non-renewable energy sources. Similarly, the coalition could enact a weakly enforced blacklist: Members of the coalition could choose not to include some transactions in their blocks – this would not immediately censor illicit network users such as ransomware purveyors, but could increase the friction and costs of illicit activities and would also ensure that the above-board mining institutions do not directly profit from anti-social uses of the network.

While this may begin as a public relations move, it could also calcify into a more coercive structure. If a majority of miners agrees to blacklist an execrable terrorist or dictator who has committed some unspeakable act, they can make this enforceable by agreeing to supplant any blocks that include the forbidden transactions. Depending on the mood or social pressure, this could be a welcome and uncontroversial response.

Again, it is important to note that what may be controversial to the economic majority of Bitcoin users in 2017 or 2022 might be not controversial at all to an economic majority in 2042. It is possible that there will be no economic majority opposed to miners enacting sanctions against a despot by censoring transactions. A panel invalidating a malicious reorg to right a chain that was disrupted by a deep reorg may be preferable to most users.

An entity interested in arrogating control over blockspace may begin to send signals that centralization has improved outcomes. It the resistance is feeble, the entity is likely to continue. If the pieces for cartelization are in place, it could happen quickly in response to a geopolitical event.

14.3 Alliances are beneficial

Alliances and coalitions are natural in repeated games. If Bitcoin mining trends strategic, it may become worthwhile for miners to form alliances for various purposes. For example, some miners may sign an agreement not to undercut each other. This is particularly useful for larger players. If two miners, each with 20 % of the hashrate, agree to not undercut one another, the expected rewards for undercutting either of these two will diminish: Any undercutter is afforded at a maximum of no more than 60 % of petty compliant miners to help them consummate the undercutting reorg. This increases the amount that each of the larger miners can safely include (recall (11.2)). The same two miners could confer on other strategic plays: If a bribe is offered and the two miners could decide to both accept or both ignore the bribe, it is certainly in their best interests to work together. Conversely, miners may also decide to communicate directly in regard to their own undercutting strategies. They can communicate or negotiate criteria upon which they will use undercutting and pledge to support each other if one undercuts, perhaps taking advantage of unaffiliated miners. Naturally, if more than one miner is coordinating with another miners about their strategies, it is only a small step to begin selfish mining together.

14.4 ESG takeovers

Alliances of miners can form for environmental, social, and governmental (ESG) reasons. Miners who follow sustainable mining practices may form an alliance to accomplish environmental objectives. If forcing less sustainable miners to become unprofitable and cease operations becomes a moral imperative, they can use their alliance to begin to selfishly mine for this explicit purpose as part of a pushout attack on the other miners. This only requires 34 % to begin: This coalitional selfish mine could snowball into a full arrogating operation in which only approved, sustainable firms are able to engage in mining.

Similarly, alliances with geopolitical objectives may form. In fact, an alliance can form based on any common enemy: Pollution-intensive mining, terrorists, ransomware purveyors, despotic regimes who violate human rights, or even miners who are threatening the network with unpredictable behaviors. If at any time 34 % of the hashrate is interested in reducing the slice of the pie taken by another group, they can begin selfish mining to accomplish this. Once they begin, they can demand that other miners choose a side: Miners that attempt to remain neutral will be victimized by selfish mining, while those who join the selfish miners (provided they meet whatever criteria are needed) will find themselves with a larger share of the pie.

Note that the ESG takeover does not need to stop at 51 % and need not be motivated by profit. It can be undertaken by any alliance of miners who would like to impose ethical standards on all miners. Any miners who can verifiably meet the standards are welcome to join the coalition.

14.5 Government KYC takeovers

Similar to ESG-motivated takeovers, miners who comply with government regulations may band together to try to force all miners to submit to similar restrictions. For example, a government may insist that only whitelisted transactions be used: Users of the network can register with the government or exchanges that use full KYC protocols. This can be used to whitelist addresses or transactions. If a government forbids miners from including transactions not on the whitelist, the miners who must adhere to this might as well enforce this edict on the rest of the network. If they comprise 34 % of the hashrate, they can begin selfish mining. Other miners who then decide to join the cartel mining only whitelisted transactions will receive an increased share of the blocks mined. This will be desirable provided that the loss in fees resulting from a smaller mempool is not too large.

Beliefs about market fragility should play a role. Any move in the direction of whitelisted mining can be seen as an attack which could massively devalue the asset. Miners who believe this would be wise to resist the government compliant miners. In fact, they may even choose to form their own one-time single-purpose alliance with the purpose of running these miners off the network.

If market fragility is not a factor, profiteering miners with no ideological qualms about whitelisting might as well comply with the whitelist. Market fragility can work the other way: If institutional use of tainting has created a market in which tainted coins are significantly less valuable, taking this principle one step further into a walled garden by implementing a whitelist could make the coins more secure and more valuable.

14.6 Overexposed bagholder exploit

Another motivation to perform a direct takeover is to harm or extort a single entity who is overexposed to Bitcoin. This can be a nation-state, a large corporation, or simply a particularly annoying billionaire. When the security budget of Bitcoin over a significant period of time becomes significantly smaller than the value of Bitcoin to a particular individual, a strategic player can use this to exploit the overexposed party.

The market value of mining equipment should be roughly the expected revenue of the equipment going forward minus the expected operating expenses. Given that most ASICs die or become obsolete in a few years, it is not difficult to imagine a scenario in which the total security stock value (recall §13.4) is much smaller than the total value of the asset held by one entity.

Example 14.1. Suppose that a single entity own 210,000 bitcoins, which is 1 % of the total supply. The security budget in the year 2037 will be 20,508 bitcoins as a block subsidy plus fees. The block subsidy for a whole year would comprise less than a tenth of this whale's holdings. Suppose that fees have matched the block subsidy in rewards share, and suppose that mining profit margins are 20 % and that the average life expectancy for the hardware is 4 years.

A back-of-the-envelope computation (recall §13.4) with these parameters suggest that the total security stock value will be about 15,000 bitcoins, which is a small fraction of the whale's holdings. The discrepancy between the exposure for the single entity and the security budget sets up an opportunity for someone who may try a direct takeover. A better description of how this exploit is beneficial requires some discussion of bargaining, so we will return to this exploit in §15.4.

14.7 Fee market and long-term viability

As noted in Chapter 11, a dwindling block subsidy could create security problems if a robust fee market does not develop. In order for a consistently high demand for blockspace to develop, blockspace bidders must be willing to pay high prices. Bidders' willingness to pay high prices day after day means that less expensive viable alternatives must not be available. In other words, if Bitcoin users are willing to pay $1,000 per transaction on a regular basis, they are doing this because this is clearly the superior option, not because of some esthetic reasons or a desire to experiment with new technology. If other blockchains or centralized private networks can accomplish the same objectives for $2 per transaction, rational agents will naturally prefer the cheaper option.

By nature, high fee rates will exclude many parties from making on chain transactions: It makes sense that many of the transactions will be paid for by larger institutions, or at least quite wealthy individuals. This would make it even easier to put together a substitute product that would arbitrage the high fee rates: If a handful of institutions are collectively spending billions of dollars annually on transactions, they may seek a cheaper solution.

In order for Bitcoin to maintain high fee rates, the Bitcoin blockspace must be something for which no substitute is readily available. This comes with a catch: If no substitute is readily available, the market fragility hypothesis should fail, and Bitcoin is vulnerable to an arrogating exploit. In the event of a takeover, there are no easy exits for Bitcoin users; their best option will be to stay put.

14.8 How much is censorship resistance worth?

According to early conventional wisdom about Bitcoin, market fragility protects against direct frontal attacks. Any takeover would spell the end of the censorship resistance narrative, and Bitcoin would become essentially worthless for many people.

But this assumption may or may not hold up when challenged. This raises the following question: How much is censorship resistance worth? Can this be modeled?

The question is slightly muddled because it involves a bit of a Keynesian beauty contest: Some market participants who have no need for censorship resistance themselves will invest in the asset because they believe that in the future, censorship-resistant money will acquire a premium. Censorship resistance is perceived to be of high value, so in some cases, a takeover would cause a sell-off not only among those who actually need it, but also among those who believed that censorship resistance gives the asset special value.

This highlights competing narratives about Bitcoin that contribute separately to the asset price and market robustness. The store-of-value narrative can be independent of censorship resistance, but need not be, and probably is not in practice. Certainly a large corporation who holds Bitcoin on their balance sheet as a store of value is not doing this in hope of one day making clandestine transactions; however, they may perceive the asset to be more in demand and hence more valuable so long as it retains this property. Another source of demand for Bitcoin is legitimate above-board use as a currency, especially when there are no obvious substitutes. This use is completely independent of censorship resistance. Corporations that use Bitcoin do not always need to hold significant amounts and may be less concerned about price movements provided they have access to the liquidity they need. We note four (non-exhaustive) categories of premiums which bring demand and value to Bitcoin:

1. censorship-indifferent store-of-value premium,
2. censorship-resistant store-of-value premium,
3. censorship-indifferent utility premium,
4. censorship-resistant utility premium.

From a security and game theory perspective, an entity who would attempt a frontal assault should estimate the contribution of premiums derived from censorship resistance. If the premium is small, or if the censorship-indifferent premiums are large, the arrogating entrepreneur may see this as a green light. We revisit this with models in Chapter 17.

14.9 A 67 % fiat attack

The most extreme form of the arrogating exploit is a 67 % or fiat attack. The goal of this attack is to gain enough control over Bitcoin's protocol to arbitrarily change rules. Taken to the extreme, changing the rules can actually change the amount of Bitcoin held by individual accounts.

The idea behind this attack is that if the attacker has 67 % of the available hashrate, they can hard-fork Bitcoin with their own rules and then split their resources between defending the hard fork and attacking the legacy chain, both battles which they are assured to win with high probability. The arrogating attack described previously in this chapter should be considered a soft fork: The miners are restricting blocks to those which are produced by a member of the cartel; all blocks remain backwards compatible. However, hard forks can change rules.

Example 14.2 (Miner-activated hard fork). Suppose that in 2031, a miner cartel with over 70 % of the hashrate decides that the halvings should no longer happen (this would enact a security proposal called *tail emission* designed to prevent the security budget from vanishing). These miners declare that at the next halving block, they will continue to produce blocks that credit to themselves the same block subsidy of 1.5625 as was credited in the previous halving period. This rule change would constitute a hard fork, and nodes following the legacy rule set would necessarily reject these blocks as invalid. The cartel would like to ensure that these hard fork blocks become the only viable Bitcoin. In order to ensure this, they spend half of their hashrate using empty-block, seesaw, edge-case, or other attacks on the legacy blockchain, rendering it completely unusable. Businesses who want to continue using Bitcoin will have to upgrade to the new software, which sees these blocks with the 1.5625 subsidy as valid.

Example 14.3. Miners can be more extreme. Suppose a 67 % cartel announces a hard fork to a more permissive rule set: Transactions that source from a published set of blacklisted UTXOs can now be spent without a valid signature, provided the output is to a published set of addresses that are associated with charitable institutions. Ideologically, this is a major departure from the original Bitcoin protocol, even if it only affects 200 bitcoins. Provided that the cartel sets aside enough hashrate to destroy any legacy chain that exists after this hard fork, they can make sure that the blockchain allowing charities to spend confiscated terrorist Bitcoin is the only viable Bitcoin.

So while it is often claimed that a 51 % attack cannot steal anyone's Bitcoin, it is true that a particularly heavy-handed attack can steal Bitcoin. However, to do this, it must be performed judiciously enough to not destroy Bitcoin completely, and it must be done in concert with attacks on competing chains. The attacker must destroy all other forms of Bitcoin without making their version of Bitcoin unpalatable.

14.10 Permissioned bypass

Consider a situation in which public-facing corporations comprise a significant amount of Bitcoin use and holding. These could be financial and technological institutions throughout the world. As the fee market begins to replace the subsidy, the fees could mushroom and create significant cost. While these corporations value Bitcoin and like the network topology and reach, they are reluctant to continue to pay heavily for fees. In response, rather than taking over the network via brute force, they simply declare a set of permissioned validators and essentially soft-fork the chain. They could follow this with a hard fork, expanding the blocksize considerably, which would have the effect of bringing down fees significantly. Any other business that wanted to remain in the same network as the majority of these corporations would continue to transact here, as the network effect would trump the decentralized esthetic. Provided other corporations are not breaking any laws or doing anything untoward, they would not be censored.

A *permissioned bypass* is a much easier transition away from proof of work compared to proof of stake, which is complicated and has its own set of game-theoretic concerns. If proof of work is being attacked or if a strong economic majority would like to transition away from proof of work, a permissioned bypass requires very little alteration of the software: Essentially only the difficulty adjustment must be recalibrated once in order to keep the chain moving, and all nodes should be programmed to reject any blocks not coming from an agreed-upon set of permissioned validators. This requires a very strong and clear economic majority and is predicated on the idea that the blockchain that has been abandoned by the bulk of corporations will become economically worthless. More generally, a *Davos attack* is a move by world financial powers to arrogate the store of value placed in Bitcoin by unilaterally agreeing to a new means of transaction validation, while dumping all coins held on the legacy chain in order to distinguish the new chain as the clear store of value.

15 Nash bargaining

Bitcoin is designed to operate peer-to-peer, globally, outside of any particular legal jurisdictions. Courts or legal powers have no easy way to wield any authority over the Bitcoin blockchain. This has some interesting consequences; existing beyond the reach of traditional authorities means necessarily also existing beyond the reach of any protections these authorities would provide.

This leads to an economic regime that may behave differently than those that are subject to traditional law and order. In particular, blockchain ecosystems are more adversarial by nature, being designed with adversarial actors in mind. This leads to different considerations: Adversarial actors may attempt to leverage destructive powers in ways that might not be possible inside of law-and-order dictated regimes.

For example, it is generally not legal in most jurisdictions to intentionally destroy or sabotage the competition. One wireless cellular carrier cannot deliberately interfere with its competitor's network in order to drive customers away; in most places this is illegal or legally actionable. This means that when a wireless carrier considers strategies for beating their competition, they do not consider directly sabotaging the competition, and they do not have to spend great resources defending against sabotage by the competition. Generally this feature of law-and-order is a very good thing.

Extortion is the use of coercive means to extract something of value and is ubiquitously illegal. One should not legally be able to demand a payment in exchange for not burning down a house. Ransomware is real, but this works only because those who create and profit from ransomware are out of the jurisdiction of their victims, by virtue of either physical location or the shield of anonymity. The cost to destroy a $400,000 home is merely the cost of matches and a few gallons of accelerant, but this does not create an arbitrage opportunity; there is not a good way for an arsonist with extortionary aspirations to extract payment from a victim in a rule-of-law setting.

Thus an attempt to appreciate the game-theoretic dynamics of Bitcoin must take into consideration how malicious actors could use any destructive means to try to coerce or extract value from other users of the ecosystem. One should attempt to account for all reasons for attacking the network. So far we have discussed security considerations in which miners attempt to exploit the network via double-spends, to attack the network for short-selling or authoritarian reasons, or to take over the network in order to extract extra profits or gain control. In this chapter we discuss the Nash bargaining solution and explore the economic regime in which nothing stops actors from behaving destructively. This is more interesting when discussed in the context of decentralized ecosystems beyond Bitcoin. Smart contract platforms afford adversarial actors access to a number of exploitative strategies that are not typically available in law-and-order regimes [27].

https://doi.org/10.1515/9783110772845-015

15.1 The bargaining problem

The bargaining problem is often given as follows [46, p. 97]:

> Two individuals have before them several possible contractual agreements. Both have interests in reaching agreement but their interests are not entirely identical. What "will be" the agreed contract, assuming that both parties behave rationally?

This age-old question has long frustrated game theorists. Without ways to enforce binding threats, negotiation can be unpredictable, and often the outcome is determined by "negotiation skills" that are difficult to quantify. Solutions to hard problems are by nature evasive, so academics usually restrict themselves to tractable models. In 1950, Nash gave axiomatic characterizations of what is now called the Nash bargaining solution. The solution itself was not presented in a procedural or dynamic way – it was not at all clear how players would arrive at this when starting negotiations from scratch. Since then, dozens of economists and game theorists (including Nash in 1953 [39]) have offered models which indicate possible strategic approaches to induce a solution. A notable paper by Rubinstein [46], in 1982, introduced time preference into the model; players are penalized as negotiation grinds on. Other papers study how negotiation proceeds under the probability of spontaneous breakdown (see for example [6]). Impatience and breakdown risk are realities that resonate with experience, so these models have been very important to the theory. A 1992 paper [47] introduced a model in which players get to create their own probability of breakdown. While this has been dismissed as unrealistic, we would like to examine models with this feature in light of current technology: In 2022 decentralized smart contracts allow for threat, probabilistic threat, and other commitment mechanisms that were not available decades ago [27].

We refer the reader to T. C. Schelling's classic Cold War era text [48] for a colorful and fascinating discussion of how bargaining theory meets the real world. A more modern reference on the theoretical side is given in the monograph [37], which includes a thorough discussion of Rubinstein type games.

15.1.1 Motivating real-world problems

Example 15.1. Mike creates $1,000,000 in value for corporation X every year, but is paid only $100,000. Because Mike is constrained by a unique skill set and geographical location, X never offers him more compensation. Mike threatens to leave if he is not given a raise, but has nowhere to go.

Example 15.2. Jane pays $13,000 for a product from manufacturer X, and the product arrives dysfunctional. Jane asks for a refund, but X says "Sorry." Jane does some calculations and realizes that the total costs of a lawsuit would probably exceed the benefits in recovering the losses, and decides not to purse such an action.

Example 15.3. Ed the Extorter determines that the total value accrued to X, a major user of a permissionless blockchain payment network, is larger than the security budget. Ed threatens to employ a 51 % attack on a blockchain (harming X) if Ed does not receive recurring ransom payments.

What is common in the above three examples is that one party is leveraging the other party's rational self-interest. While we may be morally sympathetic to Mike or Jane and less sympathetic to Ed, game theory is not. Each of these three would like to use their leverage to gain something. However, each also faces a problem; they can only threaten to act, and if the threats are not taken seriously, the status quo will remain. If Mike has a choice between being employed and being unemployed, he will choose the former. Because the ball is in Mike's court and he has to make the choice between leaving or staying, X can rest assured that Mike will stay. This default to status quo is an example of subgame perfect equilibrium.

15.1.2 Motivating game theory games

Example 15.4 (Ultimatum game). (Recall Example 3.7) Player 1 is handed \$100. He offers to split it $(x, 100 - x)$ with Player 2. Player 2 accepts this split and walks away with $100 - x$ or he rejects it and both players walk away with none.

If Player 1 offers a split of (\$80, \$20), a rational Player 2 will probably accept the offer and walk away with \$20, rather than reject it to spite Player 1. In reality, experiments have shown that sometimes Player 2 will spite Player 1 by rejecting a low offer [45].

For comparison, we point out a less rigid version of the game.

Example 15.5 (Negotiation game). Player 1 and Player 2 must come to an agreement on how to split \$100, or each gets none. Either player can walk away from the game at any time. This is the only rule.

Example 15.6 (Rubinstein type game, probabilistic version). Players 1 and 2 make alternating offers. They each get to impose a probability that the game will break down in subsequent rounds.

15.1.3 Dividing a surplus and the Nash bargaining solution

What should the fair arrangement be? Egalitarian principles dictate that the surplus should be shared, and a belief in free market competition suggests this as well: Unless one party has demonstrated it is better suited to control a resource, sharing a divisible resource promotes competition. The Nash bargaining solution [39] agrees: The surplus should be split down the middle when the players are on equal footing.

15.1.3.1 Axiomatic approach

Nash [39] studied the problem of bargaining using an axiomatic approach, and characterized equilibria. Given a set of axioms (which are probably not realistic) Nash determined that the best solution was given by the value x that maximizes the Nash product

$$(u_1(x) - d_1)(u_2(x) - d_2), \tag{15.1}$$

where u_1 and u_2 are the utilities enjoyed from a particular agreement by Players 1 and 2, respectively, over a set of possible agreements $x \in \mathbb{A}$, while d_1 and d_2 are the utilities gained if the negotiation fails. For example, in the ultimatum game, Example 15.4, the disagreement point $d = (d_1, d_2)$ is $(0, 0)$, that is, if they disagree, each player gets nothing. Nash's axioms included symmetry: Except for the very important question of who offers first, Examples 15.4, 15.5, and 15.6 are symmetric between the two players.

Nash's axioms also included translation invariance. The solution should not depend on where the disagreement point lies on the plane, but where it lies with respect to the other possible agreement points. For example, if both players are charged \$50 to play the game (making it zero sum), the negotiations should be the same: Each player should be equally averse to losing \$50 as they are keen to gaining the same.

Note however that if Player 2 could raise his own disagreement value d_2, this would alter the location where the maximum occurs in (15.1). Also, if Player 2 could cause Player 1 to feel more pain by lowering d_1, they would raise their own value in the Nash bargaining solution.

In Example 15.1, the idea of fair market value of Mike's effort is not so clear, because the market is not stepping in to narrow the surplus. If a competitor to X were to move down the street from X, Mike with his unique skill set could spark a bidding war, and find his effort is suddenly worth several times more. Even without a competitor, we can view Mike's situation either as an ultimatum game (Example 15.4), in the case that X gives Mike a "take-it-or-leave-it" offer, or as a negotiation game (Example 15.5), in the case that Mike feels empowered to negotiate. Suppose we assume that Mike, with his work ethic and skill set, could earn \$100,000 at another position in another industry if he left X. So Mike's disagreement value is \$100,000, while X's disagreement value would be the maximum of 0 or \$1,000,000 minus the cost of finding and training a suitable replacement for Mike minus any other losses incurred in the process. If we assume X's disagreement value is 0, then maximizing (15.1) one concludes that Mike could be making \$550,000.

In reality, one would not expect Mike to negotiate such a pay raise. In many industries, employees who want a raise are often best advised to walk into their manager's office with an outside offer. This leads to the inefficient practice of raise-deserving employees applying for jobs with companies at which they are not interested in working.

In comparison, the disputes in the other two examples appear to be not over a surplus, but over negative consequences. However, if we accept translation invariance of the problem, this is a matter of perspective: In Example 15.2, one can set the disagreement point to be the costs to both parties after a long, expensive and possibly public lawsuit. To explore this situation further, let us assume that both parties understand that a legal action will most likely end with Jane being refunded. Both parties also know that it will be easy for lawyers to obfuscate long enough to make the case significantly painful to prosecute. In this particular jurisdiction it is likely that the court may only award Jane a portion of legal fees. Suppose Jane calculates that she can expect to gain the \$13,000 difference, but possibly incur \$25,000 of legal fees, of which only \$10,000 will may be reimbursed. On the other hand, X will expect to pay perhaps \$30,000 for legal fees, pay a \$13,000 reimbursement, \$10,000 in Jane's legal fees, and in addition maybe lose \$50,000 in profits as sales drop due to bad publicity. So Jane stands to lose \$2,000, while X stands to lose \$100,000. The disagreement point is then (–\$2,000, –\$100,000) and the set of possible agreements $\mathbb{A}$ should be the set of partial or complete refunds given to Jane instead of going to trial. Simple calculus determines that the maximum of (15.1) occurs on the far end point: Jane should get her full refund. Intuitively, this is obvious; the incentives of coming to a settlement are clearly much more compelling for X than for Jane. If Jane threatens to sue and she can be taken credibly, X will immediately make the payment. The problem is that they have no reason to take such a threat credibly. They are certain Jane will not sue, as for her, the decision is a net negative.

15.1.4 Exogenous probability of breakdown

Suppose that two players are playing the following negotiation game over a single unit of some currency. Player 1 makes a proposal, which is then accepted or rejected. If rejected, the game continues with probability p and terminates (both players losing) with probability $(1-p)$. If we make the following assumptions, the game has a quick and easy solution:

- Players consider themselves on equal footing except for the asymmetry of who makes the first offer, that is, neither is a superior negotiator or has different risk preferences.
- A player will offer a split $(x, 1-x)$ which maximizes x among all proposals that will be accepted by the other player.
- Player j rejects any offers that are "too low." Here "too low" refers to the existence of a value (assuming $j = 2$) y such that

$$py > 1 - x \tag{15.2}$$

and

$$1 - y > px. \tag{15.3}$$

Player 2 could propose $(1-y, y)$ as a counter – (15.2) says that even after discounting by the probability of loss this offer is preferable to Player 2 (if accepted by Player 1).

Meanwhile (15.3) says that the offer of $1-y$ back to Player i will be preferable to what Player 1 proposed for themselves after discounting for the probability of breakdown.

The solution: Player 1 offers to split

$$\left(\frac{1}{1+p}, \frac{p}{1+p}\right)$$

and Player 2 accepts.

Proof. Player 1 knows from (15.2) and (15.3) that if they propose x such that there is a $y \in (\frac{1-x}{p}, 1 - px)$, the offer will be rejected. This window is open whenever $x > \frac{1}{1+p}$, so Player 1 chooses the largest possible x not in the window, namely $x = \frac{1}{1+p}$. □

A low probability p can give Player 1 a considerable advantage. As $p \to 0$ this becomes the ultimatum game, at which point Player 1 could demand any value less than 1. As $p \to 1$ the split approaches $(\frac{1}{2}, \frac{1}{2})$ and the symmetry is restored.

Suppose that there is a fixed probability of breakdown in some time interval, say 8 % each day. In this case, if offers are made once per day, p is 0.92. By allowing for rapid fire negotiation, breaking up the day into dozens or even hundreds of intervals, we can model the game with p very near 1.

15.1.4.1 Modeling time preference

The easiest way to model time preference is by using an exponential discounting factor. We consider the utility of x units of money at time t in the future to Player i to be

$$u_i(x, t) = xe^{-r_i t}. \tag{15.4}$$

The mechanics are quite similar to the exogenous breakdown scenario modeled above, but with a slight difference: The players can have different time preferences, which will lead to a different outcome of the game. Suppose that the offers are made in time intervals Δ, and we then let

$$\delta_1 = e^{-r_1\Delta},$$
$$\delta_2 = e^{-r_2\Delta},$$

which describe the discounting factor to either player between rounds.

For this model we make the following assumptions:

- Players consider themselves equal negotiators, but they may have an asymmetry as far as time preference is concerned. Player 1 makes the first offer.

- Player j rejects any offers $(x, 1-x)$ that are "too low." Here "too low" refers to the existence of a value y such that

$$\delta_j y > 1 - x \tag{15.5}$$

and

$$1 - y > \delta_i x. \tag{15.6}$$

Player j could propose $(1-y, y)$ as a counter – (15.2) says that even after discounting for time preference of the round, this offer is preferable to j (if accepted by i). Meanwhile (15.3) says that the offer of $1-y$ back to Player 1 will be preferable to what Player i proposed for themselves, after discounting.

The solution: If Player 1 offers first, they choose

$$\left(\frac{1-\delta_2}{1-\delta_1\delta_2}, \frac{\delta_2(1-\delta_1)}{1-\delta_1\delta_2}\right) \tag{15.7}$$

and this is accepted by Player 2.

Exercise 15.1. Prove that (15.7) is the unique solution to the above game.

Note that if we let the time intervals approach zero and apply L'hôpital's rule, we have

$$\lim_{\Delta\to 0} \frac{1-e^{-r_2\Delta}}{1-e^{-r_1\Delta}e^{-r_2\Delta}} = \frac{r_2}{r_1+r_2}.$$

Conclusion 15.1. With continuous bargaining but different time preferences, the bargaining should result in a split of

$$\left(\frac{r_2}{r_1+r_2}, \frac{r_1}{r_1+r_2}\right).$$

Observe the expression is scale-invariant in terms of (r_1, r_2). In particular, it does not matter which time units we are using when we compute r_1 and r_2. If the time preference of Player 1 is twice as strong as that of Player 2, they can expect to receive $\frac{1}{3}$ of the surplus in this negotiation.

15.1.5 What if players can determine the probability of breakdown?

We propose the following game. The game starts with Player 1 proposing a split $(x, 1-x)$ and a probability p_0. The game terminates with probability p_0 (even before Player 2 has a chance to accept). Player 2 can then either accept, or, if they make a counteroffer, they also include probability q_0. The counteroffer is presented to Player 1 again, but arrives only with probability p_0q_0 (with probability $1 - p_0q_0$ the game terminates). Player 1 can then accept. If they choose not to accept they propose a probability p_1 and their offer arrives for consideration to Player 2, with probability p_1q_0. The game continues with each counteroffer being accompanied by a probability of failure that will apply following the offer and the next counteroffer.

We assume this game is governed by the following rules:

- Players consider themselves on equal footing except for the asymmetry of who makes the first offer, that is, neither is a superior negotiator or has different risk or time preferences.
- Player 1 will offer a split $(x, 1 - x; p)$ which maximizes px among all proposals that will be accepted by Player 2.
- Player 2 rejects any offers that are "too low." Here "too low" refers to the existence of a value y such that

$$\begin{aligned} pqy &> 1 - x, \\ 1 - y &> pqx. \end{aligned}$$

Solution: Player 1 will offer $(\frac{1}{2}, \frac{1}{2}, 1)$, which will arrive intact to Player 2, who will accept.

Proof. First, we claim this solution will be accepted. If not, there exist y and q such that

$$\begin{aligned} qy &> \frac{1}{2}, \\ 1 - y &> \frac{1}{2}q. \end{aligned}$$

Noting that the function qy is increasing in q and y, we can look for the maximum of qy along the boundary of the second equation, namely

$$\begin{aligned} y + \frac{1}{2}q &= 1, \\ q &\le 1. \end{aligned}$$

Using Lagrange multipliers, we see that the point on the line at which qy is maximized must satisfy

$$y = \frac{q}{2}.$$

Solving for q and y we find $q = 1, y = \frac{1}{2}$, so we see that

$$\frac{1}{2} = qy > \frac{1}{2}$$

fails, thus we have a contradiction. So $(\frac{1}{2}, \frac{1}{2}; 1)$ represents a solution.

Now suppose there is some other solution offering a better outcome for Player 1, that is, $px = \frac{1}{2} + \varepsilon$ for some positive ε. Considering the constraints

$$q\left(\frac{1}{2} + \varepsilon\right) + y = 1,$$
$$q \leq 1,$$

maximize qy. This happens at

$$\frac{y}{q} = \frac{1}{2} + \varepsilon.$$

Now

$$q\left(\frac{1}{2} + \varepsilon\right) + q\left(\frac{1}{2} + \varepsilon\right) = 1,$$

so

$$q = \frac{1}{2(\frac{1}{2} + \varepsilon)} = \frac{1}{1 + 2\varepsilon},$$
$$y = \left(\frac{1}{2} + \varepsilon\right)\frac{1}{1 + 2\varepsilon}.$$

Now claim that

$$\left(\frac{1}{2} + \varepsilon\right)\left(\frac{1}{1 + 2\varepsilon}\right)^2 > \frac{1}{2} - \varepsilon$$

for all $\varepsilon > 0$, which follows by a computation. Perturbing y slightly we get a y demonstrating that the offer was too low and would be rejected by Player 2. We conclude that 1/2 is the best possible outcome for Player 1. □

Conclusion 15.2. If players are allowed to determine the probability that the negotiation will break down, then the players should come to a fair agreement very quickly.

15.2 Creating commitment

In the divide-the-dollar type games in which players are negotiating over how to split a surplus, the disagreement point and the pre-agreement are the same – each gets 0. However, in the examples of Mike, Jane, and Ed, the pre-agreement point is not the same as the disagreement point. If negotiations stall, but do not blow up into full disagreement, the status quo is preserved. For each of Mike, Jane, and Ed, they do not get to take advantage of their bargaining position until they have a way to move the disagreement point away from the status quo. This is a practical shortcoming of Nash's solution; it does not instruct players how to get unstuck from the subgame perfect equilibrium in which they must choose between status quo and an unpleasant disagreement option.

The goal of someone in Mike, Jane, or Ed's position is to move the disagreement point by creating a commitment to it. Paraphrasing Schelling's definition of commitment [48, p. 127], we say that commitment involves maneuvers that leave one in a position so that their only rational option is to follow through with a threat. The object of commitment is the credibility of a threat. Nash introduced the issues involving threats as follows [39, p. 130]:

> If one considers the process of making a threat, one sees that its elements are as follows: A threatens B by convincing B that if B does not act in compliance with A's demands, then A will follow a certain policy T. Suppose A and B to be rational beings, it is essential for the success of the threat that A be *compelled* to carry out his threat T if B fails to comply. Otherwise it will have little meaning. For, in general, to execute the threat will not be something A would want to do, just of itself. The point of this discussion is that we must assume there is an adequate mechanism for forcing players to stick to their threats and demands once made and one to enforce the bargain once agreed. Thus we need a sort of umpire, who will enforce contracts of commitments.

What Jane needs in Example 15.2 is the ability to make binding threats. Jane can threaten to sue if she does not get her money back, but nothing forces her to follow through. Now suppose that Jane is able to deposit $10,000 in escrow and makes an arrangement with a third party that if she does not get refunded and does not sue, she will forfeit her money.

Now, the tables have turned. Jane contacts X, and tells them that she is not merely threatening a lawsuit, but that it is now against her interest to fail to file the lawsuit. At this point, X is wise to assume that Jane is not bluffing. If X is rational X will pay her immediately.

The game is not completely set, however, as there are other moves that X may be able to make. They may be able to create imaginative responses, such as making the own arrangement binding them to not pay unless they receive certain concessions from Jane. If they are able to communicate this to Jane, they have created a way out for Jane, which rationally, she should accept. She could respond with a countercommitment, and this could go on. The best way for Jane to preclude counteroffers is to go off the grid and leave the escrow agent to collect the check.

Implementing a legal contract accomplishing the above appears to be somewhat problematic in many jurisdictions; enforceable contracts may require considerations from both parties. It is not clear that the third party has an enforceable claim in the case Jane does not sue.

Exercise 15.2. Draw a game tree representing the game in the above example when Jane is able to escrow money with a third party as a commitment mechanism.

15.2.1 Enter blockchains?

Smart contracts, oracles, randomness beacons, and stablecoins running on some sort of decentralized blockchain could represent a way to create such a commitment [27]. Because smart contracts do not require a court for enforcement, questions about enforceability do not arise. As legal situations often involve litigious characters, a decentralized system will be more difficult for a vexatious bully to target. If an escrow firm attempted to offer the type of contract that Jane wanted, they may face blow-back from parties that have much to lose. The typical reaction of a vexatious bully is to become even more litigious, but they will find it very difficult to punish an anonymous global network with scorched earth lawsuits.

This is an area that merits further exploration. Many smart contract platforms are designed to be quite versatile and could afford such arrangements. We will assume such devices can be constructed and leave it to the engineers to provide the actual specifications.

15.2.2 Commitment can backfire

If Jane is successful in creating a commitment, she has no way out. This is by design. However, if her adversary is stubborn, vindictive, not thinking clearly, or simply unable to comprehend the commitment device, Jane may have made a bad decision. Jane does not know ahead of time if this will happen; however, she may be able to place a probability on it. If the probability that X is going to proceed irrationally is greater than a certain threshold, Jane is wise not to proceed. In the following section we offer a way to minimize the risk when the adversary has a possibility of being irrational, while still taking advantage of available leverage.

15.3 Mixed strategies and Schelling threats

Recall that a mixed strategy is a strategy that has a random component. For example, a soccer player making penalty kicks may be more accurate kicking to her left. However, if she always kicks to her left, the goalkeeper will anticipate this. The shooter should devise a plan to kick, say 60 % of shots to her left: This will not allow the goalie to anticipate the shot, but still allows her to use her better shot more often than not. The idea goes back to Von Neumann in 1928 [40], in his proof of a famous theorem stating that all N-player games enjoy Nash equilibria, provided that players are allowed mixed strategies. Intuitively, by asserting a small probability of a move that will be damaging to the opponent, one can change the opponent's strategy entirely, even without committing to the move. By using all probabilities in between 0 and 1, one can set up an optimization problem – maximizing the opponent's exposure while minimizing one's own.

Suppose that Players 1 and 2 are playing the negotiation game (Example 15.5), but Player 1 is impatient. Player 1 decides to make an offer to split the pot down the middle and then adds that if the offer is not accepted, Player 1 will terminate the game and walk away. This threat becomes useful only if Player 2 receives the threat, believes the threat, and is willing to act rationally upon it. However, Player 1 is not 100 % confident that Player 2 will respond: The threat may not have been received credibly, or maybe it was not received at all. So if Player 1 is actually committed to the threat, they face the risk that the strategy will backfire. Player 1 wants to minimize the risk of walking away with nothing.

15.3.1 Jane's example, again

Jane would like a way to create a commitment that will hold with some probability. She takes \$10,000 in stablecoins and deposits it to an address. The contract runs until Jane terminates it. Given a source of randomness on the blockchain, the contract has a probability of triggering some irrevocable action every day; for example, it could take the \$10,000 and deposit it as a non-refundable retainer in an account held by law firm Y. Jane has signed a retainer agreement with law firm Y. Their agreement is that as soon as they receive a payment from Jane, they immediately begin filing a civil action against X and will deliver 40 hours of billed work for Jane on this case. Jane presents this arrangement to X. Taking $p = 0.05$ as an example and assuming that commencing the lawsuit will result in losses to both parties as described (\$2,000 loss to Jane, \$100,000 loss to X) each day, the expectation of cost to Jane is \$100. This is not a dear price to pay to show X that she is serious. Meanwhile the expectation of cost to X from this arrangement will be \$5,000 per day. This is much more significant, and should be taken seriously. Jane can leave this event to chance for several days or weeks – until she determines that there is no further value in doing this. Of course, in the process she may end up filing a lawsuit. But she can bound the probability of this event from above.

In practical terms, Jane estimates that there is a probability $q(p)$ that X will respond rationally to the threat, where p is the per-day probability of the trigger event. One would expect that $q(p)$ is a concave function with $q(0) = 0$ flattening out at some value $q(1) < 1$. She wants a probability p that maximizes

$$13{,}000 \times q(p) - (1 - q(p)) \times p \times 2{,}000.$$

For example, we may take

$$q(p) = \frac{1}{\pi} \arctan(20p),$$

which assigns a probability of near 1/2 for p near 1. Also note $q(0.05) = 1/4$. Maximizing

$$13{,}000\frac{1}{\pi} \arctan(20p) - 2{,}000\left(1 - \frac{1}{\pi} \arctan(20p)\right)p$$

suggests a value of p slightly above 0.45 should be optimal for Jane.

In comparison, consider the following hypothetical situation discussed by Schelling [48, pg 196]: A hitchhiker pulls a gun on a driver. The driver responds that he will crash the vehicle, killing both, unless the hitchhiker tosses the gun out the window. The hitchhiker finds this threat irrational and continues to make demands on the driver. So the driver begins to accelerate, reasoning that the credibility of the threat increases with speed. At 120 miles per hour, the hitchhiker must realize that the probability of a fatal accident is now very much a positive number.

The hitchhiker situation is similar in some aspects, but economically much different. Indeed, both the hitchhiker and the driver stand to lose their lives. The driver (who may have a respectable job and a family) likely has more to lose than a hitchhiker who is robbing kind strangers. If the hitchhiker realizes this, the hitchhiker is smart to keep the weapon cocked. The driver is making an irrational decision to drive recklessly and will eventually slow the car. On the other hand, Jane has much less to lose. Company X, which is (in theory) motivated by money, will not be interested in playing this game for long. This the major element that she is leveraging, and this is why her strategy should work if X is rational.

Schelling's point is summarized as "the only way to becoming physically committed to an action is to initiate it." When Jane posts and allows a contract to be executed, she is doing precisely that. We define a *Schelling threat* as a threat which one can commit themselves to by initiating it, but the consequences of which depend of some random event beyond the control of the party making the threat.

Schelling describes mixed strategies as a tool to "scale down" the threat [48, pp. 178–179]. This accounts for the possibility that a threat may be ignored: the threat was not communicated, the threat was miscommunicated, the adversary may have a nature which ignores the threat (e. g., the scorpion who stings the frog, ushering in his own watery death), the adversary is playing a "long game" and is trying to create an illusion of irrationality, or other unforeseen reasons. In fact, if one has a reasonable idea of the probability that a threat will fail for unknown reasons and also has an idea of how high the probability needs to be to form an effective threat, one can compute a range of probabilities that will be useful. Based on the calculus in [48, p. 180], it may be that the only threats worth making involve probabilities in a small interval.

15.4 Extortion and griefing

In a permissionless network with anonymous actors, one cannot preclude actors from performing or threatening to perform actions that would be illegal in law-and-order settings.

Consider some good G. Valeria values G and it brings her value v. Ed has the means to destroy G, at cost d.

Proposition 15.1 (Extortion principle). *If the value v of G to Valeria is less than the cost to Ed to destroy G, then Ed can extort Valeria. The Nash bargaining solution suggests a ransom of $\frac{v-d}{2}$.*

While an extortion arrangement is not legal, we are interested in situations where nothing is in place to prevent it. All Ed needs is a means to commit to destroying G when Valeria does not pay up. The modeling is simple, so we present it here.

Proof. Assume $d < v$. Assume that Valeria responds by negotiating with Ed in a rational way. The Nash bargaining solution can be determined as follows. The disagreement point is $(-d, 0)$: Ed loses d and Valeria ends up with 0. Thus the function to maximize is

$$f(r) = (r + d)(v - r),$$

where r is the ransom that Ed and Valeria agree upon. This is easily seen to be maximized at

$$r = \frac{v - d}{2}.$$

Note what happens if Ed demands a higher ransom, say

$$r = v - \varepsilon.$$

If Valeria is savvy, she will signal to Ed that she is committed to engage in probabilistic bargaining, only paying such a ransom with some probability p chosen to make more extreme threats dangerous for Ed. □

Example 15.7. Ed estimates that the monthly security budget for a proof-of-work blockchain is \$100 million and that this blockchain brings \$200 million of value per month to corporation X. Ed sets up a smart contract escrowing \$150 million. If the smart contract is triggered by a random event, the contract offers \$150 million over the next month to wreak havoc on the blockchain by bribing miners to attack. Ed sets the probability of triggering at 5 % per day and presents the contract to X. Corporation X reviews the contract, finds it credible, and commences to negotiate a monthly ransom payment for Ed to disable the contract.

The disagreement point here is (–\$150 million, –\$200 million), so the Nash bargaining solution suggests that X should pay \$25 million to Ed. If we reset the disagreement point to $(0, 0)$ the set of divisions of the surplus is of the form $(x, 350 - x)$ (in millions). Suppose that Ed proposes \$50 million, represented by the agreement $(200, 150)$. There certainly exist values y such that

$$(1 - p)y > 150,$$
$$350 - y > 200(1 - p),$$

so X rejects the offer, perhaps waiting a short time period. They can continue to reject such offers, knowing that Ed is facing considerable exposure, until Ed offers to accept a $25 million ransom, which represents the Nash bargaining solution.

Example 15.8. We consider a situation where Ed would like to destroy a network X, which brings values $v_1, \dots, v_N$ to N different participants. First note that if the cost to destroy the network $d < v_i$, then Ed can extort participant i as in the previous example. More likely, however, is that the value of the entire network exceeds d, while the value enjoyed by any individual participant does not, that is,

$$v = \sum_{i=1}^{N} v_i > d,$$
$$v_i < d, \quad \forall i \in \{1, \dots, N\}.$$

Nash bargaining in this case is similar, but multivariate. Let r_i be the ransom participant i pays to Ed. The disagreement point is $(0, \dots, 0, -d)$, so the function in this $(N+1)$-player situation to maximize is

$$\left(d + \sum_{i=1}^{N} r_i \right) \times \prod_{i=1}^{N} (v_i - r_i).$$

It is easier to maximize the logarithm:

$$\ln\left(d + \sum_{i=1}^{N} r_i \right) + \sum_{i=1}^{N} \log(v_i - r_i).$$

Taking the derivative with respect to r_j and setting to 0, we have

$$\frac{1}{d + \sum_{i=1}^{N} r_i} - \frac{1}{v_j - r_j} = 0$$

for each i. Solving

$$\sum_{i=1}^{N} r_i + d = v_j - r_j,$$
$$2r_j = v_j - d - \sum_{i \neq j}^{N} r_i < 0.$$

We have demonstrated the following.

Proposition 15.2. *Suppose the value of the network enjoyed by any given participant is less than the cost to destroy the network. The Nash bargaining solution suggests it will be irrational to attempt to extort value from the network.*

Example 15.9. Suppose that in addition to Ed, Fred also has the means and ability to destroy the network and begins threatening corporations who value the network. In fact, Fred is negotiating at the same time with Ed. If Ed or Fred were to trigger the destructive event, they would destroy the leverage of the other. This Nash bargaining solution would be determined by maximizing a Nash product involving more variables.

15.4.1 Schelling attacks as an inexpensive way to destroy a network

Aside from extortion, one may have an interest in attacking a blockchain for other reasons. If the blockchain is useful, being used and built upon, there may be a non-linear relationship between the negative effects of probabilistic attack and the probability p as negative network effects can compound.

Example 15.10. Greg the Griefer has some interest in damaging blockchain X. Greg designs a contract (presumably on another blockchain Y) that is activated with probability p and when activated pays miners for garbage blocks at a somewhat lower difficulty and at a higher rate than the block reward: If anyone submits a block to the contract and that block is the immediate child of a stale block and has a hash within a factor of 100 necessary to mine a block, Greg's contract pays them 1.2 % of the block reward. Now any miner who is able will opt to pursue not only the more sure reward but also the 20 % higher payoff. In the meantime the targeted blockchain X will languish, becoming slow and less secure. If Greg has a significant amount of money, he could cause this to go on for a long period of time. Now Greg can also choose p very small. Suppose Greg chooses the value small enough, so that each year there is a 40 % chance the blockchain will become attacked for a period of 14 days. The expected cost to Greg would be only about 2 % of the security budget of X. For an enterprise wishing to operate on the blockchain, a 40 % chance of a 14-day downtime in any given year might be a complete deal-breaker, driving builders and investors away.

If Greg is a heavy holder of tokens for a competing blockchain Z, which is designed to not have this vulnerability, this attack would be a cheap way to drive adoption away from the target blockchain X and towards Greg's blockchain Z. If Greg is interested in shorting X, this attack could be an opportunity to send X into a death spiral. The point is that Greg may not need to spend even a single dollar attacking, provided he has enough backing to enact a probabilistic threat. Dangling the probability of an attack may create sufficient FUD to damage the network. A well-funded venture capitalist that trades tokens or perhaps a nation-state with a zero-sum game mentality may find such an attack attractive.

15.5 Wars of attrition

Wars of attrition between irrational parties or parties that have completely different beliefs about their strengths or their opponent's strengths may be hard to model. Humans do struggle with rational approaches to such games. In the *dollar auction game*, players continue to pay non-refundable amounts to attempt to outbid opponents to acquire a dollar: Experiments have shown that people will pay much more than a dollar to obtain a dollar [50]. Wars of attrition can fall into this dynamic if players are well matched; however, if one side is clearly favored, the Nash equilibrium (after bothering to set up an appropriate game) should be that the weaker side defaults immediately.

For one-shot games, this follows by a quick backwards induction argument. The best time to give up a losing fight is almost always now. In general, if a player does not want to be victimized by stronger bullies, they can commit to engaging in a war of attrition for some time, or if they can manage it, with some probability. This may discourage other bullies from attacking or may result in changing the game to a Nash bargaining.

Example 15.11. Consider a war-of-attrition game: The winner is given a pot of \$30. Jack has a bank of \$10 while Jill has a bank of \$50. Each round Jack must expend (never to be recovered) \$1 to continue, while Jill must expend \$2. This can behave somewhat like Nash bargaining; Jane must expend \$20 to gain the \$30, in which case Jack expends \$10 to get nothing. The disagreement point is (−\$10, \$10) while the surplus (suppose Jack or Jill defaults immediately) is \$30. If there were only two options for Jack – play to the end or not play to the end – Jack may attempt to use a Nash bargaining to offer to default in exchange for \$5.

By further engaging in a war of attrition round-by-round, a doomed player can demonstrate commitment to the next round. Of course, there are mental games and elements of poker that can be difficult to model but a Nash bargaining perspective suggests that a player who is destined to lose (due to being not as well resourced) but is spending less round-by-round may be able to extract concessions in order to drop out of the war. If the less-resourced player is spending more, they have no negotiating power (unless they are credibly and convincingly irrational) and should drop out immediately.

If there is no way for the eventual loser to extract concessions, in a one-shot game the loser is always wise to give up immediately.

16 Coalitional game theory

Cooperative and non-cooperative game theory are distinct. In the former, players are assumed to have the ability to negotiate, form coalitions, and enter into binding agreements with other players. This leads to different dynamics: The idea of a Nash equilibrium is no longer applicable; a player should consider not only the effects of changing their own strategy but also what would happen if they changed their strategy in concert with other players. Immediately the dynamics of games such as stag hunt and the prisoners' dilemma change. If two players in a prisoners' dilemma can agree to be bound to cooperate, certainly they will cooperate; this is preferred over each defaulting to the less optimal Nash equilibrium (the question of how such binding agreements are enforced in practice is a separate question).

Cooperative game theory studies the behavior of players not only as individuals but also potentially as coalitions. Typically, this means we have $n \geq 3$ players; otherwise, it is more of a straightforward bargaining problem (which are rarely "straightforward"). These games naturally possess considerably more complexity. In this section we will offer some solution concepts that play roles similar to the Nash equilibrium. Cooperative game theory is a huge topic, so we will tailor discussions to the game closest to our interests: coalitional mining. This will form a basis for security discussions in the next chapter. Much of the material (including the examples) is discussed in greater length in [28] and [41, Chapters 13 and 14].

16.1 Motivating examples

Example 16.1 (Odd man out). Instead of the negotiation game (Example 15.5), suppose that three players are told they can split $100,000, provided two of them agree to the split. One possible solution would be to split it in three equal proportions; this is easy and obvious. However, any two of the players can agree to exclude the other and increase their take by 50 %. So the three-way split lacks a Nash equilibrium type property: A subset of the players could break off from the three-way split and create an outcome that favors both of them. The problem with this jumps out immediately: If Player A offers Player B to share in a 50/50 split, Player C can offer Player A a 60/40 split, which improves upon the agreement with B, at which point B can offer C a 50/50 split that excludes A. There is no end to such negotiations.

Example 16.2 (Piano movers). Suppose that the job of moving a piano pays $1,000 and requires one large, beefy person and one thin, wiry person. There are three potential movers; Mover A is large and beefy, and Movers B and C are thin and wiry. Clearly Mover A is necessary for the job. It is easy to see that by making a series of alternating offers, Player A can choose a partner that concedes almost 100 % of the takehome pay to A. Movers B and C can only undercut each other and cannot form a team exclud-

https://doi.org/10.1515/9783110772845-016

ing A. However, the game is more subtle: B and C can collude to negotiate; as a union they have equal negotiating power as A and as such may demand no less than their share of the two-player Nash bargaining solution. Alternatively, they could agree to flip a coin; the loser drops out and the remaining player is left on equal footing with Player A. With this power dynamic in mind, one may suggest a 50/25/25 split is natural.

This power dynamic is what causes labor unions to form. While workers themselves may not have the ability to form a productive business without the resources provided by the capitalist, the union is a bargaining coalition which seeks to take advantage of the fact that the capitalist cannot form a productive business without cooperation with a least some subset of workers.

Example 16.3 (Weighted majority games). There are N players each with positive weights $\omega_1, \ldots, \omega_N$ which sum to 1. Any winning coalition takes home the pot, with a winning coalition being any collection of players $i \in I$ such that

$$\sum_{i \in I} \omega_i > \frac{1}{2}$$

(more generally, the value $\frac{1}{2}$ can be replaced with any $t \in (\frac{1}{2}, 1)$ to create a different game).

Example 16.4 (Coalitional Bitcoin mining game). Let $\mathcal{C}$ (for "centralized") represent the profits following from controlling a centralized Bitcoin network and let $\mathcal{D}$ (for "decentralized") represent the total profits to miners under decentralized mining. There are N miners each with positive weights $\omega_1, \ldots, \omega_N$ which sum to 1, representing their share of the hashrate. Any winning coalition I takes home the maximum value

$$\max\left\{\mathcal{C}, \left(\sum_{i \in I} \omega_i\right)\mathcal{D}\right\}$$

with a winning coalition being any collection of players $i \in I$ such that

$$\sum_{i \in I} \omega_i > \frac{1}{2}.$$

On the other hand, if either (1) no winning coalition forms or (2) a winning coalition forms but prefers to stay decentralized, that is,

$$\mathcal{C} \leq \left(\sum_{i \in I} \omega_i\right)\mathcal{D},$$

then every coalition I takes home $(\sum_{i \in I} \omega_i)\mathcal{D}$.

We have to make a few remarks about the coalitional Bitcoin mining game (CBMG):

- The values $\mathcal{C}, \mathcal{D}$ require more unpacking; this will happen in §16.6.4 and Chapter 17.

- This game varies slightly from the traditional setup that is used for coalitional games, as the payoff depends on whether a majority coalition forms (see §16.3).
- If $\mathcal{C} > \mathcal{D}$, the reward to every majority coalition that forms is the same value $\mathcal{C}$. A game where every winning coalition claims the same reward is referred to as a *simple game* (each of the preceding examples was a simple game). In general, CB-MGs need not be simple when $\mathcal{C} < \mathcal{D}$, as different winning coalitions may collect different rewards.

16.2 Characteristic form and payoff configurations

A *coalitional game in characteristic form* is as follows. There are n players, denoted A, B, C, etc., and a function v which describes the payoff to any coalition that forms. For example, in odd man out (Example 16.1) with three players A, B, C, the function v is defined as

$$\begin{aligned} v(A) &= v(B) = v(C) = 0, \\ v(AB) &= v(BC) = v(AC) = \$100{,}000, \\ v(ABC) &= \$100{,}000. \end{aligned}$$

We will make several remarks:
- By convention, we will always assume that $v(\emptyset) = 0$.
- Players can only join one coalition, but many coalitions can form.
- We use S, T to denote typical coalitions, while we use N to denote the *grand coalition*, that is, the coalition of all players.
- The way the payoff is split among the coalition is not specified in the characteristic form of the game.
- The fact that the payoff to each player is not specified is often described as a game with “side payments” or with “transferable payments.” Side payments allow players to transfer money to players in the same coalition to ensure their cooperation.
- A game is *cohesive* if for every partition into coalitions

$$v(N) \geq \sum_{S_i \text{ partitioning } N} v(S_i).$$

- It is convenient to assume games are cohesive; in this case side payments apply to the grand coalition.
- While this is a nice framework, it does not capture all the dynamics; for example, the piano mover game is more interesting than the game specified by the characteristic form; there is a useful alliance that can be formed between Players B and C and one would expect side payments to be allowed to flow between B and C even though this will not be the final winning coalition. The “power to exit” wielded by Players B and C is not explicitly accounted for by the characteristic expression.

- We assume the unit is some divisible money, and as always throughout this text we assume that players believe that "money is good and more is better."
- The payoff to a coalition does not depend on the formation of other coalitions in the complement. In this sense, it is different from other strategic games in that if Player A forms a coalition with Player B, both know the payoff and do not need to speculate about C possibly forming coalitions with D or E.
- The CBMG in general cannot be expressed in characteristic form: The payoff to a miner mining alone or with a minority coalition depends on whether or not a majority coalition exists.

A *payoff configuration* (PC) is a selection of winning coalitions together with payoffs to each of the individual members. The payoffs to each of the players must be feasible given the coalitions form (more cannot be paid out than is collected). For example, for odd man out, possible PCs include $[\{AB\}, (50, 50, 0)]$ and $[\{AC\}, (70, 0, 30)]$. We use notation [winning coalition(s), payoff vector], where the payoff vector denotes the payoff to each player ordered in the obvious way beginning with A.

Given a payoff vector x and a coalition S, we denote

$$x(S) = \sum_{i \in S} x_i.$$

Note that we often will be using S to denote a subset of N, without insisting that the coalition is forming to take some action in the context of the game. We always assume payoff vectors have non-negative entries.

An *imputation* x is a PC of the grand coalition, which is N-feasible, that is,

$$x(N) = v(N),$$

and will be denoted by a payoff vector of the form $x = (x_1, \ldots, x_n)$.

16.3 Standard coalitional game theory and the Bitcoin mining game

While the set of cooperative games one could encounter is quite large, there are two types that have been studied extensively in the literature. One type of game is a coalitional game with side payments as described above, which allows the coalition to split winnings however they choose among there members. This is amenable to a number of tools from calculus and convex analysis, so it lends itself to rigorous study by mathematicians and game theorists. On the other end of the spectrum are coalitional games without side payments – the set of payoffs is specified. The study of these games requires a different set of tools leading to different types of results and theory.

The CBMG can be modeled in a few ways, none of which fits nicely into either category. Several issues to consider in the CBMG are the following:

- If no coalition forms, each miner takes home their share of the pie. If a majority coalition forms, those not in the coalition receive nothing. In standard games of characteristic form the payoff to a coalition should not depend on the actions of other coalitions, but in this case, clearly they do.
- Coalitions should be allowed to make side payments (i. e., share profits). If a cartel forms, any imputation which distributes the value $\mathcal{C}$ should be allowed.
- It makes sense to allow for a grand decentralized coalition. This is indeed the default mode for bitcoin mining. This is interpreted as "the majority agree to remain decentralized by default." This absence of a cartel has different properties than other cartels; in particular, the winnings are not redistributed via a negotiation, but rather they are apportioned precisely according to the hashrate each miner puts in. The sum total of the grand decentralized coalition is still $\mathcal{D}$. The fact that no side payments exist means that the CBMG is not a game with side payments.
- It also makes sense to allow for a grand centralized coalition in which all players accept an imputation which distributes the value $\mathcal{C}$. This allows more flexibility than the grand decentralized coalition and sometimes will appear in the solution concepts.
- One way to summarize the above two observations is that the set of payoff vectors afforded to the grand coalition is either arbitrary splits of $\mathcal{C}$ or a very specific split of $\mathcal{D}$. This suggests that we should define imputations slightly differently: those y with $y(N) = \mathcal{C}$ together with the decentralized set of hashrates dividing up $\mathcal{D}$. This will require that we shoehorn this definition into the solution concepts as they arise, leading to slight modifications of the theory.
- If we interpret "no winning coalition forms" as defaulting to "a grand decentralized coalition forms," we can assume that any individual coalition receives zero rewards. This salvages part of the theory (now the payout does not depend on other coalitions' behavior) but the payout to the grand coalition is now undetermined; it could be $\mathcal{C}$ or $\mathcal{D}$.
- We are going to try to adapt the definitions of the solution concepts as closely as we can. The theory is not quite the same, so we cannot simply cite theorems that have been proven in the literature and we have to check things by hand for the CBMG. For example, an important solution concept in coalitional game theory with side payments is the nucleolus, which has the very gratifying property that it is not only non-empty, but also unique. This is not immediate for CBMGs.

For convenience, we can normalize so that $\mathcal{D} = 1$ and call this a *normalized CBMG*.

16.4 Solution concepts

We discuss solution concepts for coalitional game theory while keeping our eyes on generalizations involving the Bitcoin mining problem.

Game theorists can only talk about solution concepts. Solution concepts present possible solutions that are driven by some mathematical considerations, but for many of these games there is subtlety and unpredictability of human behavior which cannot be ignored when attempting to connect to reality. Social norms, ideas of fairness, and a desire to have things settled likely play a decisive role when the solution concepts do not precisely describe an obvious solution. Ickiness of certain bargaining strategies may make certain types of solutions unattractive to humans. Coalitional game theory solution concepts tend to factor in an aversion to endless negotiation. While solutions are unlikely to be optimal for all parties, they may be acceptable due to the lack of a clear solution.

For coalition games, there are a number of proposed solution concepts. These might seem unsatisfying at first, but in view of how perplexing these games can be, the fact that solution concepts exists will in the end be somewhat reassuring. A more satisfying solution concept for a particular game will suggest a limited number of reasonable solutions. Some solution concepts can suggest a set so big that it becomes meaningless. Others are so restrictive that they suggest no solutions at all.

16.4.1 The core

The core is the closest thing in cooperative games to a Nash equilibrium. A PC lies in the core if there exists no coalition that could decide (as a coalition) to opt for a PC which is superior to all members of the coalition. Note that with side payments, there is not a distinction between one and all members of the coalition being better off: If a new PC can be arranged so that all players are no worse off and one player is better off, that player can "ε-share the wealth" so that every player is strictly better off.

Many games have empty core. For odd man out, it is easy to see that if we fix a PC, the player who receives the smallest payout can partner with another player to create a more favorable PC for both of them. Thus the core is empty.

However, if we go back to the two-player negotiation game, any split will be in the core: Changing the PC always comes at the expense of one player, so will not be favorable for both. In this case the core is a 1D set.

The piano mover game does have a non-empty core: Both $[\{AB\}, (50, 0, 0)]$ and $[\{AC\}, (50, 0, 0)]$ (which are essentially the same solution) cannot be improved upon. Player C cannot entice Player A by any means, as A is already receiving the maximum reward. Nor can Player C entice B with anything: The coalition $\{BC\}$ has no payout.

Exercise 16.1. Prove that a weighted majority game with no sum of weights summing to exactly $\frac{1}{2}$ and no player with weight over $\frac{1}{2}$ will have no core. Construct by hand a set of weights for which there is a core when sums of weights are allowed to sum to exactly $\frac{1}{2}$.

Proposition 16.1. *In the CBMG with no majority miners, the core is non-empty if and only if*

$$\mathcal{C} \leq \left(\sum_{i\in S} \omega_i\right)\mathcal{D} \tag{16.1}$$

for all winning coalitions S, in which case the core includes the decentralized PC:

$$\{N; (\omega_1\mathcal{D}, \omega_2\mathcal{D}, \ldots, \omega_n\mathcal{D})\}.$$

Proof. Suppose that (16.1) holds. Any coalition that forms which would be superior for all members must split the payoff $\mathcal{C}$ amongst its members so that each gets strictly more than their share $\omega_i\mathcal{D}$. If the decentralized PC is not in the core, then there is some winning coalition for which $\mathcal{C} > (\sum_{i\in S}\omega_i)\mathcal{D}$, contradicting (16.1).

On the other hand, if $\mathcal{C} > (\sum_{i\in S}\omega_i)\mathcal{D}$ for some winning coalition, the decentralized solution cannot be in the core as S can clearly choose a payout vector in which all members of S are strictly better off. So $\{N; (\omega_1\mathcal{D}, \omega_2\mathcal{D}, \ldots, \omega_n\mathcal{D})\}$ is not in the core. Now consider any PC (S, x) involving a winning coalition S: At least one of the members (say i) must have a positive payoff in this PC, $x_i > 0$. By removing this member i and replacing them with members outside the coalitions $N\backslash S$, the positive payoff can be redistributed to everybody to strictly increase their payoff. This is strictly preferred by the winning coalition $N\backslash\{i\}$ (we are assuming that i is not itself a majority). So (S, x) cannot be in the core either. Thus the core is empty. □

The core, when it exists, offers somewhat stable and satisfying solutions for a game, much like the Nash equilibrium. But for many interesting games, the core will be empty. Thus we need more subtle solution concepts. The common idea behind these other solution concepts is one common to many problems in real life: While a solution may seem suboptimal for one party, a proposed better solution will come with its own set of problems and failures to be optimal, making the alternatives no more stable. This reflects the principle that negotiations and lack of stability have a cost: A coalition may choose to forgo efforts at increasing their position in order to maintain a social order. For odd man out, for any coalition that forms, there is always a way for some party to propose a preferable solution to some coalition. To avoid this game traveling the circuit ad infinitum, players may just agree to split the prize in three equal proportions and move on with their lives.

This reasoning is important to the stability of decentralized Bitcoin. Even when it may be apparently better for a majority coalition to form, a member of that would-be coalition may be hesitant to propose such a cartel. Any serious suggestion that a coalition forms to centralize Bitcoin may set off a round of renegotiating and horse trading that need not terminate in a favorable way for any given player.

16.4.2 The stable set

In many of the solution concepts that follow, we use the idea of an *objection*. Roughly speaking, given a proposed solution x, an objection to x will be a claim by some coalition that they can do better (for themselves) than x. There are a few different types of these objections.

We say that an imputation x is an *objection of the coalition S to the imputation y* if

$$x_i > y_i, \quad \forall i \in S$$

and the imputation x is feasible for the coalition S, that is,

$$\sum_{i \in S} x_i \leq v(S).$$

Essentially, S is making the claim that y is suboptimal from their perspective. They can each improve their payoffs by coalescing and agreeing to pay out x.

Definition 16.1. A*stable set* is a set of imputations X with the properties that:
1. For every $y \notin X$, there is an objection x of the coalition S to the imputation y for some S and $x \in X$.
2. If $x \in X$, then there is no other $y \in X$ and coalition S such that (S, y) is an objection of the coalition S to x.

Example 16.5. The set

$$X = \{(50, 50, 0), (50, 0, 50), (0, 50, 50)\}$$

in odd man out is a stable set. Given any imputation y outside X, if some set of two of the players are together receiving less than 50, these will prefer one of the options in the stable set, so it satisfies the first condition. However, for any x in the stable set, there is no other way for both members to strictly improve via another member of the stable set. Note that this does not preclude any objections outside the stable set: For example $(0, 40, 60)$ is a objection to $(50, 0, 50)$ by the coalition $\{BC\}$, while $(50, 50, 0)$ is an objection to $(0, 40, 60)$.

There can be many stable sets, and in general the set of stable sets can be quite large. Often this set will be non-empty; in fact, it was conjectured in the 1940s that it should always be non-empty. However, a counterexample of a 10-person game with no stable set was found in 1968 [33]. The core, if it exists, will be part of any stable set.

For CBMGs, consider the set of minimal winning coalitions $\mathbb{W}^m$. A *minimal winning coalition* is a coalition which commands a majority, but where no member can be removed while maintaining the majority. For a given minimal winning coalition S, define

$$S^* = \text{imputations } x \text{ such that } x_i = 0 \text{ for } i \notin S \text{ and } \sum x_i = C.$$

Proposition 16.2. *In the mining game, suppose that* (16.1) *fails. Let $\mathbb{W}^m$ be the set of minimal winning coalitions. The union*

$$X = \bigcup_{S \in \mathbb{W}^m} (S^*)$$

is a stable set.

Exercise 16.2. Prove Proposition 16.2.

16.4.3 The bargaining set

The bargaining set follows up on some of the ideas behind the stable set, but pairs players against one another directly.

We say that a PC (S, y) is an *objection of* i *against* j *to the imputation* x if:

- S includes i,
- S does not include j, and
- for all $k \in S$, $y_k > x_k$.

In other words, Player i is making a claim that a coalition excluding j can do better for themselves than the proposed x.

Now Player j can view this objection and aim to propose a new PC involving a different coalition T. If there is a (T, z) such that

- T includes j,
- T excludes i,
- $z_k \geq x_k$ for all $k \in T \backslash S$, and
- $z_k \geq y_k$ for all $k \in T \cap S$,

we call this a *counterobjection to the objection* (S, y) *of* i *against* j. In other words, Player j is responding by making a claim (directed at i) that if S were to move towards the proposal by i, Player j could then make another proposal to T that would end up causing i to be excluded. The *bargaining set* is the set of imputations x with the property that for each objection (S, y) of i against j, Player j can make a counterobjection to (S, y).

Example 16.6. In odd man out, $[\{AB\}, (50, 50, 0)]$ is not in the bargaining set. Player C may propose an objection, $[\{AC\}, (50 + \varepsilon, 0, 50 - \varepsilon)]$, which excludes B, but B cannot propose a counterobjection. Anything B proposes to A must improve upon A's take of $50 + \varepsilon$, which means than B has no hope of reaching 50. The counters to C's proposal will be strictly worse for B than the original PC $[\{AB\}, (50, 50, 0)]$.

There is, however, a unique member of the bargaining set: $[\{ABC\}, (\frac{100}{3}, \frac{100}{3}, \frac{100}{3})]$.

This is more satisfying, at least in odd man out. The bargaining set appears equitable.

Exercise 16.3. Prove that $[\{ABC\}, (\frac{100}{3}, \frac{100}{3}, \frac{100}{3})]$ is the unique element in the bargaining set in Example 16.6.

Exercise 16.4. Consider a mining game with miner weights $\{0.4, 0.2, 0.2, 0.2\}$ and $\mathcal{C} = \mathcal{D} = 1$ (this is equivalent to a game also known as "My aunt and I"). Argue that the bargaining set is of the form

$$\left\{(1-3a, a, a, a) \mid a \in \left[\frac{1}{7}, \frac{1}{5}\right]\right\}.$$

The bargaining set for coalitional games with side payments is always non-empty, as we will see. As the above two examples show, the bargaining can be a singleton, but need not be.

16.4.4 The kernel

For the kernel we consider a different variety of objection and counterobjection. First we define the excess. Given an imputation x, the *excess* of S is defined

$$e(S, x) = v(S) - x(S).$$

The excess measures how much the coalition would be sacrificing by accepting an imputation x. Positive excess means that the coalition could form and claim $v(S)$ and this could be distributed to members of S so that each improves their payoff. However, S may forgo this, as this could lead to more negotiation and disruption of the social order. If the core is empty, one would not expect to find an imputation x for which no coalition has positive excess.

Definition 16.2. Given an imputation x, a coalition S is an *objection of i against j to x* if:
- S includes i,
- S excludes j, and
- $x_j > v(\{j\})$.

A coalition T is a *counterobjection to the objection S of i against j* if:
- T includes j,
- T excludes i, and
- $e(T, x) \geq e(S, x)$.

The *kernel* is defined as the set of imputations x such that every objection of i against j to x has a counterobjection.

This definition of the kernel here is equivalent to the original definition (cf. [12] and [28, §6.2]). However, the definition in terms of objections and counterobjections puts this in a consistent framework with the bargaining set and nucleolus, with the unfortunate side effect of incrementing the list of definitions of objection and counterobjection.

Recall that for CBMGs we are assuming $v(\{j\}) = 0$ for any singleton, so the condition $x_j > v(\{j\})$ requires only positivity of x_j.

Exercise 16.5. Recalling Exercise 16.4, show that the kernel is the singleton $\{(\frac{2}{5}, \frac{1}{5}, \frac{1}{5}, \frac{1}{5})\}$.

16.4.5 The nucleolus

The nucleolus requires yet another type of objection and counterobjection.

Definition 16.3. A pair (S,y) consisting of a coalition S and an imputation y is an *objection to* x if

$$e(S,x) > e(S,y).$$

A coalition T is a *counterobjection to the objection* (S,y) if

$$e(T,y) > e(T,x)$$

and

$$e(T,y) \geq e(S,x).$$

The *nucleolus* is the set of imputations x with the property that for every objection (S,y) to x, there is a counterobjection to (S,y).

Definition 16.3 is not the standard one for nucleolus (although it is the one given in [41]).

Exercise 16.6. Show that $[\{ABC\}, (\frac{1}{3}, \frac{1}{3}, \frac{1}{3})]$ is in the nucleolus for odd man out.

The following are some useful statements about coalitional games with side payments. We will not prove these, as they do not directly apply to the CBMG which we are most interested in (proofs can be found in more expansive textbooks on game theory; see for example [41]).

- The nucleolus is non-empty.
- The nucleolus is a singleton.
- The nucleolus is a subset of the kernel (hence the kernel is non-empty).
- The kernel is a subset of the bargaining set (hence the bargaining set is non-empty).

The nucleolus provides a solution that both exists and is unique and is nested within two broader solution concepts. This also means that if the kernel or the bargaining set is a singleton, the nucleolus must be as well. Depending on the game, one variety of a objection may be easier to work with.

As we will see, the containment relationships listed above need not hold for CMBGs.

16.5 Analysis of solution concepts for weighted majority games and CBMGs

For weighted majority games, we say the game is *constant-sum* (in some places, *zero-sum*) if

$$v(S) + v(N\backslash S) = v(N) \quad \text{for all } S \subset N.$$

Barring trivial or contrived instances, the CBMG is constant-sum if and only if $\mathcal{C} \geq \mathcal{D}$.

For weighted majority games, a player is called a *dummy* if there is no coalition that becomes winning by adding this player. A weighted majority game contains no dummies if every player is part of a minimal winning coalition.

In weighted majority games, the players can be ranked into different classes according to power levels. To see this, we define a relation. We say

$$i >_P j$$

whenever there exists a coalition T such that $T \cup \{i\}$ is a winning coalition, but $T \cup \{j\}$ is not. We say that

$$i \sim_P j$$

if neither $i >_P j$ nor $j >_P i$. For weighted majority games, it is not hard to show that this relation $\sim_P$ is transitive, reflexive, and symmetric, so it gives an equivalence relation on the set of players (see Exercise 16.7). It is also clear that the ordering cannot reverse the order of weights, that is, one cannot have $\omega_i > \omega_j$ but $i <_P j$. Often this will collapse players with slightly different weights into the same power classes.

Example 16.7. Consider the vector of weights $(0.28, 0.23, 0.19, 0.16, 0.14)$. We have

$$\mathbb{W}^m = \left\{ \begin{array}{c} \{1,2\}, \\ \{1,3,4\}, \{1,3,5\}, \{1,4,5\}, \\ \{2,3,4\}, \{2,3,5\}, \{2,4,5\} \end{array} \right\}.$$

We can see that there is no situation in which adding 1 would be more beneficial than adding 2, so these are in the same equivalence class. Similarly, 3, 4, and 5 have no way to distinguish one another in power. However, clearly $1 >_P 3$ because $\{2\} \cup \{1\}$ is a winner whereas $\{2\} \cup \{3\}$ is not. So despite the fact that each player commands a distinct amount of hashrate, there are only two equivalence classes.

A weighted majority game is *homogeneous* if every minimal winning coalition has the same total weight, that is,

$$\sum_{i \in S} \omega_i = q$$

for all $S \in \mathbb{W}^m$. The value q is called the *quota*. Generically, CBMGs will not be homogeneous, but the homogeneous ones are easier to study.

A weighted majority game with weights $\{\omega_i\}$ can have a homogeneous representation if there exists a set of weights $\{\omega_i'\}$ such that the set of winning coalitions for the weighted majority game is the same and the weights $\{\omega_i'\}$ define a homogeneous weighted majority game.

Example 16.8. Consider weights $(\frac{2}{5}, \frac{1}{5}, \frac{1}{5}, \frac{1}{5})$. This is homogeneous, with the quota $q = \frac{3}{5}$.

Example 16.9. Consider weights $(\frac{2.1}{5}, \frac{1.3}{5}, \frac{0.8}{5}, \frac{0.8}{5})$. This is not homogeneous; $\{1, 2\}$ is a minimal winning coalition with total weight 0.68, while $\{1, 3\}$ is also minimal, but with total weight 0.58. However, the weights $(\frac{2}{5}, \frac{1}{5}, \frac{1}{5}, \frac{1}{5})$ provide a homogeneous representation. The sets of winning coalitions are the same.

Example 16.10. Consider weights $(\frac{2}{9}, \frac{2}{9}, \frac{2}{9}, \frac{1}{9}, \frac{1}{9}, \frac{1}{9})$. This is not homogeneous; the coalition $\{1, 2, 3\}$ is minimal and has total weight 2/3, whereas $\{3, 4, 5, 6\}$ is minimal with weight 5/9. Moreover, this game does not admit a homogeneous representation.

The theory for homogeneous games is somewhat nicer. Example 16.10 shows that it will not always apply.

Example 16.11. *Finding the homogeneous representative.* Consider again Example 16.7. For a homogeneous game, the weights should be the same across the equivalence class. Hence we expect the homogeneous representation to look like

$$(a, a, b, b, b)$$

with

$$2a + 3b = 1.$$

Now for each type (up to equivalence class) of minimal winning coalitions, we compute the equation required for homogeneity:

$$\begin{aligned} 2a &= q, \\ a + 2b &= q. \end{aligned}$$

In this case, there are only two types of minimal winning coalition. The three equations above give us the system

$$\begin{aligned} 2a + 3b &= 1, \\ 2a - q &= 0, \\ a + 2b - q &= 0, \end{aligned}$$

which we can solve:

$$a = \frac{2}{7}, \quad b = \frac{1}{7}, \quad q = \frac{4}{7}.$$

So the homogeneous representation of the game is

$$\left(\frac{2}{7}, \frac{2}{7}, \frac{1}{7}, \frac{1}{7}, \frac{1}{7}\right).$$

Example 16.12. Consider a normalized CBMG with weights $(\frac{2}{5}, \frac{1}{5}, \frac{1}{5}, \frac{1}{5})$ and $C \in (\frac{3}{5}, 1)$. This is homogeneous, as every minimal winning coalition has weight $q = \frac{3}{5}$.

There is no core. The set of imputations for the grand coalition is

$$\left(\frac{2}{5}, \frac{1}{5}, \frac{1}{5}, \frac{1}{5}\right) \cup \left\{(x_1, x_2, x_3, x_4) \mid \sum_{i=1}^{4} x_i = C, x_i \in [0,1]\right\}.$$

The kernel contains the decentralized distribution $(\frac{2}{5}, \frac{1}{5}, \frac{1}{5}, \frac{1}{5})$. To see this, let

$$x = \left(\frac{2}{5}, \frac{1}{5}, \frac{1}{5}, \frac{1}{5}\right).$$

Note that any majority coalition that excludes any miner could be an objection (Definition 16.2). If S is any coalition that includes i but not j, it is clear that there is a minimal winning coalition T that includes j but not i. Now

$$e(T, x) = C - q > 0.$$

On the other hand, if S is winning but not the grand coalition, then

$$e(S, x) = C - \sum_{i \in S} x_i \leq e(T, x),$$

so T must be a counterobjection.

The kernel contains also the redistributed distribution

$$c = \left(\frac{2}{5}C, \frac{1}{5}C, \frac{1}{5}C, \frac{1}{5}C\right)$$

for precisely the same reasons: Any objection is simply a coalition S excluding one member. In this case we still have

$$e(T, c) = C - qC \geq e(S, c).$$

The nucleolus contains the decentralized imputation but not the centralized redistribution. To see this, recall Definition 16.3: The objections defining the nucleolus do not require anyone is excluded. So we can dismiss c by presenting the decentralized imputation $(N, y = (\frac{2}{5}, \frac{1}{5}, \frac{1}{5}, \frac{1}{5}))$ as an objection. Checking that it is an objection,

$$e(N, c) = 1 - C > 0 = e(N, y).$$

Any counterobjection T must necessarily satisfy

$$e(T, y) > e(T, c),$$

which is impossible because y dominates c. Since there is an objection to c with no counterobjection, c is not in the nucleolus.

However, the decentralized imputation is as follows: Any objection to (N, x) is some (S, y) with $e(S, x) > e(S, y)$, namely

$$y(S) > x(S).$$

Next we show that there exists a T with

$$x(T) = x(S)$$

and

$$y(T) < x(T) = x(S) < y(S).$$

Instead of doing this by hand, we offer a trick using linear algebra. Considering the vectors

$$\begin{aligned}\mathbf{1}_{\{1,2\}} &= (1, 1, 0, 0),\\ \mathbf{1}_{\{2,3,4\}} &= (0, 1, 1, 1),\\ &\text{etc.,}\end{aligned}$$

note that

$$\mathbf{1}_{\{1,2\}} + \mathbf{1}_{\{1,3\}} + \mathbf{1}_{\{1,4\}} + 2 \times \mathbf{1}_{\{2,3,4\}} = 3 \times \mathbf{1}_{\{1,2,3,4\}}. \tag{16.2}$$

Now

$$y(S) = y \cdot \mathbf{1}_S,$$

so from (16.2),

$$\sum_{T \in W^m} y(T) = 3y(N).$$

Now if $y(T) \geq \frac{3}{5}$ for every $T \in W^m$ and $y(S) > \frac{3}{5}$, it is easy to see from (16.2) that

$$\sum_{T \in W^m} y(T) > 5 \times \frac{3}{5} \geq 3y(N),$$

which is a contradiction. So there must be some T which satisfies $y(T) < \frac{3}{5}$. For this T, we have

$$e(T,y) > e(T,x)$$

and

$$e(T,y) \geq e(S,x).$$

So T is a counterobjection.

Finally, we claim the nucleolus is unique. For any y, if there exists (S,x) with $y(S) < \frac{3}{5}$ for $S \in W^m$, this will be an objection with no counterobjection. Certainly, $x(S) = \frac{3}{5} > y(S)$, but for there to be a counterobjection T we would need

$$e(T,x) \geq e(S,y),$$

that is,

$$y(S) \geq x(T) \geq \frac{3}{5},$$

which contradicts our choice of S. It follows that we must have $y(S) \geq \frac{3}{5}$ for any y in the nucleolus and $S \in W^m$. Arguing as above, we get $y(S) = \frac{3}{5}$ for each $S \in W^m$. This y satisfies a linear system of equations, namely

$$\begin{bmatrix} \mathbf{1}_{\{1,2\}} \\ \mathbf{1}_{\{1,3\}} \\ \mathbf{1}_{\{1,4\}} \\ \mathbf{1}_{\{2,3,4\}} \end{bmatrix} y = \begin{bmatrix} \frac{3}{5} \\ \frac{3}{5} \\ \frac{3}{5} \\ \frac{3}{5} \end{bmatrix}.$$

The rows of the 4×4 matrix are linearly independent, so this system has a unique solution, namely x. So $y = x$.

For normalized homogeneous constant-sum majority games without dummies, it is known that the nucleolus is the homogeneous representation [34, Theorem 20.54]. We do not prove the full generalization of the above example to CBMGs (which are not constant-sum) but the same reasoning suggests the conclusion that for a homogeneous game with $\mathcal{C} < \mathcal{D}$ the nucleolus should be the decentralized solution. Of course, this is a lucky situation and only serves as a model. In reality, one would not expect sums of hashrates to add up exactly.

In general, the lack of ability to redistribute the decentralized solution (make side payments) creates some technical issues in discussing the solution concepts.

Exercise 16.7. A relation $\mathcal{R}$ on a set Z is a subset of ordered pairs of elements of Z. We say a relation is transitive if $(x,y) \in \mathcal{R}$ and $(y,z) \in \mathcal{R} \implies (x,z) \in \mathcal{R}$. A relation is reflexive if $(x,x) \in \mathcal{R}$ for all $x \in Z$. We say that a relation is symmetric if $(x,y) \in R \implies (y,x) \in \mathcal{R}$.

A relation that is transitive, reflexive, and symmetric is called an *equivalence relation*. Show that the relation $(x, y) \in \mathcal{R}$ when $x \sim_P y$ is an equivalence relation.

Exercise 16.8. Show that the game in Example 16.10 does not admit a homogeneous representation.

16.6 The three-player case

Consider a CBMG with weights $(\omega_1, \omega_2, \omega_3) \neq (\frac{1}{3}, \frac{1}{3}, \frac{1}{3})$ and $C \in (m, 1)$, where

$$m = \min\{\omega_1 + \omega_2, \omega_1 + \omega_3, \omega_2 + \omega_3\},$$

and assume that each $\omega_i < \frac{1}{2}$. This is not homogeneous, as the minimal coalitions will take different weights. The set of distributions for the grand coalition is

$$(\omega_1, \omega_2, \omega_3) \cup \{(x_1, x_2, x_3) \mid x_1 + x_2 + x_3 = C, x_i \in [0, 1]\}.$$

The core is empty; recall Proposition 16.1.

16.6.1 Kernel

Claim 16.1. *The decentralized PC is not in the kernel.*

Consider $x = (\omega_1, \omega_2, \omega_3)$. This admits an objection $S = \{2, 3\}$ of 3 against 1 (we may assume that $\omega_1 > \omega_3$). To find a counterobjection, we have the only option $T = \{1, 2\}$ and this must satisfy $e(T, x) \geq e(S, x)$, which fails:

$$e(T, x) = C - (\omega_1 + \omega_2) < C - (\omega_2 + \omega_3) = e(S, x),$$

so T is not a valid counterobjection. So the decentralized PC is not in the kernel.

Claim 16.2. *The centralized redistribution $c = (\frac{C}{3}, \frac{C}{3}, \frac{C}{3})$ is in the kernel.*

For any objection of i against j, which is just some $S \in \mathbb{W}^m$, there will exist some $T \in \mathbb{W}^m$ which includes j and excludes i. This T will be a valid counterobjection because

$$e(T, c) = e(S, c) = C - \frac{2}{3}$$

for any $S, T \in \mathbb{W}^m$.

Remark 16.1. This is a slightly unexpected outcome, considering that the nucleolus may contain either the centralized or decentralized solution, depending on how m is related to C. This is probably a quirk of shoehorning the CBMG into solution concepts designed for characteristic games with side payments. The results for the nucleolus are more intuitively satisfying, as we will see.

16.6.2 Nucleolus

Claim 16.3. *If*

$$m \geq \frac{2}{3}C, \tag{16.3}$$

the decentralized solution is in the nucleolus.

Let (S, y) be any objection. Then

$$y(S) > x(S) \geq m. \tag{16.4}$$

Now suppose (to be contradicted) that neither of the other two coalitions S', S'' are counterobjections. If S' is not a counterobjection, then either

$$y(S') \geq x(S') \geq m$$

or

$$y(S') > x(S) \geq m.$$

Both of these conditions imply that $y(S') \geq m$. The same applies for S'', giving $y(S'') \geq m$. Summing with (16.4),

$$2y(N) = y(S) + y(S') + y(S'') > 3m,$$

contradicting (16.3). It follows that at least one of S', S'' is a counterobjection. Thus x is in the nucleolus.

Claim 16.4. *If*

$$m < \frac{2}{3}C, \tag{16.5}$$

the decentralized solution is not in the nucleolus.

Take $y = (\frac{C}{3}, \frac{C}{3}, \frac{C}{3})$ and let S be a coalition such that $x(S) = m$. We claim that (S, y) is an objection (to the decentralized solution x) with no counterobjection. Note that

$$y(T) = \frac{2}{3}C > x(S) \quad \text{for all } T \in \mathrm{W}^m. \tag{16.6}$$

In particular,

$$y(S) > x(S),$$

so (S, y) is an objection. Now (16.6) rules out any counterobjection: Any counterobjection would require

$$x(S) \geq y(T) = \frac{2}{3}\mathcal{C}.$$

So (S, y) is an objection with no counterobjections, and thus x is not in the nucleolus.

Claim 16.5. *If*

$$m \geq \frac{2}{3}\mathcal{C}, \tag{16.7}$$

the centralized redistribution $c = (\frac{\mathcal{C}}{3}, \frac{\mathcal{C}}{3}, \frac{\mathcal{C}}{3})$ *is not in the nucleolus.*

We claim the decentralized PC (N, x) is an objection with no counterobjections. First, note that

$$e(N, x) = 0,$$
$$e(N, c) = \mathcal{D} - \mathcal{C} > 0,$$

so this is an objection. There can be no counterobjections: If T is a counterobjection, then

$$x(T) < c(T) = \frac{2}{3}\mathcal{C}. \tag{16.8}$$

But

$$x(T) \geq m \geq \frac{2}{3}\mathcal{C} \quad \text{for all } T \in \mathrm{W}^m,$$

contradicting (16.8).

Claim 16.6. *If*

$$m < \frac{2}{3}\mathcal{C}, \tag{16.9}$$

the centralized redistribution $c = (\frac{\mathcal{C}}{3}, \frac{\mathcal{C}}{3}, \frac{\mathcal{C}}{3})$ *is in the nucleolus.*

Suppose that (S, y) is a centralized objection to c. Then

$$y(S) > c(S) = \frac{2}{3}\mathcal{C}.$$

Now suppose (to be contradicted) that neither of the other two coalitions S', S'' are counterobjections. If S' is not a counterobjection, then either

$$y(S') \geq c(S') = \frac{2}{3}C$$

or

$$y(S') > c(S) = \frac{2}{3}C.$$

In either case, $y(S') \geq \frac{2}{3}C$. Summing up,

$$2y(N) > 2C,$$

with is a contradiction because for the centralized PC (S, y), $y(N) = C$.

The decentralized PC (N, x) is still an objection, but we can find a counterobjection. In particular, if we choose T such that

$$x(T) = m,$$

then we can check that

$$e(T, x) = C - m > e(T, c) = C - \frac{2}{3}C$$

by virtue of $m < \frac{2}{3}C$, and also

$$e(T, x) \geq e(N, c) = 0.$$

So for either the centralized objections or the decentralized objections, there exist counterobjections. It follows that c is in the nucleolus.

Summary 1. If $m < \frac{2}{3}C$, the centralized redistribution is the nucleolus, while if $m \geq \frac{2}{3}C$, the decentralized imputation is the nucleolus.

Corollary 16.1. *Suppose that $C < \frac{3}{4}$. Then the decentralized solution is the nucleolus.*

Example 16.13. Suppose $C = 0.84$, $(\omega_1, \omega_2, \omega_3) = (0.4, 0.37, 0.23)$. Then

$$m = 0.6,$$
$$\frac{2}{3}C = 0.56 < m.$$

So the decentralized imputation $(0.4, 0.37, 0.23)$ is the nucleolus.

16.6.3 Beyond $n = 3$

For a number of technical reasons we will not attempt to state and prove a generalization. The theory may be delicate, especially for non-homogeneous hashpower arrangements. In the $n = 3$ case, we are not forced to contend with a non-minimal winning coalition attempting to enforce decentralization, which could happen for larger numbers of players.

Example 16.14. Consider a game with weights $(0.23, 0.20, 0.19, 0.19, 0.19)$ and $\mathcal{C} = 0.8$. While the minimal winning coalitions prefer centralizing, the coalition $\{1, 2, 3, 4\}$ will prefer decentralization and the coalition $\{2, 3, 4, 5\}$ will prefer centralization.

Based on the result for $n = 3$, one may conjecture that for games with homogeneous representation the condition

$$m = q\mathcal{C}$$

serves as a cutoff: For values of $\mathcal{C}$ larger than m/q there is incentive to centralize and redistribute, while for smaller values of $\mathcal{C}$ the system is more likely to remain decentralized.

16.6.4 Notes on $\mathcal{C}$ and $\mathcal{D}$

The ratio of $\mathcal{C}$ to $\mathcal{D}$ could take a large range of positive values. It is important to emphasize that $\mathcal{D}$ represents the value accrued to miners during decentralized operation, not the value of decentralized network itself. No value of Bitcoin as a public good is considered, as miners have no way to divvy this up. If profits are near zero, this value can be quite low. On the other hand, centralized mining can result in much larger profit margins, even though the value of the network itself may have decreased. In general, lower miner profit margins and muted market fragility suggest $\mathcal{C} > \mathcal{D}$, while higher miner profit margins and substantial market fragility would lead to $\mathcal{C} < \mathcal{D}$.

16.7 Discussion

In two-player bargaining, the cost of negotiation was quantified in Rubinstein bargaining, confirming the Nash bargaining solution. With more players, the concept is similar; one wants to avoid going in circles. In the real world there is cost to going in circles, so if players have equitable ways to terminate this process without any player forfeiting any advantage, they may be induced to accept such an agreement.

If the nucleolus is already being enacted by decentralization, it would be dangerous for players to cartelize. Even though they may see an opportunity to achieve a higher payoff, they open the door to a round of negotiation which may or may not improve their situation. So the concepts of objection and counterobjection are germane to Bitcoin. In fact, the principles apply to many aspects of social organization. Many parties prefer to make small sacrifices in order to avoid being constantly engaged in hardball moves or negotiation tactics. Social norms can re-enforce this.

On the other hand, overhead is required not only for negotiation (as is somewhat accounted for in Rubinstein's model) but also administration of agreements. This is hard to be explicit about in game-theoretic models, but is somewhat common sense: A near majority player may choose to collude with the second largest player rather than dozens of smaller players as this makes things simple. The costs of administration are lower and the number of parties needed to trust in an agreement is smaller.

We also are not considering that some players may have affinities that facilitate collusion. One may expect that a handful of mining corporations co-located in Texas are more likely to form a coalition than, say, a coalition of two corporations in Texas together with two corporations in Kazakhstan.

We are also ignoring the fact that some miners may refuse to join any coalition whatsoever out of principle. Perhaps paradoxically, there are situations in which this can increase the likelihood of a cartel forming: The notions of stability underpinned by the concept of counterobjection require that all possible cartels must be willing to form. If some cartels are not open to this, the set of counterobjections is smaller, allowing for more objections with no counterobjections, leading to more destabilizing pressure.

In reality, the acting coalition may just be deciding to behave opportunistically, especially if other players are not eager to form coalitions. It is not clear when the game has moved from a clearly decentralized regime to a regime where bargaining and negotiating strategically is appropriate. The idea of splitting the proceeds from a grand coalition makes more sense in a laboratory setting. If each player in the room is prepared to negotiate, one would expect an equitable solution. Bitcoin is unlikely to behave in this way: If a cartel formed (by analogy, an objection is asserted) the counterobjection (some other cartel consisting of diverse players throughout the world) may not form as easily as it would in a laboratory. The cartel may see this as a time-dependent opportunity and choose to act before there is competition.

Remark 16.2. We are not discussing the Shapley value. The Shapley value is commonly regarded as one of the best ways to apportion value to players in a coalitional game. This suggests the "fair value" of playing a coalitional game. While Shapley values are interesting to compute, we are more concerned with one of the driving questions in the game theory of Bitcoin: whether centralization is a likely outcome or if decentralization will continue.

17 Stock, flow, and economic considerations

Bitcoin's *fundamental economic security assumption* is the following:

$$\text{total cost of attack} > \text{reward for an attack}, \tag{17.1}$$

for each agent or collection of agents who possess the ability to launch a successful attack.

In terms of coalitional game theory, the above condition is a statement that decentralized operation is in the core, that is, no coalition has incentive to attack the network (recall §16.4). If (17.1) fails, some stability may persist, but this relies on the objection–counterobjection reasoning underpinning solution concepts discussed in §16.4. In this chapter we explore the fundamental question of when (17.1) fails. If there is a way one coalition of miners can attack the network improving payoff to all its members, one should conclude that the fundamental security assumption is no longer in place. While any attack can be thought of as an objection to decentralization to which there may be a counterobjection, we would not want to rely on this nuance for security.

First, it is important to point out that the set of collections of agents who could successfully attack the network is always non-empty. This follows from the mathematically obvious fact that 51 % < 100 %, as any collection of miners who together marshal over 51 % of the hashrate is a potential attacker. The condition (17.1) is meant to hold not just for some bogeyman or malignity outside the ecosystem who wants to attack, but most importantly, also for groups of miners actively maintaining the security. Majority collections of miners always exist, whether or not they are actively collaborating. Indeed, this was discussed in the whitepaper [38, §6]:

> If a greedy attacker is able to assemble more CPU power than all the honest nodes, he would have to choose between using it to defraud people by stealing back his payments, or using it to generate new coins. He ought to find it more profitable to play by the rules, such rules that favour him with more new coins than everyone else combined, than to undermine the system and the validity of his own wealth.

The set of coalitions who could potentially launch an attack depends heavily on the hash deployment ratio (recall §12.1). If the HDR is near 100 % and no hashrate is available for rent, the only coalitions with the ability to attack need to have 50 % of the hashrate (or 34 %, if one includes selfish mining in the set of possible attacks). On the other hand, if the HDR is low and the availability of hashrate for rent is large, there can be quite a large handful of coalitions that could attack the network, possibly even including actors who have no investment in the network at all.

Example 17.1. Consider the coalition Bitcoin mining game (Example 16.4) with $\mathcal{C} = 100$ and $\mathcal{D} = 20$ and three miners each with 1/3 of the hashrate. This represents a situation where the market fragility hypothesis fails and mining is expensive relative to revenue;

https://doi.org/10.1515/9783110772845-017

there is much to be gained by cartelizing. This behaves like the odd man out game (Example 16.1). The core is empty, while the kernel and nucleolus consist of the centralized solution obtained by centralizing as a grand coalition and then dividing the profit. In practice, if any two miners share some affinity, they may go ahead and collude, excluding the third.

Defining the left- and right-hand sides of (17.1) is a perplexing problem, laden with imprecision, especially when coalitions are involved. In addition to containing parameters that can only be guessed, it also should be interpreted as a risk-adjusted expectation. In order to get a single number on both sides, one would assign probabilities and utilities to each of the outcomes of an attack to compute an expectation.

To begin attempting to make (17.1) precise, we should point out that there is more than one possible type of unit. We distinguish between stock and flow type units. A *stock* value is the value of an asset or network as computed at a particular time and does not have a per-time unit associated to it. For example, if one says "the cost of attacking Bitcoin is $5 billion, which is less than the $10 billion payoff of big short play" this is a statement in terms of stock units: comparing one-time expenditures to one-time payoffs. On the other hand, one might say that "the cost to attack the network is twice the security budget." This is a statement about *flow*, which has a cost-per-time unit. Note that the integral of flow over a time interval will represent a stock unit; in particular, the flow cost of attack over time multiplied by the time interval required to complete the attack will give a stock cost involved in the attack. This relationship as it pertains to (17.1) is explored in a somewhat well-known paper of Budish [9]. For a miner participating in proof of work, the flow represents the income derived from mining rewards, while stock can represent the value of a one-off attack on the network, or also the value of precious mining equipment that may be coveted: The value of this equipment is naturally tied to the brand name of Bitcoin and to the efficiency of competing equipment.

Perhaps the largest variable in this discussion is the availability of additional hashrate, for sale or for rent. In normal operations we assume that the marginal cost of creating hashrate for each miner will be no more than the expected revenue; this is the standard zero-profit condition. Most miners use all of their equipment if all of their equipment is efficient. If some is less efficient, miners may be using only a fraction of their equipment. If a miner has access to cheap electricity, one would expect them to obtain less efficient ASICs and use these to expand operations. There is less competition for the less efficient equipment, so miners with cheap electricity should be able to purchase this at advantageous prices (see [53]).

At any given time, there should be some amount of equipment that is not profitable, for anybody, given mining rewards and costs. This equipment could be sitting idle in a warehouse next to active ASICs, or it could be available for rent or for sale. Transportation costs are non-trivial, so even if someone in Brazil has access to a limited amount of very cheap electricity, it need not be the case that they acquire unprofitable equipment left in a warehouse in Poland.

A given miner making the decision to attack or not will consider the cost of their own equipment (some of which is perhaps profitable) as well as the cost of renting or purchasing other equipment that is being sold on the market. The costs of purchase are one-time stock costs, while the costs of electricity and renting additional hashrate are flow costs.

With this in mind, we offer a list of values that will play a role in our analysis of (17.1).

I	some agent or collection of agents
$\Delta_{\text{attack},I}$	factor by which the value of Bitcoin declines after attack by I, due to market fragility
$V_{\text{DS},I}$	maximum value of low-trust transactions that are double-spendable to agents I
S_I	value of one-time purchase required to obtain a majority of the hashrate
U_I	unrecoverable purchase costs in obtaining stock S_I
$C_{\text{DS},I}$	double-spend flow costs, including possible rental or electricity costs to maintain a majority attack
$C_{\text{Sab},I}$	sabotage flow costs, including possible rental or electricity costs
T_I	total time needed to execute a majority attack
A_I	market value of I's ASICs stock
R_I	probability that a community coalition overrides the attack or blacklists proceeds
s_I	total amount of short interest available to I
E_I	authoritarian bribe or value of sabotage not accounted for by shorting

As far as motivations for an attack, we are grouping these into three categories:

1. The double-spend exploit, as alluded to in the whitepaper.
2. Speculative profit obtained from movements the financial market would make in response to an attack (Goldfinger attack).
3. Authoritarian or adversarial reasons to attack: A nation may not like Bitcoin, or a competing blockchain or financial institution may be damaged by Bitcoin's operations.

In all the above quantities, for simplicity we consider these values as being in excess of the rewards and costs of obtaining the blocks themselves: By assuming a zero-profit condition we can avoid adding the same values to both sides.

17.1 Double-spend exploits

Anyone wishing to use a double-spend exploit should calculate the maximum value that can be obtained by such an attack, the escrow time necessary to take possession of the initial spend, and the cost of acquiring and maintaining more than 51 % of the hashrate. Other risks can be more difficult to quantify. The community could override the attack via either suspending Nakamoto consensus or blacklisting or tainting the stolen funds; or the value of the double-spend could be reduced if this triggered a market fragility response.

Note that the total value accrued during a double-spend relies completely on the opportunities to participate in *low-trust* transactions. While the original goal of Bitcoin was to create a space where low-trust transactions could take place, in practice, there may be only a small percentage of transactions that are legitimately low-trust. A low-trust transaction is a transaction that the spender would have a clear incentive to reverse at some point in the near future, with no recourse for the recipient. A Bitcoin transaction can be like handing someone a check. When you pay a local business to install HVAC equipment in your home via a check, there is little risk to the business of a double-spend: While you could empty out your checking account, they know who you are and where you live. Handing them a check is not a low-trust transaction. On the other hand, purchasing bicycle parts in the parking lot of a Walmart in response to a Craigslist advertisement is a low-trust transaction, and the seller is wise not to accept a check.

An often-conjectured way to perform a large-value double-spend is to cheat an exchange in a different jurisdiction: deposit a large amount, then withdraw (perhaps via another cryptocurrency), and then perform the reorg, or use the funds to bid up thin markets, laundering money via exchange markets into other accounts.

The following model equation offers a basic criterion for when an attack by a collection of agents I is incentivized, in which case (17.1) fails:

$$(1 - R_I)(1 - \Delta_{\text{attack},I}) \times V_{\text{DS},I} > C_{\text{DS},I} \times T_I + \Delta_{\text{attack},I} \times (A_I + S_I) + U_I. \tag{17.2}$$

Several remarks are in order.

- These values are not well-defined or deterministic: Other miners could fight back, the market fragility effect is unknown, and the miner might be unsure which values T_I and S_I are most appropriate (larger S_I means smaller T_I, usually).
- We are only discounting the stock purchased when the attack devalues Bitcoin: If additional hashrate is purchased, we assume that it maintains its market value, and hence it is not considered a "loss," unless Bitcoin actually loses value. Transaction costs incurred obtaining the hashrate can be lumped into U_I.
- The values of all parameters necessarily depend on I. For example, if a very large cartel of miners I colluded to attack, S_I would be zero, but $\Delta_{\text{attack},I}$ would likely be closer to 1.
- The value $V_{\text{DS},I}$ appears to be overestimated in some literature. For example, in [9, §2.1] the number of transactions per block is multiplied by "highest-value transactions that are possible on the Bitcoin system"; certainly this is an overestimate by orders of magnitude, especially when most transactions are not low-trust.
- $V_{\text{DS},I}$ could be driven higher if the agents I are actively collecting (and keeping secret) a pool of transactions of which they are taking a cut. For example, an individual wishing to double-spend may reach out to a malicious attacker, directly sending them, say, 60 % of the transactions while directing 40 % back to themselves. The attacker may also perform a wide and shallow double-spend as described in [29].

- If $\Delta_{\text{attack},I}$ is not small, the feasibility of a double-spend attack is extremely low, that is, the market fragility hypothesis renders double-spends unlikely.
- Values of R_I near 1 also greatly quash the incentive to attack. The strong expectation of tainting or blacklisting renders double-spends unlikely.
- If the amount of available unprofitable hashrate is quite large, S_I may tend towards 0 (although U_I need not).
- The value A_I may also be quite small if the hashrate market is competitive and ASICs are widely available.
- While T_I or S_I should be finite, these could be ridiculously large (for instance, certainly $17 trillion would be enough to purchase enough hashrate for a majority attack, even if there are some reluctant sellers).
- For coalitions of public-facing entities, $V_{\text{DS},I}$ would be near zero. Branded Bitcoin corporations with ongoing contractual relationships would rarely find occasion to double-spend (there are always imaginable circumstances; the officers of a corporation going belly-up may want to steal clients' funds on the way out, or law enforcement demands an exchange claw back a withdrawal from a bad actor).
- If a majority of all available hashrate is controlled by public-facing entities with no desire to participate in a double-spend, there is no risk.
- The terms $C_{\text{Sab},I} \times T_I$ and $(A_I + S_I)$ should grow superlinearly as functions of the security budget. Higher security payments to honest miners means it costs more to obtain and deploy equipment required to perform an attack.

All things considered, isolated double-spend attacks appear to be unlikely in the near future and are only incentivized in very narrow sets of circumstances. The left-hand side of (17.2) should likely be quite low, while all three terms on the right hand side should be quite high. In the early days of Bitcoin, when a much higher percentage of transactions where illicit and before ASICs raised the stock costs of being involved in mining, double-spending was more of a concern.

17.1.1 Finality

The concept of *probabilistic finality* was introduced in the whitepaper [38, §11], but remains hard to precisely quantify and is poorly understood. The probabilistic nature means that a given transaction that is on the chain today has a high probability (near but not equal to 1) of being part of the heaviest chain at some date in the future. Paying enough to obtain blockspace in a given block is pointless if the block is to be supplanted in the near future.

There are different notions of finality. In the whitepaper, it was assumed that an attacker has a fixed amount of hashpower and that this hashpower is less than 50 %. Then, as more blocks are mined on top of the transaction, the finality approaches 1 over time (a more precise computation is given in [20]). This idea continues to be a part of the common understanding of finality. Provided the attacker has less than 50 % of the hashrate, the finality always approaches 1 very quickly. An attacker with 10 % of the hashrate attempting to create a reorg 200 blocks deep would have better odds simply attempting a brute-force guess of Satoshi's private keys. While the probability that an attacker will overcome such odds should be ignored, the probability that an attacker will come to obtain more than 50 % of hashpower at some point in the future is not infinitesimal.

This soft understanding has led to some flawed assumptions about blockchain security. One common misunderstanding is that if the security budget becomes quite small after the subsidy declines, those accepting payments can simply wait longer to consider the payment final. However, this is not mathematically justified. The non-linear nature of the finality computation depends delicately on how close the attacker is to 50 %. If an attacker is near enough to 50 % that a payment six blocks deep is not deep enough to be considered final, this is a security concern that waiting until 12 or 15 blocks have passed will not remedy. While an attack with 48 % of the hashrate is very likely to fail in attempting a 1,000-block reorg, the attacker is all but guaranteed (eventually) to succeed by gaining 3 % of the hashrate to become 51 % of the network.

Bayesian reasoning is appropriate here: It is natural to have beliefs about the probability that different hashrates are being used against you. If your prior belief is that nobody is going to attack your transaction and that there is not a lot of latent hashrate laying around, you have little to worry about. Bayesian reasoning is particularly important when observing the number of blocks that are being produced.

Example 17.2. Suppose you are paid a large low-trust transaction and have reason to believe the spender may attempt to reorg the chain to reclaim the payment. If five confirmations appear in 25 minutes, the event that the chain is being attacked is unlikely: If around half of the hashrate was working on a secret reorg, producing blocks with the frequency you observe is unlikely. On the other hand, if you wait 2 hours and see only six confirmations, this is reason to be suspicious as this slowed block production is not an unlikely occurrence given that an attack is happening.

A more important notion is the idea of *economic finality*. We refer to the definition provided in [3, p. 11]: *"a cryptocurrency payment can be considered as final once it is certain that from a certain moment of time onwards, it will never be profitable to undo the payment via a double-spending attack."*

This definition is more robust in that it does not make assumptions about the hashpower of parties who would be willing to attack. The hashpower of attackers can change dynamically as coalitions can form or be disbanded in an instant. This definition asks a better question: When do the incentives for a coalition to attack exist at all? If there is no incentive to attack, the payment should be considered final. While not explicitly spelled out in the above definition, the profits from overturning the particular payment may be computed while considering attacks that may be primarily targeting other transactions. For example, a 0.1-bitcoin transaction that is six blocks deep may be considered secure on its own; however, if there is also a low-trust 10-bitcoin transaction that is seven blocks deep, it may be profitable for an attacker to go after the larger transaction and in the process also accept a double-spend for the smaller transaction. If they are offered 0.05 bitcoin in a double-spend routing the other half back to the sender, it is indeed profitable to undo the payment, provided they are already going after the larger transaction.

Time (measured by addition of blocks) should increase finality, but not necessarily linearly: If an attacker has the ability to marshal 60 % of the hashrate, they may be doing so at additional cost, perhaps renting inefficient equipment. Longer reorgs will take much more time and require much more reward to become profitable. Deeper reorgs are also more damaging to Bitcoin's brand name, so Δ_{attack} can be large: A 500-block reorg which is not forked away could significantly damage the value.

One interesting case is the situation to be described in §17.4.2: Large institutions who remain unaffected by deep reorgs would be unharmed by a double-spend. In this case notions of finality are more important only to those dealing with low-trust counterparties. In this regime there would be a limit to the amount of illicit transactions that could safely take place over time – if recent blocks become heavy with illicit transactions, the incentive to reorg becomes quite large. Miners processing such transactions could take this into consideration, prioritizing transactions that are from known parties and demanding higher fees for larger transactions sourcing from suspicious parties. This makes for some interesting game-theoretic calculations by miners deciding to confirm low-trust transactions.

Bidding up the fee market can increase security for recent past transactions if it drives the blockchain forward, but will not provide security against future attacks if the fee market were to implode. In fact, a precipitous dropoff in fees presents an incentive to reorg: Miners may find it more profitable to go back in time and reclaim old fees when the current mempool is unappealing. In particular, paying a fee larger than required for inclusion into the next block does not increase finality or reduce the time until a transaction can be considered final, as stronger finality requires persistent block rewards into the future.

When Bitcoin users are creating transactions with fees, they are typically not paying for security; rather, they are paying for blockspace. Because there are thousands of transactions per block, individual transactors would not be considering the marginal security benefits by offering the miners a higher fee; usually they just wish to pay the minimum fee required for the transaction to clear. The security budget over time is an aggregate of all fees plus the subsidy. Users would not willingly donate to this cause any more than an individual US citizen would willingly donate to the US Department of Defense, which is already amply funded. Transactors have no inducement to overpay.

As the block subsidy dwindles, the security budget will depend largely on the utility of the network as a means of making peer-to-peer transactions. There is no way to extract a security tax from those who use Bitcoin as a store of value. In this sense, there is a free-rider problem: Long-term holders of Bitcoin are benefiting but not paying for ongoing finality. There is more value being enjoyed than what is being paid for.

17.2 Sabotage (Goldfinger) attack

As above, we can model the incentive condition for sabotage via

$$\Delta_{\text{attack},I} \times s_I + E_I > C_{\text{Sab},I} \times T_I + \Delta_{\text{attack},I} \times (A_I + S_I) + U_I. \tag{17.3}$$

We make several remarks:

- We use the term $\Delta_{\text{attack},I} \times s_I$ to denote the value obtainable by shorting. We could construct a similar expression if the mechanism for profiting from collapse is buying put options.
- The term s_I can be somewhat large and is likely to scale linearly with the total market cap. If the total market cap of Bitcoin is trillions of dollars, it may not be unreasonable to obtain billions of dollars of short interest.
- Included in the factor s_1 should be any ETFs which are tied to Bitcoin or any publicly traded companies that hold Bitcoin on their balance sheet. Any publicly traded corporations or funds that are leveraged long into Bitcoin can contribute in a nonlinear way.
- Also included in the short factor is the possibility of shorting Bitcoin with respect to another decentralized currency which gains if Bitcoin was sabotaged. For example, a cryptocurrency using proof of stake that would not be subject to the same attack could stand to gain from a sabotage attack on Bitcoin.
- These values are not deterministic. For example, T_I should have some range and a probability distribution over that range: The point of a sabotage is not to steal specific funds but to destroy faith in the ecosystem, so it is unclear from the start how long this effort would take. A single 10-block reorg may do the trick, but could also require a protracted attack over a period of days.
- The modeler gets to choose if E_I is a function of $\Delta_{\text{attack},I}$. Here we assume the value of an attack to the hostile actor is independent of the drop in price; however, one may choose to model this term to depend on the drop in price.
- We are proposing one attempt at modeling (17.1). There are many variations on this theme. See also [9, §2.1], [3], and [25], which offer similar analyses beginning from (17.1).

A sabotage attack is likely to remain a serious concern into the future. A definitive enough attack is certain to damage the price of Bitcoin, and as Bitcoin continues to permeate traditional financial markets, the available short interest is only going to increase. If Bitcoin grows, the reasons why actors could become actively hostile also grows, increasing E_I. On the other hand, dwindling security budget will suppress the right-hand side.

17.3 Determining the stock value of ASICs

Under a standard assumption of free entry for miners, profit margins will compress to zero and the marginal cost of mining will be the marginal value of reward. This is only approximate, as miners face both operating and capital expenses, which may vary over time and from miner to miner, and receive income that is variable and depends on fee markets, the hashrate, and the market value of Bitcoin. We express the zero-profit condition to a particular miner as

$$\chi_{\text{hash}} = d + e,$$

with

$$\begin{aligned} \chi_{\text{hash}} &:= \text{monetary reward for a unit of hash,} \\ d &:= \text{amortization rate of equipment while producing a unit of hash,} \\ e &:= \text{electricity cost required to produce a unit of hash.} \end{aligned}$$

One can approximate d by dividing the cost of the equipment by the lifespan (lifespan should be given in hashes as units must agree). In other words,

$$\frac{\text{price of an ASIC}}{\text{hash units produced over lifespan}} = \chi_{\text{hash}} - e.$$

As long as there is a significant discrepancy between the reward for producing hashes and the electricity cost, the value of machines should remain positive, and more ASICs will continue to be produced in factories. However, if the price or fee market drops significantly relative to the hashrate, the hashrate will readjust (via difficulty adjustment) to the point where $\chi_{\text{hash}} \leq e$ for many ASICs. In this case, only the most efficient machines will have positive productive value. The value of the machines will then be determined by the possibility that they may become valuable to produce Bitcoin in the future or by their usefulness in either waging or defending an expensive hash war. If the hash reward of new machines being produced does not exceed electricity costs by an amount significant enough to justify production, new ASICs will not continue to be built.

Over time, as new machines become more efficient, unless the price and fee rewards continue to grow fast enough to outpace the production of new equipment, the stock value of older machines will trend towards zero. Any stock value will arise as a hedge against extreme market situations or attacks. This can create a somewhat dangerous situation: The values of A_I and S_I may become quite small: $C_{\text{Sab},I}$ or $C_{\text{DS},I}$ will become larger but may not become large enough to prevent the attack incentive from occurring.

Returning to the marginal cost and value functions described in §12.3, we can determine how much of the total stock of ASICs will have a value near zero. In Figure 12.2 there is a point where the curves meet, a hashrate h_* such that $\text{mc}(h_*) = v(h_*)$. As the dependent axis is the cumulative hashrate of machines in action, it follows that h_* represents hash produced by the machines with non-negative value: This many machines can be used immediately to return profits on energy. In practice, the curve $\mathrm{mc}(h)$ will probably have a vertical asymptote at some value $H_{\max}$, which represents the maximum amount of hashrate that could be brought on board at any price. The amount of hashrate that would have a small or near zero value would be $H_{\max} - h_*$, as this represents hashrate that is not profitable to be turned on at current prices. In this case the hash deployment ratio is given by

$$\mathrm{HDR} = \frac{h_*}{H_{\max}}.$$

17.3.1 Moore's law

It is unclear to what extent Moore's law holds for ASIC production, as some evidence suggests that ASIC efficiency increases may be approaching a limit. However, we can perform some simple computations based on the assumption that the total amount of hashrate capacity produced grows exponentially. (Side note: Koomey's law governs efficiency, whereas the better-known Moore's law governs processing power. Koomey's law says that the energy efficiency of technology is also doubling roughly every year and a half.)

Proposition 17.1. *Suppose that the amount of hashrate that is produced per year is given by $H_0 b^t$. Consider the fraction of the total hashrate which has been produced within the last x time units. Then as time increases, this fraction approaches*

$$1 - b^{-x}.$$

Exercise 17.1. Prove Proposition 17.1.

The significance of this is that if the hashrate production grows exponentially, one may attempt to acquire a majority by purchasing only the newest equipment available. The amount of time required to do so remains roughly fixed: It should take

$$x = \frac{\ln 2}{\ln b}$$

units of time to acquire a majority of all hashrate. Thus Moore's law has double-sided implications for security: While control of the majority of hashrate may be susceptible to one party making large purchases over a short period, on the flip side, acquiring decades worth of obsolesced equipment does not allow one to gain a majority, as long as growth continues at the same rate.

If Moore's law is out-sized in effect, then the stock values of ASICs behave more like flow costs: Because the depreciation of equipment would be quite rapid, miners must continually buy new equipment to produce competitive hash.

17.3.2 Flow-based miner cost model

It is interesting to consider a pure flow-based economic model where capital expenses and operating expenses are not distinguished. This could be true if hashrate is available for rent or if new equipment is becoming efficient fast enough that it is burned through and obsolesced shortly after it is purchased. In this case we can consider a simplified zero-profit condition: The hashrate over a period is equal to the reward over that period divided by the market cost of producing hash.

In this case the double-spend and sabotage attack incentive conditions become, respectively,

$$(1 - R_I)(1 - \Delta_{\text{attack}}) \times V_{\text{DS},I} > C_{\text{attack}} \times T_{\text{attack}}, \tag{17.4}$$

$$\Delta_{\text{attack}} \times s_I + E_I > C_{\text{attack}} \times T_{\text{attack}}, \tag{17.5}$$

where C_{attack} is the cost of producing a majority of hashrate and T_{attack} is the time necessary to complete this particular attack. In comparison to (17.2) and (17.3), these conditions have become more feasible after removing the stock terms. Looking at these conditions, one can also observe that Δ_{attack} is a "pick-your-poison" parameter [9, §2.2]: If near 1, the incentives to sabotage are large, whereas if near 0, the incentives to double-spend may become significant. However, the indeterminacy of Δ_{attack} may provide some protection against an attack from a party looking to accomplish either one goal or the other. If the goal is entirely to sabotage the network, the attacker must reckon with the probability that Δ_{attack} is smaller than expected. On the other hand, an agent wishing to use the double-spend exploit must be careful not to underestimate Δ_{attack}. A clever and well-resourced attacker may try to combine both as a hedge: By trying to take advantage of portions $\mu_1, \mu_2 \in (0,1)$ of each of the proceeds, one can consider a more general condition that will remain true even for a larger range of values Δ_{attack}. We have

$$\mu_1(1 - R_I)(1 - \Delta_{\text{attack}}) \times V_{\text{DS},I} + \mu_2\Delta_{\text{attack}} \times s_I + E_I > C_{\text{attack}} \times T_{\text{attack}}. \tag{17.6}$$

Based on this inequality we can make an observation: If R_I is close to 1, it will be very difficult to guarantee that an attack will pay off, especially if one is assuming that Δ_{attack} is small.

Conclusion 17.1. If the soft-forkability hypothesis holds, Bitcoin remains economically secure against attacks, provided that market fragility is not triggered by a soft fork.

While explicitly deferring to governing entities (see §13.8) represents an anathema to Bitcoin true believers, it would provide significant incentive against both double-spend and sabotage attacks.

17.3.3 Stock value of latent hashrate as a threat

ASICs that are not profitable for mining have at least two reasons to maintain positive value. One is as a type of call option – ASICs that are currently unprofitable give the holder of the machines an option to create bitcoins at a price depending on electricity costs. This would be fruitful to exercise if Bitcoin's value were to spike above that price. The ASICs value derived from this optionality should not exceed the market value of financial derivatives with the same function. Another is as an extortionary weapon, as the following example demonstrates.

Example 17.3. Suppose there are four firms who hold all ASICs. Firms A, B, and C each have ASICs which produce 10 units of hash at a cost of 1 unit of value per unit of hash. Firm D has ASICs which would produce 40 units of hash, but at a cost of 4 units of value per unit of hash. Suppose that the block reward per unit time is 105 units of value.

If each of the firms A, B, and C are producing at maximum rate, the total production is 30 units, so that the return on a unit of hash is $105/30 = 3.5$ value/hash, which puts each of these firms clearly in profit, as their costs are 1. Firm D, however, would not be making profit as their cost for creating a unit of hash is 4. However, firm D makes the following observation: If they are to produce 31 units of hash in order to collect all mining profits, their cost would be 123 and the return would be 105, for a loss of 18 which is less than the loss to A of 25. Because the loss to firm A is greater than the value lost to firm D via this attack, firm D can attempt to claim their portion of the Nash bargaining solution by extorting A. They could demand that A subsidize D, paying them 3.5 units of value to not hash at all, or demand that A directly purchase enough of their stock to obviate any future attacks (note we are not assuming market fragility, which would clearly change the calculation).

17.4 Stable regimes

If the conditions (17.2) and (17.3) both fail, there could be stable outcomes where not only the market cap is large but also Bitcoin is reasonably secure. We consider a couple of scenarios. These visions of Bitcoin differ immensely from those of the early community.

17.4.1 Strong expectation of community reorg, no major adversaries

One scenario is that there is a strong expectation that a community reorg will occur in response to any attack.

- R_1 must be near 1: Any major attack will have a near certainty of being soft-forked away by the community, rendering attacks no more than a waste of resources.
- Δ_{attack} should be very small. This may follow from R_1 being near 1: If the ecosystem is confident that Bitcoin will fork away any attacks, attacks will not shock the market.

- E_I should be small. No major nation-state should be willing to spend military-budget amounts of money to attack Bitcoin. Note that the reason for this could be precisely the soft-forkability hypothesis: As much as a superpower may want to destroy Bitcoin by overpowering the network, if there is a community mechanism for deflecting such attacks, the attacks are not incentivized.

17.4.2 Attacks happen, but only affect a minority of users

Another regime which would allow for some security assurances is one in which attacks may happen, but they do not affect an economic majority. In this case, both sides of equation (17.2) could be small.

- $V_{DS,I}$ should not be large. This may arise if the overwhelming majority of Bitcoin transactions are not low-trust, but between financial or other institutions that maintain semi-trustful relationships with counterparties. While drug dealers, child-pornographers, ransomware attackers, or terrorists may require the irreversibility and censorship-resistant properties of the network, most users do not, and hence very few transactions will be at risk of double-spends.
- Similarly, if tainting is a common practice, the value of a double-spent output may be near zero.
- Δ_{attack} should be small. If published transactions between semi-trusting institutions that have been posted to the mempool can be considered somewhat reliable, a large reorg need not affect the accounting. The transactions would still be expected to make their way onto whatever longest chain persists, so attacks would not be disruptive for most users.
- Here E_I can be significant, but would occur in the double-spend incentive equation rather than the sabotage incentives. A government or international coalition of governments who would like to discourage low-trust use of the network may encourage an occasional purge attack: As long as important institutions ignore the opportunity to double-spend (perhaps double-spending could even be declared illegal), the only users negatively affected will be those dealing with untrustworthy counterparties. In this case, purge pools may operate very much in the open and even compete against one another. KYC and anti-money-laundering procedures would become self-enforcing: An exchange that chose to deal with unsavory characters or used lax KYC procedures would be putting themselves at significant risk.

17.5 Efficiency, security, and decentralization

We now explore security through a slightly different lens, based on splitting the possibilities into two sets of scenarios:

1. If decentralized Bitcoin is inefficient, then improvements can be made by centralization and decentralization is unlikely to persist.
2. If decentralized Bitcoin is efficient, a risk arises if the surplus efficiency created by the network can be leveraged by a malicious attacker.

In a system where security operations are performed by unknown economic free agents, the same economic free agents can be outbid by actors who are either hostile or simply profiteering. Any efficient system must be creating some value, otherwise it would not exist. If the system is creating value, the security budget (the amount paid to the economic free agents that are providing security) should be less than the benefit conferred to users. If there is a gap between the security budget and the value created, any outside actor, or perhaps even the free agents themselves, may hope to use this gap as an arbitrage opportunity. If the network would continue to be valuable despite centralization, the free agents have an incentive to cartelize in order to seize the surplus profits for themselves. If the system would fail to be valuable under centralization, a well-resourced profiteer seeks to use this failure mode as leverage in order to extract value. If the value created is part of a zero-sum game, that is, the value is created at the expense of another party, that other party will find it more costly to leave the network in place than to destroy it.

Ignoring motives for direct hostility, the key to Bitcoin longevity as a decentralized network is the following: The network must create excess value for its users via decentralization, without creating a situation such that any individual users could be directly leveraged.

Example 17.4. Consider a set of 100 users who form a decentralized network. The network benefits each user by 1 unit of value and the total security budget is 30 units. If the network were to fall into centralization it would be useless to all users. Thus there is a surplus of 70 units created by the network, with no incentive to centralize. The system remains secure and efficient.

17.5.1 Efficiency

In economics, a system is efficient if it uses resources optimally. Much of economic activity falls into this category, almost by tautology: An ecosystem made up of rational agents would not choose an inefficient system if there was a better option that is easily available. Generally speaking, inefficient economic activity faces annihilating pressures. A decentralized system will be efficient when the added cost due to the *price of anarchy* (PoA) is less than the value that would be destroyed if the system where to be centralized. The PoA is the ratio of the cost of the least efficient Nash equilibrium arrangement to the optimal arrangement that would be chosen by a central planner with a socialized objective in mind. A related notion is the *price of stability* (PoS), which compares the best Nash equilibrium to the central planner's solution. By definition, PoA bounds PoS.

When Bitcoin is functioning as intended, the Nash equilibrium in which all miners mine honestly is roughly the best and also the worst: Barring strategic mining, most deviations would be less than optimal for most miners. So the PoA and the PoS should be practically the same. Of course, both should be quite large. It would be more "efficient" were a small handful of entities to oversee the UTXO confirmation process using Amazon Web Services. Obviously, there is some inherent value in decentralization.

Looking solely at transactions, we can assume that the fees provided in each transaction must exceed the value of making the transaction; otherwise, the fees would not be offered. Sending the payments via Bitcoin and paying the fees is the most efficient arrangement available to the transactors. The block subsidy itself serves dual functions: distributing new coins as well as providing temporary security while bootstrapping the network. So while it is not completely accurate to lump the entire subsidy into security costs, it is a cost that does get paid when miners create hashes. We avoid this wrinkle entirely if we look to the future when the subsidy plays a diminished role.

The major cost of Bitcoin in the future is likely to be the security budget provided by the fee market. This is where computations of PoA become more relevant. Today, spending $2 to send thousands or million dollars is a negligible fee to those using Bitcoin for making transactions, but this would probably only provide around $1 million per day in security budget, if transaction fees were to be the sole source of revenue for miners. If the costs of Bitcoin were limited to today's fees, Bitcoin would be quite efficient as a way to transfer, but a security budget at this level would be very enticing for an attacker.

Put another way, the value of the Bitcoin network must exceed the fee payments made to miners. This is not saying much: Most days in early 2022, total transaction fees were less than $500,000 per day. Part of the value of Bitcoin lies purely in the peer-to-peer payments functionality, which can be evaluated as one would evaluate the value of the Visa network. However, there is also value in the network as a store of value over time: One can place funds in the network and know that months or years into the future, the funds will still be there to spend. When the block subsidy exists, it can be thought of as security budget which comes from this store of value; the inflation of the currency to compensate the miners directly devalues the Bitcoin of current holders by reducing the scarcity.

To remain efficient, the surplus value created by decentralization should be at least the costs created by proof of work, less the cost if the transactions were to be processed by a small handful of entities.

Note that this statement says absolutely nothing about the total stock value of the Bitcoin network itself, which is a significant component of Bitcoin's value: Anyone who has the capacity and willingness to transact in Bitcoin is part of the network. Centralizing blockchain confirmation itself need not change the topology for most of the peer-to-peer network. As Bitcoin grows and the network becomes larger, the value of the network itself could become quite large. Indeed, Metcalfe's law suggests that the value of a network grows quadratically with its users (this is probably overstating the usefulness, but network value should certainly grow superlinearly). The economic value of the Bitcoin network is sometimes dismissed; "append-only ledgers are nothing new" is a common criticism, but the network effect combined with billions of dollars of infrastructure (much of it open source) built to plug into the network can create something novel and quite useful. The value of this network need not be significantly diminished by centralized block confirmation. Thus, Bitcoin would become inefficient the moment that it would become more valuable centralized than decentralized. This inefficiency is itself a security weakness: The moment Bitcoin becomes inefficient (by becoming more valuable as a centralized institution), anyone willing to take over the network by overwhelming security has an arbitrage opportunity.

Letting DB denote "decentralized Bitcoin" and CB denote "centralized Bitcoin," this insecurity condition becomes

$$\text{Value(DB)} - \text{Cost(DB)} < \text{Value(CB)} - \text{Cost(CB)}. \tag{17.7}$$

One may interpret this as flow or stock values.

Example 17.5. Suppose that in 20 years, the fee market has replaced the subsidy market (as denominated in Bitcoin) and the value of Bitcoin has increased significantly. Transaction fees are on the order of $1,500, naturally limiting most transactions to larger institutions dealing with large sums of money. Many of the institutions are public-facing, law-abiding corporations which like Bitcoin for its store-of-value and medium-of-exchange qualities, but are not particularly eager to collectively spend $100 billion per year out of allegiance to the cypherpunk esthetic. These institutions could collectively decide to form their own permissioned version of Bitcoin (recall §14.10), which would be validated according to some agreement they come up with, designed to minimize costs. With the network in place, this cartel could make other modifications, such as increasing the blocksize or decreasing the interval between blocks, based entirely on their own assessment of efficiency. As a peer-to-peer network, most of the network topology would remain intact, excluding only those who rage-quit following centralization. On the other hand, previously hesitant users may join, reassured not only that the network is operating with higher energy efficiency, but also that it is now a walled garden free from unsavory characters.

Again, value itself and the value of storing value are not the same: It is unclear what relationship these two have to one another. The market cap could be extremely large, but if markets are thin and few coins are changing hands, the value is restricted to the value provided by a tenuous store-of-value feature for a few large holders. On the other hand, instead of this low-velocity situation, consider a high-velocity situation in which Bitcoin is being used regularly and markets are active and robust, but most users consider it only as a medium of exchange. In this case the value of the network can be quite large.

Inspecting (17.7) we see that if Cost(DB) becomes large enough, this fail-to-be-efficient condition will be satisfied unless the surplus value of decentralized Bitcoin grows commensurately. Thus, for Bitcoin to remain efficient, the size of the security budget should be bounded above by the ongoing extra value flowing from decentralization. This leads to the following observation.

Conclusion 17.2. If Bitcoin adoption grows to the point at which Bitcoin becomes a preferred payment method for a significant part of the global population, decentralization will become inefficient and centralization will become inevitable.

This underpins a central and delicate paradoxical conflict inherent to Bitcoin. In order to function, it should be useful as a currency or as a store of value. On the other hand, in order to remain decentralized, it cannot be the most compellingly useful currency to people who do not care about centralization (recall the premiums discussed in §14.8). This is simply a restatement of the fact that if Bitcoin gains widespread adoption, the market fragility hypothesis is likely to fail. The inevitability of centralization follows from the following claim.

Claim 17.1. *Inefficient Bitcoin is insecure.*

Proof. Suppose that (17.7) holds. Any cartel of miners making up over 50 % of the hashrate can declare themselves the validators of the chain. As they would win any war of attrition with any other cartel, they can make this claim without serious challenge (even with coalitional game theory in mind, they could simply agree to split profits with any serious challengers to maintain social order among miners). As sole controllers of the chain, they can then try to extract up to Value(CB) in rent from Bitcoin users who would still appreciate the centralized chain, at cost Cost(CB). Since the left-hand side was assumed to be non-negative, the right-hand side should be strictly positive. As this positive incentive is available for any 50 % cartel of mining power, Bitcoin should be considered insecure against centralization. □

This is a flow-based discussion. We are largely ignoring stock value and cost of commandeering the network, which may be somewhat different.

17.6 Theoretical pushout and epsilon-attacks

In proof-of-work, proof-of-stake, or any other "proof-of-resource-X" protocols which delegate validation to economic free agents, a security budget is paid out to those holding or expending some scarce resource or combination of resources. Whenever there is a fixed security budget being doled out according to the proofs provided by validators, a pushout attack becomes achievable if the markets for all the ingredients are well developed. By expending more than is economically sensible in the short term, any participant can increase their share of rewards.

For Bitcoin, ingredients in proofs of work are somewhat heterogeneous: Each ASIC exists in a specific place and time and consumes electricity that is provided at a cost determined by local market conditions. In particular, while hashes are essentially fungible, the resources expended in a proof of work are not fungible. It is possible to assign market value to the hashes according to the block reward and difficulty adjustment, but it is not possible to universally determine the market value of the inputs.

On the other hand, proof of stake requires staked tokens, which are essentially fungible. As a result the staked token has a clearly defined flow value determined by the security budget. There may be competing options for flow value, as the token can be lent out or tied up in smart contracts. If the token's flow value stemming from validation exceeds alternatives provided by the market, tokens will be staked. However, if the market determines a flow value of a token that exceeds the returns from staking, the dynamics become interesting, as the rational token holder will move away from staking in pursuit of higher rewards.

Such market conditions need not arise organically; they can easily be orchestrated in an *epsilon-attack*. While the mechanics depend on the protocol and the market, the economics underlying this attack are rooted in a simple principle: If I can pay you a flow value of $(1 + \varepsilon)$ for something which generates a flow value of 1, this is an offer you will generally agree to. Thus in a completely liquid market for the scarce resource, the attacker should incur a cost of ε.

18 Bitcoin as it matures

In this chapter we tread into speculative territory, asking how Bitcoin will play out if it continues to grow in reach. There are many open questions. Will Bitcoin be adopted, used as a geopolitical weapon, used as a treasury reserve, or directly supported by nation-states? Is hyperbitcoinization feasible? Are sovereign currencies on the decline? Can the 21 million supply cap be altered? Who will protect Bitcoin if it starts to falter? In public discourse these questions seem to attract an unlimited diversity of answers. Attempting to provide thorough answers would require expertise in areas beyond the scope of this book, including macroeconomics, international relations, domestic and international politics, and philosophy of money. Even with such expertise, making confident predictions in these areas would be a wild goose chase.

Rather than an attempt to spear into these areas, we do our best to set up a framework which identifies different assumptions and considerations and let the reader decide which are reasonable or salient as history develops.

18.1 Hyperbitcoinization

We begin by discussing the idea of hyperbitcoinization, a state of affairs theorized by some of Bitcoin's most optimistic advocates in which Bitcoin is ubiquitously accepted as a currency. These aspirations cover a lot of territory, so we divide this into two rough categories. We define *subjective hyperbitcoinization* as a state in which Bitcoin is widely regarded as a reliable unit of account. We define *coercive hyperbitcoinization* as an economic regime in which any parties who opt out of, or are excluded from, the Bitcoin network are placed at an economic disadvantage. These two concepts should not be used interchangeably in game-theoretic discussions; the implications are quite different.

Subjective hyperbitcoinization is something like Bitcoin replacing and going a bit beyond the "gold standard." Bitcoin would be accepted by many, but not all retailers and financial institutions, and people would think of it as a relatively stable investment (at least as much as gold) and a reliable method to transfer money overseas. This does not require a fantastical imagination; it could be the result of decades of Bitcoin slowly gaining prominence and acceptance through fits and starts, it need not be disruptive to the social order.

Coercive hyperbitcoinization, on the other hand, is quite different: The concept is rooted in the idea that sovereign currency will decline in prominence to become essentially worthless, not only in per-unit terms but in aggregate. One must be careful not to conflate the observation of inflation of a currency with the decline of a currency. Inflation is happening. Historically, inflation is a fact and it is expected to continue in the future, but major currencies such as the dollar are not in any way declining. A single $1 bill may be worth less than it was 20 years ago, but the capital-D Dollar is more powerful than ever – it grows with the dollar-denominated economy, which reaches throughout

https://doi.org/10.1515/9783110772845-018

the globe. Sovereign currencies show no signs of "going to zero," despite the fact that individual units lose buying power. The rise of Bitcoin as a global currency or unit of account (subjective hyperbitcoinization) does not require sovereign currencies to stop being accepted. Coercive hyperbitcoinization becomes possible only if sovereign nations either choose to cede their power to create and maintain sovereign currencies or, for some reason, lose their ability to issue currency.

Even in this case, the current Bitcoin blocksize only allows for a limited number of on-chain transactions to occur each day. The number of Lightning channels to be created or closed is bounded by the set of on-chain transactions. The population of humans grows faster than the number of transactions on the blockchain, so for coercive hyperbitcoinization to occur, the financial system would need to accommodate billions of users by adding expansive layers upon layers. It is unclear at this point how this could happen without financial institutions or sovereign nations issuing digital (or perhaps even paper) IOUs that are essentially fiat.

Coercive hyperbitcoinization requires that the US, China, the European Union, and other powerful economic entities would be powerless or uninterested in intervening while hundreds of trillions of dollars of value is usurped by a leaderless currency with no physical defense mechanisms. A rapid collapse of sovereign currency is a major disruption of social order; sovereign nations who seek to maintain order would be in dereliction of their duty to permit this to happen. A slower collapse also faces serious obstacles (see §18.6).

18.1.1 Decentralized hyperbitcoinization

Under decentralized hyperbitcoinization, the following properties of decentralization would be in place as they are today:

1. Miners are decentralized; no cartels form to monopolize the network.
2. Miners choose the highest-fee rate transactions and do not engage in censorship.

However, with subjective hyperbitcoinization, we would also expect the following:

3. Volatility is reduced. In order to become a reliable unit of account, volatility must have decreased in terms of real buying power. Otherwise people would not use it as unit of account.
4. A healthy portion of corporations will be using it. The network would be well established and wide-reaching.
5. The total value should be much larger than today. If everybody is using it all the time, the demand would be incredible.
6. The total fee market should be significant. If the fee market is not much larger, the network would be insecure. A frustrated villain in some corner of the world may try to commandeer Bitcoin or launch a devastating attack to cause pain for the developed world which has adopted it. If the security budget is small enough, an offshore

hedge fund would find short-selling while launching an inexpensive attack too good to pass up. Businesses would not rely on Bitcoin if it was not secure.

7. Mining should be a low-profit undertaking: If the fee markets are consistently strong, mining infrastructure will remain a prudent capital investment until margins compress significantly.
8. The market fragility hypothesis should be true only to a minimal extent. This follows from the fact that corporations are using it. If Bitcoin's reach is dense enough, most users will not care who is processing their transactions, provided that the transactions are being processed.

In coercive hyperbitcoinization, we also expect that:

9. Sovereign currency has not only declined in per-unit terms, but also in total purchasing power. Very few people will choose to use sovereign money.
10. Market fragility should be large when gauged in any other competing currency (whose values are near zero) but small when gauged in purchasing power.

Claim 18.1. *Subjective hyperbitcoinization is unstable. Coercive hyperbitcoinization is extremely unstable.*

Proof. Recalling Example 16.4 and Proposition 16.1, decentralization is in the core when the mining profits resulting from centralization are less than half the profits accruing to miners during decentralization. This should fail drastically under hyperbitcoinization. To argue that it does not, one would need to argue that the global unit of account Schelling point presented by Bitcoin is significantly devalued by centralization. However, in the case of centralization, central validators would be careful to avoid making any disruptive changes to the network, so the network itself would retain value for most users (this is just game theory, as we will argue further in §18.1.2). Decentralized miners would be making small profits, even while the revenues for mining will be quite high. The market fragility is minimal, so even after centralization, the revenues for mining would remain high; after all, Bitcoin functions well for many throughout the world. By centralizing and limiting their own hashrate, miners can reduce mining costs by orders of magnitude and share the wealth amongst members of their coalition.

Nation-states or collections of nation-states would also want to step in and perform a direct frontal takeover. By arrogating control of the network, they can prevent malign actors such as rogue dictators from benefiting from the network and dangle or exercise power directly to sanction nations who misbehave. They would do this while promising not to disrupt the prosocial economic uses of the network.

If fees are quite large and miner profits are quite high, miners may be reluctant to make drastic changes. In this case there is still another option for corporations or nation-states who do not want to foot the capital to perform a brute-force takeover: They could simply perform a permissioned bypass (recall §14.10). If the network effect provided by the network of the world's largest corporations is more compelling than the advantages of an unpermissioned network, users engaged in commerce could stick with the corporate fork, which could be rather stable. Future global corporations would have plenty of data, research, and social influence and would not perform such a takeover on a whim.

If censorship-resistant money remains in high demand, holders of a competing currency using a different consensus algorithm may have significant motivation to sabotage: The demand must go somewhere. If Bitcoin has been attacked or centralized, close competitors will stand to benefit. In fact, if Bitcoin encompasses a large enough portion of buying power, a traditional sovereign currency backed by staid central banks and stable governments could be huge winners in a flight to safety. Speculative attacks are a sword that cuts both ways.

In the coercive setting, whoever controls Bitcoin would literally control the entire global economy; naturally coalitions would form to make a grab for this power, a power previously unimaginable in the history of humankind. This could be a cabal of corporations or a group of economically powerful allies. Further, any coalition who sets up a private or nation-state currency has an extremely strong incentive to sabotage Bitcoin. They would be in line to take the lion's share of global buying power and economic power. This would be simply irresistible for the power hungry and a moral necessity for anyone hoping to front-run an evil power-seeking cabal. With coercive hyperbitcoinization, anyone excluded from the Bitcoin network is at an extreme disadvantage: Their adversaries or competitors will be aware of this and strive for a governing role in Bitcoin's network.

If Bitcoin is being relied upon in economies throughout the world, whoever controls the blockchain has immense economic power; the incentives to coalesce and perform a takeover remain extremely high. If a corporate takeover were to cause large-scale abandonment of the network, the world would not be properly hyperbitcoinized. Even if a state of decentralization is temporarily in balance, there always exist incentives to form coalitions. Such a balance is therefore delicate and unstable.

The idea that coercive hyperbitcoinization will lead to a persistent, hard-fought, but never-quite-won hash war is fantastical and not well founded in game theory. Rather than spending billions or trillions of dollars of energy so that each player maintains their own position in a decentralized network that necessarily includes global adversaries and pariahs, a majority coalition of powerful nations or corporations would form quickly in order to preserve resources while excluding bad actors; this is a win–win situation. Any hash war would be quite short: Once the losers realize they are not going to win a hash war, they would back down, as there is no point to continue. Even if they secretly mined a longer chain and foisted it, the majority coalition would simply fall back on brute force. If a majority coalition has already succeeded in performing a takeover, they may even declare their Bitcoin to be the true one and simply ignore any attempts by the excluded parties, even those with more chainwork. The best the excluded parties can do is create their own permissioned fork. This would be preferable to expending resources in an extended hash war and would allow economic activity to continue. □

18.1.2 Centralized hyperbitcoinization

Claim 18.2. *Centralized Bitcoin is stable if Bitcoin is an efficient means of commerce.*

Even after a centralization event, one would expect a version of market fragility to hold the central validator to certain standards of behavior. If the central validator became capricious, heavy-handed, censorial, or extortionary, Bitcoin would suffer a decline in value, as users would be forced to develop or seek alternative payment or store-of-value methods. The central validator would attempt to maintain as much continuity and openness in the network as possible, so as not to destroy their revenue source. They would also like to withhold their censorial power for use in more extreme situations. If the central validator is under a nation's jurisdiction, the nation would be prudent to save exercising any use of power only as an extreme measure. Unless the central validator is exercising totalitarian authority, alternatives to Bitcoin should exist and will become attractive if the validators become excessively abusive.

One of the philosophical underpinnings of the market fragility hypothesis is the assumption that a centralized takeover of Bitcoin is a watershed moment at which point abuse of power becomes inevitable. Once abuse of power is inevitable, the currency becomes less attractive and demand will begin to decline. However, it is not clear why abuse of power is not subject to the same type of reasoning: The central planner still has the opportunity to convince the public that the powers are never to be abused; the best way to do this is simply to never abuse power. In this way centralized Bitcoin could continue to function without intervention for decades; the central planner has no interest in destroying their cash cow, so they refrain from making waves, except in situations that are deemed necessary by the economic majority of users of the network.

An often-voiced concern is that centralization will lead to profligate money-printing; the central validator will not respect the 21 million supply cap, allowing for an increased supply to flow to associates in a corrupt fashion. This falls under the same principle: The instant a central validator deviates from the emissions trajectory, this would be cause for alarm by anyone hoping to use Bitcoin as a long-term store of value; they would be motivated to seek more reliable alternatives, unless, again, there is widespread agreement that increasing the supply is a prudent means to achieve some greater goal and does not represent a departure from sound principles.

Another concern is that if Bitcoin were to be centralized, it would fall into a different regulatory category. We can follow familiar backwards induction arguments: Would governments suddenly enforce disruptive regulations when the currency is enmeshed in the economy? The will to do so may be small, especially if Bitcoin has become central to the status quo. Too Big to Fail may mean Too Big to Regulate.

Overall, if the network functions well, allowing corporations and individuals throughout the globe to settle balances reliably and quickly, there is little reason to believe that the network would be abandoned: If Bitcoin makes for good money, then it makes for good money.

The more well suited Bitcoin is as an international currency, the more a coalition of business and political leaders are induced to coalesce and negotiate a permissioned version with prosocial governance procedures.

On the other hand, if Bitcoin is not particular efficient, centralized Bitcoin still could remain a stable (in terms of governance) but possibly volatile currency. Bitcoin could be viewed as a clunky currency, not well suited for most transactions, but still retain speculative value as a hedge against inflation or as asset diversifying an investment portfolio. In this case, centralized Bitcoin could still remain a Schelling point for this type of asset. Fund managers need not care that the asset is censorship-resistant; in fact, knowing that the blockchain for the asset is reliably maintained by staid financial institutions or nation-states could be reassuring.

As mentioned in §14.8, it is important to disambiguate the value of Bitcoin as a Schelling point for value storage from the value of Bitcoin as censorship-resistant Schelling point for value storage. If there are no censorship-resistant Schelling points for value storage, anyone who is not in danger of having their assets seized should be comfortable with the former. If the central validator resides in a jurisdiction governed by liberal democratic principles, the fear of confiscation should be low. If Bitcoin is centralized under a cartel of Wall Street institutions, a dissident in Myanmar, Saudi Arabia, or Tigray who stores wealth in Bitcoin only has to fear that these institutions are somehow beholden to powerful figures whom they are criticizing. Unfortunately, due to global enmeshment of powerful politicians and business leaders, this concern is ongoing, but if the Wall Street validators have stuck to a well-observed policy of liberal principles, it would require a very strong twist of leverage to break from this policy. Certainly, we are not attempting to claim that Wall Street institutions have a spotless record of respecting liberal principles, but rather that many people throughout the globe could make a risk assessment and find the risk of their own Bitcoin being meddled with to be acceptably small, at least in view of alternative savings vehicles.

One of the most interesting and important debates surrounding Bitcoin is whether it resists or advances illiberal uses of power. Bitcoin presents a viable alternative seizure-free store-of-value asset (inasmuch as it functions as a store of value) for many who live in illiberal regimes throughout the globe (see [17]). If Bitcoin were to become centralized under capitalist institutions who favor liberal principles, it could still remain a powerful force of resistance in these regions. When discussing Bitcoin as "censorship-resistant," one must ask "for whom and from whom?" Corporate- or state-controlled subjective hyperbitcoinization is probably not much qualitatively different from recent economic regimes in liberal democracies, which function but are far from equitable. However, corporate-controlled coercive hyperbitcoinization would represent a dystopian imbalance of power. With no viable alternatives, a cabal of corporations could wield financial power over the entire population; such a cabal could print money at will or censor whoever they choose to, likely with the help of machine learning algorithms. We leave further exploration of this possibility to writers of dystopian science fiction.

18.1.2.1 Clash of chains

Suppose a confederacy of governments and corporations makes a move to arrogate control over Bitcoin via a permissioned fork. This would be ideologically unpalatable to a subset of Bitcoin users. Historically, chain splits usually end with a clear winner being determined rather quickly; the majority of economic activity abandons one chain for the other. Notably, this was the outcome for Ethereum versus Ethereum Classic and for Bitcoin versus Bitcoin Cash. If an arrogating attack is launched, a minority of true believers would face an uphill battle against a corporate and government cartel who control various aspects of traditional and social media. Convincing businesses to oppose what appears to be the economic majority is very hard-sell. The winner could be decided by a Keynesian beauty contest as most businesses simply want to be on the winning side.

18.1.3 Sovereign currency will not die

We claim that coercive hyperbitcoinization is a chimera. This claim rests on an assumption that appears quite reasonable: At any point in the future, at least one sovereign jurisdiction will retain a currency strong enough that it will not evaporate during a speculative frenzy.

The so-called *speculative attack* has been proposed as a possible path towards coercive hyperbitcoinization. It is important to understand how far-fetched this is. Bitcoin itself, being backed by only beliefs about future demand, is much more likely to become the victim of a speculative attack. Bitcoin's price relative to the dollar has surged many times in the past, and it has an established pattern of falling back by over 70 % after the frenzy. The expectation that *the next* short-term bull run will be the one that does not pull back before destroying fiat currency is simply not a viable one; this would require a sort of "mass formation psychosis" in which businesses, families, nations, non-profits, fishermen, bartenders, automobile makers, bail bondsmen, dentists, social workers, and great-aunts and -uncles all became so enraptured in an all-consuming speculative mania that they abandoned sovereign currencies, even as Bitcoin continues to achieve all time high prices and despite a well-demonstrated history of Bitcoin price repeatedly succumbing to gravity. Rational business and political leaders will understand the risks and will continue to operate in sovereign currencies, and opportunists will find asymmetric risk in moving back into sovereign currencies. At some point the speculative attack against Bitcoin will become so lucrative that it will become inevitable. Risk seekers will shift away from betting on Bitcoin to betting against it; this will inevitably lead to gravity catching up and sending the market crashing in the other direction. To buy into a coercive hyperbitcoinization thesis one must believe that the economic totality of the population is suggestible to such a speculative frenzy; also one must assume that the majority of corporations are not only in the business of currency speculation but will be unwilling to speculate the other direction. Most businesses avoid currency speculation, instead focusing on their own business model.

Historical instances of hyperinflation are not germane. While smaller currencies backed by unstable regimes are subject to collapse, global currencies backed by hegemons simply are not. Bitcoin cannot win a speculative war, whereas a sovereign currency can. Bitcoin can continue to defy gravity, but cannot eliminate it, whereas sovereign nations waging a battle with Bitcoin possess many advantages outside meme warfare.

Decentralized Bitcoin will always face a fundamental problem as a Schelling point taking the place of the gold standard: Decentralized Bitcoin can always be destroyed via majority attack. This vulnerability is a fundamental cornerstone in the design. Demand for Bitcoin as a solid store of value will always be curtailed by the risk of attack.

Major nation-states will not simply abandon their sovereign currency; this would represent massive abdication of power. We can use backwards induction from there: If you believe the US dollar will never be destroyed, then the US dollar will always be a vehicle for a speculative attack against Bitcoin. The US dollar will always possess good buying power, and anyone can denominate shorts of Bitcoin in US dollars. This is asymmetric: Bitcoin can only undermine US dollar by FUD and memes, whereas an attacker of Bitcoin can use not only FUD and memes, but actual attacks on the network. Continued increase of Bitcoin's market share against the dollar cannot be ruled out, but the probability of a reverse attack is non-trivial; in fact, it becomes more likely as the currency increases in value. If the price is high and transaction fees are small, a sabotage attack becomes more likely. If transaction fees are high, a permissioned bypass or centralized frontal takeover becomes appealing to the economic majority. Certainly there is a limit to transaction fees; as the total market capitalization of Bitcoin becomes extremely large, even large transaction fees will not protect it. High values against the default currency create an incredibly lucrative opportunity in the default currency to attack Bitcoin.

18.1.3.1 Thier's law, Gresham's law

In the near future, the dollar will always remain a competitor to Bitcoin and probably a superior one. Barring consummately idiotic behavior by the Federal Reserve (not merely unpopular, self-serving, tone-deaf, and short-sighted, but unprecedentedly imbecilic behavior) the dollar will be more stable and should serve as a better temporary store of value. Note that the US dollar is considered a hard currency: The supply is managed well enough that much of the world's debt continues to be denominated in dollars.

Gresham's law is roughly stated as "bad money drives out good money." This refers to the tendency of "bad money" to remain in circulation: People are always happy to part with their bad money and not so eager to part with their good money; thus, the bad money has higher velocity and makes the rounds more frequently. In contrast, Thier's law presents a reverse suggestion: People will demand payment in good money, and hence bad money will cease to persist in circulation. In a functioning circular economy, one would expect that most participants would prefer the better money. Obviously, one or both of these “laws” must fail to be a universal law, as they are principles in conflict with another. We can conjecture which principles are most applicable to Bitcoin: A consumer who obtains Bitcoin because they believe the value will rapidly appreciate against the default currency will be unlikely to spend them. On the other hand, a business who is not interested in speculation will be unlikely to accept them unless this increases their market reach. If most consumers are not interested in spending, there is little chance accepting Bitcoin increases the accessible market. This can be easily framed as both sets of parties maximizing a utility function that accounts for risk preference or risk aversion.

In many nations throughout the world, dollarization is happening and is evidence of Thier’s law; being stable and widely available makes the dollar attractive to accept and to spend. People are not reluctant to spend their dollars as dollars are not expected to shoot skyward in buying power. Businesses are happy to receive dollars: The value of dollars is unlikely to crash.

Central bank digital currencies (CBDCs) are expected to play a large role in the coming decades. These would have some of the properties of Bitcoin but would be more stable. The downside would be primarily to users who are worried about surveillance or censorship by the issuers. One would hope that in liberal democracies such as the US these would be issued in a way such that civil liberties and privacy are valued: Only in extreme cases would these principles be breached. China and the US, who are in some competition for global market share, have every reason to make their CBDC attractive and user-friendly (for a much deeper dive into this topic, see [42]).

The dynamic balance of global currencies will be interesting in the coming decades. As the economy becomes more global, we may see some consolidations of currencies; a small unstable nation with an unstable currency is poor competition for the US dollar or RMB, if these are available. China could benefit from this shift, taking a larger share of economy, and a stable CBDC could be very useful towards this end. The US would like to compete and should respond. Bitcoin may play a role, but without becoming a reliable store of value, it may struggle against more stable currencies.

18.2 Will nation-states support Bitcoin?

Bitcoin is often touted as a public good. From the idea that Bitcoin is a public good, the idea that some nation-states would make efforts to support the network follows. It is worth taking the time to examine this idea and isolate the assumptions under which this would or would not take place. There is a wide spectrum of reasons for a nation-state to exhibit Bitcoin friendliness. Light-touch regulation and favorable taxation would send a welcome signal to Bitcoin, but a nation could go even further, supporting the Bitcoin infrastructure directly or via tax credits, investing in Bitcoin as a treasury reserve, or even mining Bitcoin with public resources.

When describing the various decision makers' motives and goals, we should try to disambiguate between the objectives of the policy makers themselves and those of the people they represent as these are not always in complete alignment. Policy makers may have limited power to force businesses or individuals away from Bitcoin, and any heavy-handed attempts to do so may not be welcomed by their constituents. It is difficult to construct a universal theory here, so we offer a (very much non-exhaustive) series of observations. Decentralized and centralized bitcoins have different implications for each sovereign jurisdiction, so we separate these assumptions.

18.2.1 Reasons why a nation would support Bitcoin use (centralized or decentralized)

1. The policy makers believe that Bitcoin facilitates smoother commerce. Perhaps this is because it offers a superior method for settling accounts with overseas trade partners or offers a wider network of trade partners. In most domestic situations Bitcoin does not add a significant advantage over the local currency.
2. The nation would like to make themselves attractive to economic development by Bitcoin-embracing corporations or wealthy individuals.
3. The policy makers believe that Bitcoin will rise dramatically in price and would like their nation to benefit from this rise. This is just a form of speculation: Obviously the concomitant downside risk must be accounted for. Even if this works over some time period, it is not the same as building a pattern of sustainable growth.
4. The nation does not have a truly sovereign currency and is at the mercy of central banks under outside control (for example, consider the West African and Central African CFA currencies [17, Chapter VI]).
5. A nation with liberal, democratic, and capitalist principles views Bitcoin as a subversive means for introducing their values into nations with illiberal leadership [17, Chapter XI].
6. Policy makers are corrupt or self-serving, and they could see Bitcoin as a means to enrich themselves: They could time their own purchases or sales to front-run policy changes that may affect the demand. For example, if a nation-state decides to acquire Bitcoin for the treasury, self-serving officials could acquire massive amounts in the days leading up to this announcement. Corrupt officials could also look at Bitcoin as a reliable option to accept bribes from foreign influence peddlers.

18.2.2 Reasons why a nation would oppose Bitcoin use (centralized or decentralized)

1. The policy makers believe that Bitcoin is little more than a form of gambling that drains resources from productive economic development.
2. Policy makers believe that the speculative cycle could leave a wake of toxic debt which would do considerable harm to the economy. While many market crashes are due to straightforward herding dynamics, a nation-state should also consider

the possibility that an adversary might orchestrate a rugpull maliciously. There is good reason to reduce exposure to this.

3. The local sovereign currency is struggling with inflation and the addition of a fixed supply competitor could exacerbate the inflationary pressures. Generally, economic policy makers are pursing pro-growth policies which maximize real growth and are not simply trying to protect the value of their currency. However, if the value of the currency is decreasing to the point where inflation hampers economic activity, the policy makers would be wise to take steps to combat this. This could backfire; a ban of Bitcoin may be seen as a flailing attempt to save a perishing currency and push more citizens into holding Bitcoin.
4. The nation-state would rather the economic growth go into their currency instead of sharing wealth with international holders. They see their economy as more productive relative to other nations' economies. If domestic economic growth happens in a Bitcoin economy, buying power of the network grows, enriching other parties who may or may not have contributed. The free-rider dynamic is not conducive to protecting one's own interests.
5. The nation-state enjoys advantages from their own currency being used as an international reserve. Encouraging the growth of Bitcoin risks ceding this advantage (many would argue the US have been in such a position for quite some time [17, Chapter XII]).

18.2.3 Reasons for a nation to maintain Bitcoin specifically as a decentralized network

1. The nation believes that a fixed supply international currency is a boon to a nation's economy in decentralized form, but this value would be diminished in centralized form. Perhaps the nation-state expects to be poorly treated by any party who would arrogate control over the currency. This can be due to simple imbalances in power or due to the nation's tendency to run afoul of international governing bodies' wishes and face economic sanctions.
2. The nation-state believes the market fragility hypothesis, and significant Bitcoin is held in their treasury or by nationals.
3. The nation-state has strategic adversity with parties who would be able to take over the network.
4. The nation-state may want to preserve the decentralization as an exploitable vulnerability that they can take advantage of at some future time.

18.2.4 Reasons for a nation to discourage use of decentralized Bitcoin

1. The nation-state is at a disadvantage in a centralized scenario, which they feel is imminent or likely. Any nationals who denominate their wealth or business in Bitcoin

are putting themselves at risk: A centralized takeover could leave these businesses in a bad place if the central board imposes sanctions on the nation. Effectively, allowing the nation's economy to become enmeshed with Bitcoin would hand anyone capable of taking over the network an economic weapon.

2. Policy makers believe that decentralized Bitcoin promotes illicit activities.
3. The policy makers are heavy-handed, and they would like to retain the option to seize assets of their citizens.

18.2.5 Indifference to centralization and the soft-forkability hypothesis

Corporations also face similar considerations. Some will be indifferent to centralization, provided the leader is friendly. If a corporation feels that their Bitcoin-related business model will function equally with or without centralization, they are considered *centralization-indifferent*. A corporation may prefer Bitcoin for many reasons; for example, they could find transactions in Bitcoin less expensive, either due to fees charged by financial institutions or the loss incurred in chargebacks. Any increases in profits must be weighed against the risk that they will be caught in the cross-fire of international economic disputes that culminate with a blockchain attack, keeping in mind that their own nation-state or some other friendly governing organization may respond with a permissioned chain. If the corporation believes the permissioned chain is likely to occur and will not harm business, they may believe they would face only minimal risk from an attack. Centralization indifference can be thought of as the opposing force to market fragility under a centralization exploit: If centralization indifference is high, market fragility will be low.

On the other hand, an attack that damages the business model and cannot be fixed via a soft fork does pose a serious risk. These risks might be present with traditional payment systems such as Visa and Mastercard, who participated in Russia sanctions in 2022. For a corporation domiciled in a jurisdiction that may come under sanctions, storing value in Bitcoin incurs risk. A takeover of Bitcoin by a coalition of nations for the purpose of sanctioning other nations is not a trivial undertaking and would not happen casually, but should always remain on the table for strategically thinking nations.

Nations themselves each have their own levels of vulnerability to centralization. Even under an attack, they may be confident that a friendly soft fork will arise that will allow them to continue to use the chain. If any bifurcations caused by a soft fork do not damage relations with significant trading partners, the nation could also be considered centralization-indifferent.

Not all soft forks are created equal; a soft fork that is open to only a small portion of the network is less valuable than one that walls in an economic predominance of the full network.

18.2.6 Bitcoin mining

When considering if Bitcoin mining itself stands to be a major source of prosperity, one can and should make quantitative observations. Bitcoin's total annual security budget (as of 2022) is around $10 billion. This is revenue; profits should be considerably smaller. While the price may increase, most of the security budget is derived from the block subsidy, which is exponentially decreasing over time. To appreciate the scale, the annual oil market is around $5 trillion, or 500 times as large. The US GDP is around $20 trillion. The GDP of Ireland (often compared to Bitcoin in terms of total energy use) is around $450 billion. The state of Texas contributes around $2 trillion to the GDP. So for most countries, profits to miners have very little chance of creating a transformative amount of income. Nations who set up policies to encourage mining can enjoy the upside risk by allowing mining corporations to place a bet on Bitcoin (mining is like a call option). The corporations may stand to do quite well during a bull run and some profits may end up in the local economy or become tax revenue. Downside risks are not insignificant: If mining is not as profitable as hoped, the corporations may leave a mess behind, environmentally and financially (it is worth noting that mining returns have historically not been significantly better than simply speculating on Bitcoin).

Any arguments about how Bitcoin will affect an industry (positively or negatively) should be framed quantitatively. Again, this requires an appreciation of scale. The natural gas market in the US is around $165 billion in 2022. Nuclear, coal, and hydroelectric energy all produce orders of magnitudes more energy than Bitcoin consumes. The Inflation Reduction Act of 2022 adds approximately $100 billion to existing loan programs for financing production of electric vehicles. Bitcoin advocates often claim that concerns about consumption are overblown: The total amount of energy use remains insignificant in comparison to many other industries. At these quantities, Bitcoin mining seems unlikely to revolutionize industries or economies. A jurisdiction considering encouraging or allowing Bitcoin mining should weigh the upside risks and the downside risk. The benefit mining will bring in the optimistic case should be compared with the harm (economic and environmental) done if the asset does not perform.

Arguments are also made that Bitcoin mining can help stabilize electricity grids. This is feasible, given certain ranges of parameters (the expected local demand curve, local energy costs, the availability of intermittent resources, the price of Bitcoin, the price of ASICs, the total hashrate, energy derivative markets, and how contracts are negotiated with private corporations, among others), but the argument is delicate and heavily quantitative. It requires some work and domain knowledge to model. Again, the impact is limited by the revenue that can be brought in by mining, and the risk of significant market downturn needs to be considered.

Even if the price were to increase by orders of magnitude one must consider the increased likelihood of attack or takeover. A Bitcoin with a $20 trillion market cap and a $50 billion security budget is much more likely to be attacked than Bitcoin in 2022. If attacked by short speculators, investments in mining could be devalued when the currency exhibits fragility, and if attacked by direct takeover, the payoff depends on whether or not the miners are included in the winning coalition. If so, the investment could be lucrative; if not, the mining investment is instantly devalued.

18.3 Non-profitable mining

One important question is whether powerful entities, either large corporations or nation-states, would step up and mine Bitcoin at a loss in the situation that the security budget is low or the network is under direct attack. Such an arrangement is possible, but has a number of problems and if protracted will likely result in de facto centralization of the network.

In order to be fully secure against a takeover, over 50 % of the hashpower would be required to remain committed to decentralized mining. The attacker would likely be motivated by a stronger reward (everything) than what motivates the defenders (thin profits from a decentralized competition). If the attacker were to challenge the defender in a hashwar, it would involve heavy costs to both, but give no significant payoff to the defender. Further, all of the defenders could be required to altruistically participate in defending at a loss. If some defenders are not willing to lose money defending the network, defections could cascade and the network could quickly fall into centralization.

The motivation for defending would have to be clear and convincing for miners to volunteer resources to such a fight. If the miners are nationalized using public resources, it must be clear that decentralized Bitcoin is a public good that needs defending. It should also be clear what the end-game is, if the situation being defended against is temporary or if the nation will end up footing the bill for the next 20 years.

If a nation-state has built up an arsenal in preparation for a hashwar, they are likely to follow the approach usually taken in prepping for conflict, namely, to overprepare. If this undertaking is done on a military scale, the nation-state could easily develop a hash capacity that could dwarf the network. In this case, the nation could take over Bitcoin whenever they choose. Possessing the ability to shut down the network at any time, they enjoy de facto veto power over the network.

When multiple nations have an incentive to defend, a free-rider problem may arise. If one nation steps up and takes the lead, others are not incentivized to join them, unless the leader nation demands some help. If there is an agreement that each nation contributes proportionally to a hash force to protect the network, this coalition has effectively become a governing authority controlling the network. Any agreement to partner in protecting the network would necessarily be paired with agreements on how governance decisions would be enacted if necessary.

There is only a small difference between (i) a nation acquiring the hashpower necessary to take over a network and people trusting them to leave the network intact and

(ii) a nation that actually takes over the network and people trust them to not operate the network in an untoward way. The same applies to a coalition. A governing coalition that has the power to, but has not yet attacked the network essentially controls the network. If they one day drop the charade of decentralization and explicitly take over the network, the same amount of trust must be placed in them for Bitcoin to function. Any entity with the power to protect the network from a 51 % attack also has the power to perform the 51 % attack.

There is a possibility that several nations could each develop some amount of hashing power, keeping this on reserve against an attack, knowing that they would be joined in defense and knowing that this would give them certain negotiating power. In this case, any hashwars would be fought at the negotiation table. If several nations who keep some amount of hashpower on reserve threatened to make a move to take over the network in order to sanction or censor another nation, that nation would only be able to defend themselves if they could marshal enough hashrate to defend themselves.

There are parameters under which a nation could develop hashrate that would be effective in defending a network but not taking it over. For example, if 30 % of the hashrate is dedicated to default decentralized mining and cannot be bought in any circumstances, three nations who each acquire 10 % of the global hashrate could combine forces to help the decentralized miners fend off an attack. They must be careful to maintain incentives for the decentralized miners, unless these are also committed to mining at a loss. If the hashrate becomes more expensive than the block reward, security issues could follow as some miners drop out. Any nation who thinks Bitcoin is good for their nation and consider it worth to develop a hash force must be also take steps to ensure they do not end up on the wrong side of a 51 % majority, as there is nothing they can do in this case. Unlike traditional wars, there is no home-field advantage. Sending troops across the world to invade a foreign country is fraught with asymmetric perils, which is why a nation with a somewhat small army can defend themselves against a more powerful invader. This is not true with hashrate; a weak hashrate does nothing against an international coalition determined to enact economic sanctions.

The likely outcome of a build-up of hash forces is that coalitions of nations could effectively steer the network away from unsavory uses by simply making others aware of the power they hold. In particular, if there is a clear majority of hashing power held by nice and well-behaved nation-states, Bitcoin would cease to be an open network for nations that may be thought of as on the naughty list.

The type of hashrate that nations build up is also important. Nations could buy up old equipment that is no longer profitable, or they could buy new equipment. If buying new equipment, they would be motivated to mine themselves. If this practice becomes widespread, there would be little room for free-agent miners on the network. An arrangement of nation-state miners mining for profit and to keep the network neutral could exist in equilibrium for some time, but this could deteriorate rapidly if there is ever a global conflict.

18.3.1 Will corporations protect their bags?

A corporation who has overexposed themselves to Bitcoin is unlikely to mine at a loss in order protect their value. Such a flex would not inspire confidence and could backfire. If Bitcoin has a security budget issue so serious that corporations are considering stepping up to protect the network, it would appear the incentives are broken. The Schelling point of the belief in a self-enforcing economic mechanism would be dead. A corporation vying to do so would have to explain to their shareholders why exactly they are doing this, how long they expect to do it, and why they think it would be effective.

Anytime a party mines beyond what is profitable, they are likely to increase the hashrate, making mining even less profitable for other miners. This could induce honest profit-seeking miners to drop out, perhaps permanently, generating further security problems for the network. Eventually, this leads to the entity who set out to protect the network essentially controlling the network, a death knell to a network whose narrative is already struggling. In addition, if a single entity has the power to take over a network, they may be in the unique position to be asked to exercise this power by authorities if some transactions are found offensive.

It has been argued that Bitcoin bears some analogies to free or open source projects, such as Linux or Wikipedia. Corporations and individuals around the world contribute to open source projects without any guarantee of monetization of their efforts. However, the game theory is extremely different. Because of the binary win-or-lose nature of a hashwar, any contribution that falls short of protecting the network is simply a waste of efforts. This does not happen with Linux. One does not need to forge an agreement with developers throughout the world to defend Linux from attackers. Nobody can attack Wikipedia and run off with it, and nobody can attack the internet and extract rent. Nor can any party render Linux an unusable operating system. These projects simply do not exist in adversarial domains.

18.4 Geopolitical considerations in adversarial conditions

International relations is a game based on conflicts of interest. Nations are always seeking to maximize their influence and well-being. This can be done by forming strategic alliances that exclude adversaries or by developing means to harm the adversary and using these as leverage. Conflict is inevitable and brutal.

One may hope that future wars are less brutal, perhaps more economic. The Cold War involved decades of tension and brinkmanship, but in practice much was ground out on the level of economic expansion and proxy wars. This was preferable to full-scale nuclear war. In 2022, we saw a large-scale economic attack on Russia following the invasion of Ukraine. It remains unclear how well sanctions work, especially when personalities are involved (the game theory here explained by Schelling [48] is interesting; it is often advantageous for nation-states to convince other nations that their leader is an irrational and unpredictable lunatic). Regardless of efficacy, economic weapons will continue to be used in conflicts, as these are preferable to more violent options.

In this framework, nation-states are interested in maximizing the leverage they have over adversaries and minimizing the leverage wielded by their adversaries. Nations can allow some enmeshment with allies, provided they are not exposing themselves to an exploitable risk.

Nations who benefit from the status quo are generally risk-averse. Nations who benefit from the status quo are typically going to hold a larger share of economic power, which they use to preserve the status quo; this is the nature of the status quo. So economically powerful nations (the US and China immediately come to mind) are going to be very reluctant to embrace a new system that will be potentially disruptive. It should surprise nobody that the world's second largest economy (China) has banned Bitcoin, while one of the world's poorest countries (Central African Republic, with a gross national income per capita of $510) has embraced it. China is expanding their reach with the Belt and Road Initiative. If Bitcoin becomes enmeshed in dollarized economies, the effects of a Bitcoin attack are extremely asymmetric.

This underscores a major headwind that Bitcoin faces for widespread acceptance. While China and the US may be rivals as the two largest economies, they are both beneficiaries of the status quo. So if either of the two is to embrace Bitcoin, the benefits must be clearly resounding in the face of all that can be lost if the status quo is disrupted.

An even more serious concern is the catastrophic risk to one's economy that is brought in when policy makers allow their economy to rely on Bitcoin. A situation in which many corporations in the US came to rely on Bitcoin is something to be salivated over by US adversaries: A hostile attack on Bitcoin could cause massive disruption of the US, leading to a reduction in international power. In particular when the security budget is relatively low (compared to say, the Department of Defense budget), an adversary could use resources as part of their defense budget to build up the means to launch a 51 % attack. A well-resourced dictator could direct all national resources towards the production of ASICs and energy: For billions of dollars, such a nation could create infrastructure capable of causing huge damage to the US. The US would be forced to abandon Bitcoin, champion a permissioned fork, or seek out hashrate to fight back. If they already possess the hashrate to fight back, such an attack is thwarted, but this plan does not leave Bitcoin unscathed: A major power stepping up to mine Bitcoin with their own resources poses a number of problems (see §18.3). The most likely outcome is centralization of the mining.

Fully analyzing the value of making a threat requires modeling market fragility and what would be lost if the US was forced to settle for a permissioned fork. (An obvious follow-up question: If the answer is "relatively little," then why not just perform the permissioned bypass already and save billions of dollars worth of electricity?)

If the US were to wield power internationally (by say, offering military support to a country which might be facing invasion), the adversary could retaliate by launching an attack on Bitcoin meant to damage the US and allies. Of course, the US would anticipate this and have their own plan. A cynical analyst may suggest this provides a convenient cloak-and-dagger scheme; a major power such as the US could stage a "false-flag" attack on the network as a premise to arrogate control over the network.

The safety mechanism is almost always centralization: The economically powerful have this option; they can bind together and form an economic predominance that attracts the most activity. With this in mind, the notion of nations entrenching in a hash-war becomes almost silly: The victor will win a network secured by expensive hashrate, the consolation prize is a network that must be secured by some other form of consensus or agreement, perhaps a permissioned set of validators, which is not only cheaper to operate but also easier to control.

This underscores a fundamental reality of money. Money is a negotiation over value and buying power. The goal of any individual is to maximize their buying power. An effective currency will attract the most value in order to maximize its users' buying power. Creating a Schelling point for value has an immaculate conception problem, but once created, major players can negotiate over where the locus of economic power lies. Forking to a new consensus agreement can be a strong and advantageous move if the fork captures most of the economic value in the network. If the permissioned fork has more buying power, it will be more attractive and continue to grow in value. Majority control over the hashrate does not give one control over the power wielded by money; it gives one control over a Schelling point that can be easily abandoned if a predominance of the powerful players choose to.

So the havoc that can be wreaked by an attack is limited provided that the object of the attack can simply fork the network and continue operating. This complicates the risk and reward analysis that businesses in a jurisdiction make when enmeshing themselves with a Bitcoin economy. The upsides need to outweigh the downsides, but the downside need not be as catastrophic: The heaviest-chain rule is not necessarily a single point of failure. No powerful nation would welcome putting $500 billion into a sector that could be destroyed for $15 billion, but if the $500 billion worth of value knows that it can be salvaged by ignoring Nakamoto consensus in order to keep operating, the motivations for sabotage attack are diminished.

We should distinguish between exposure to stock values of Bitcoin (that is, market value of holdings of assets) and flow value (income facilitated by using the network) of enmeshment. In terms of buying power, speculation is purely a zero-sum game; no value is created or lost when one currency changes value with respect to another, only buying powers have changed. If a nation has considerable stock exposure, they put themselves at risk of orchestrated rugpulls. These can be enacted by a simple market dump or follow as major players abandon the network. A massive abandonment of the Bitcoin network can destroy trillions of dollars of value for a nation, crippling a nation that is overexposed. There is no sovereignty and nowhere to hide if this type of attack takes place. During an escalation of economic hostilities, the nation with the most exposure is at a disadvantage.

For example, suppose that Bitcoin continues to increase in price and reach. North Korea, still holding huge sums acquired from hacking in the early days, now possesses a national treasury worth tens or hundreds of billions of dollars and begins to use this in a way that the international community finds unnerving. At an international meeting of diplomats and economists, the decision is made to centralize Bitcoin under permissioned validators. Corporations, individual investors, and national treasuries then agree that the old Bitcoin is no longer a Schelling point, and sell their tokens on the old chain, keeping only their new permissioned Bitcoin. The value of North Korea's holdings (who are obviously denied the opportunity to use their new Bitcoin) plummets and their plans to become an international nuclear superpower are thwarted.

Excluding an international pariah does not require a Davos attack on the validation process itself. If the UTXOs held by the pariah are identified, they can be declared tainted by a handful of major economic players. The tainted Bitcoin held by the pariah is no longer fungible and loses value as the unwillingness to accept it cascades out to all market participants.

Note that while centralization is always a foil against a sabotage attack, there is no obvious foil for a market dump. Unlike with gold, the ability to instantly transfer Bitcoin is a sword that cuts both ways. A rugpull can be executed in minutes; once it has happened it cannot be fixed. This can pit nations in a timing game against one another; once the value starts to approach a target price or unsustainable levels, larger holders will begin to guess when others are likely to sell. This Hobbesian trap leads to a situation where large holders could dump preemptively, to avoid getting dumped upon. The act of positioning oneself to be closest to the exit quickly progresses into a race to the door.

Recall that a nation does not need to invest heavily in ASICs in order to attack. Certainly, spending a few orders of magnitude beyond the security budget to bribe miners into denial of service would be a simple solution. For any nations with a large defense budget, this is an easy option. If the US would like to exclude a major adversary from the network, but has not built up the hashrate to do so, they can simply pay miners to denial-of-service attack the network, while promoting their own permissioned chain. The adversary then has no choice but to form their own permissioned chain; if they are economically isolated already, such a chain could be worth very little.

In order to minimize risk, most nations would like to retain the option to insulate or isolate during a global crisis of distrust. Exposure to Bitcoin may make that difficult. The economic fallout of trying to divorce oneself from the Bitcoin network can be unpredictable; naturally there are formidable technical details involved in setting up a permissioned fork. If Bitcoin is the common currency, there is no golden bridge of retreat back to safety. Historically the ability to fall back and isolate has been quite important; nations are not forced into a fight to death if each can retreat to a place where their sovereignty is respected.

Powerful nations would be quite wary to expose themselves in a game where they have much to lose, without having much to gain. It is not wise to allow yourself to be unequally yoked to a lesser player, nor to get oneself caught in the cross-fire of international disputes, if this can be avoided. There has to be some benefit in forming a coalition, either the decentralized coalition represented by decentralized Bitcoin or any proposed centralized coalition which controls a version of Bitcoin.

18.4.1 Nice and naughty nations

The ability to centralize, abandon, or attack the Bitcoin network by major players has important international ramifications. If naughty-list nations such as North Korea continue to build up Bitcoin reserves as the reach and value of Bitcoin increase, the stakes begin to change. If the US decides it is in their best interest to counter North Korea on an economic front, any weapons (direct attack, permissioned bypass, tainting, or abandonment) are on the table. At some point, this tension may become clear to leaders in rogue nations, who may begin to reduce exposure to Bitcoin in favor of more traditional hard assets. While going all-in on Bitcoin may be a brilliant gambit for some nations, they must be able to follow this up by reducing exposure in a timely way, before it becomes clear that it is in more powerful nations' interest to attack. There is a limit to the upside for such players; too much value stored in Bitcoin raises the specter of a takeover.

If at any point the US started to procure hashpower, naughty-list nations may preemptively begin to sell. The isolated nations have much more to lose if they were to be cut off and may seek to maximize their payoff by selling at the ripest opportunity.

On the other hand, if nice nations see centralization to be imminent they may also begin to allow or encourage enmeshment. If Canada sees the US as an economic ally and believes that centralization under US control is inevitable, Canadians may proceed with little risk in mind.

If there is an economic preponderance of nation-states who want to preserve the principles and ideals of Bitcoin, paradoxically, they may be inclined to perform a takeover with the explicit purposes of keeping Bitcoin accessible. This may appear to be the best option; it is a departure from the original intention, but it provides powerful nations with an opportunity to use their power to enforce international peace. While sanctioning neighbor-invading or nuclear-stockpiling nations was not the original purpose of Bitcoin, as long as most of its users see this as acceptable, there is nothing to stop it.

Again this boils down to coalitional game theory. Each coalition of players with the capacity to control the network has a payoff if that coalition forms: This can be thought of as a coalitional game in characteristic form. Naively, one would expect that a coalition of nice, well-behaved, liberal democracies with thriving economies would be most interested in forming a large-tent coalition dedicated to keeping Bitcoin open for free market operators. A United Nations type coalition to control Bitcoin preempts possible attacks by malign actors, gives the coalition economic control over dictators or other peace-disruptors who would want to use the network, and may also reduce energy consumption (which is likely to continue to be a political issue, whether one agrees it should be or not). If corporations are hesitant to become enmeshed with Bitcoin due to uncertainty, a move by respectable nation-states to give the network a foundation into the future may suffice to beckon them off the fence.

Further, by forming an explicit coalition, players could negotiate away the rigidity and unequal yoking provided by the default decentralization arrangement. Negotiation and control allows for a wide spectrum of economic arrangement between allies, while agreeing on core principles. Of course, economic alliances are never without troubles; the Euro experiment has been far from perfect.

It follows that hyperbitcoinization may be pursued by or of interest to more powerful nations who may use it as another tool for control. Rogue nations who fear a crackdown probably will be careful to avoid exposure. If they benefit too much while flaunting sanctions, they could only be inviting a takeover that would be devastating for their economy. If there is a beneficial role for a global fixed supply currency, it makes the most sense that the powerful nations would want to exercise as much control over this currency as possible. A centralized version is as unassailable as sovereign currency (at least from denial of service) so allies can involve themselves with little fear.

18.4.2 The free-rider problem with a finite supply currency

We leave it to the reader to form opinions on Menger, Mises, Rothbard, Keynes, and modern monetary theory, avoiding a larger discussion of the virtues of a fixed supply currency. While fiat currencies are often derided as debased or worthless, it should be repeated that the US dollar is considered a hard currency, that is, it functions as a currency in which long-term debts are denominated. A finite supply currency is a different type of beast.

While fiat currency supply can be tweaked to approximately track the rate of expansion of an economy, finite supply currencies have no such lever. If someone holds a portion of the finite supply currency, they hold the same portion of the buying power ascribed to that currency. While the economy grows, the buying power accrues even to those who do nothing other than hold the currency. Inflation, on the other hand, creates a slow-moving treadmill that pushes the economically inactive holder off into oblivion. We also leave it to the reader to engage in a normative debate comparing the negatives of the *Cantillon effect* against the negatives created by the free-rider issue described here, but in terms of game theory, the free-rider dynamic can become important (the Cantillon effect can be described as the ability of those nearest to the money spigot to unfairly profit under inflationary regimes). A nation who considers themselves more productive than other nations' lackluster economies may be reluctant to share growth with these nations. Economic growth poured into a Bitcoin economy benefits the entire Bitcoin economy; this can be unfair and unattractive to the larger contributors.

Rather than watch their nation's growth converted to buying power for free-riders, a nation may prefer to discourage enmeshment with Bitcoin. The nation may prefer that those with capital be using creativity and insight to direct their money to productive and fertile domestic markets, not simply parking it as a non-productive asset hoping the asset increases in value. If the supply of investment capital were to significantly be diverted to Bitcoin, it could be a vicious spiral for the economy and other economies that rely upon that nation's growth.

If the US were to allow much of the economy to become enmeshed with Bitcoin, other nation-states could simply coat-tail on the US' economic growth by acquiring a Bitcoin stack. So while some nations (say El Salvador) may embrace Bitcoin hoping that larger nations follow, many larger nations remain somewhat reluctant.

18.5 Can the supply be increased?

Perhaps the most obvious method for avoiding security problems that may occur when the block subsidy decreases is to simply halt the decreasing of the block reward. This is actually a very simple proposal. For example, instead of the block subsidy decreasing from 1.562 to 0.781 in around 2032, it could remain the same. This sort of modification to the block reward is called a *tail emission*. It requires minimal change to the code base and would only slightly modify the short-term trajectory of the supply cap, but would mean that there is no limiting upper bound to the number created as time goes to infinity. However, these sorts of proposals are extremely unpopular. There are several reasons why these proposals remain unpopular. One is the claim that the "pure meme value" of the number 21 million is incredibly powerful. Even a small departure from the well-known value would drastically erode one of Bitcoin's prime selling points. To some, the number 21 million is a non-negotiable selling point; anything else is heretical to Bitcoin. Another argument is that if this important touchstone of Bitcoin could be modified once, it could be modified again, perhaps quicker and with greater ease. This would put power back in the hands of centralized decision makers, an apostasy to Bitcoin's core principles. This argument is based on a slippery slope; if after years of discourse, the block reward decay could be halted, who is to say that a group of miners could then decide that the block reward should be increased at some point in the future, using the previous departure as a precedent? Not all slippery-slope arguments are fallacious. But one must keep in mind that along the slippery slope, there could be many lines drawn in the sand; users could continue to make threats to abandon the network. These threats become more salient if the supply changes become less cautious.

This is a touchy issue. Many Bitcoiners would like to believe that they have full veto power over changes to the network, but it is unclear how this would play out. A coalition of miners could cite security issues and declare an intent to maintain the block subsidy. Their concerns could be justified; there could be a point where miner margins are small and a halving of the block subsidy could render the network quite insecure. Then an interesting game of chicken ensues. If a wide coalition of miners declare their intentions and distributes the (minor) software patch to major financial institutions, they could set up a situation where everyone must go along. Financial institutions may share the miners' concerns; lower security could mean unpredictability and loss of business. Many users would declare this miner-activated hard fork a heresy and threaten to sell all of the coins mined on the apocryphal blockchain.

A significant event in Bitcoin's short history was in 2017, when a user-activated soft fork (UASF) was threatened against a coalition of miners. The UASF was an effective tactic; miners were not willing to engage in this game of chicken and relented. Following this, major exchanges took sides against the coalition of miners, who ended up taking a decisive loss in the blockwars of 2016–2017. However, there is no guarantee that the dynamics or demographics will be the same in 10 or 15 years; the power centers may have shifted, reliability of the chain could loom larger (less than, say, the value of the asset), and security may also play a more urgent role in the discussions. A coalition of miners in 2032 could be led by CEOs who are superior negotiators and make no missteps. The blocksize wars were complicated, but the portrayal in [5] of the coalition of miners suggests that they made a number of poorly calculated moves. One could imagine that a shrewder set of operators may win such a clash if it played out in the backdrop of an impending security crisis.

Again, we should apply backwards induction: If a coalition of businesses and mining corporations opted to modify the emission schedule, would there be a race to the exit that would crash the value and security of the network? The answer depends on how important one believes the number 21 million is to the economic majority of Bitcoin users.

Exercise 18.1. Suppose that the community widely begins to treat Bitcoin as fungible, so that proceeds controlled by different UTXOs may have different market values. Explain a way to effectively create tail emission without a hard fork.

18.6 Limits as a decentralized store of value

The principle of backwards induction allows us to start discussions from various possible end-states. One can come up with no shortage of paths that could meander us from our current situation, but any decisive movement towards an end-state needs to make sense. It is possible that Bitcoin could remain in a state of limbo for decades, but eventually, expectations will begin to congeal into realities.

One must ask whether Bitcoin will actually challenge the global power structures that exist, and if so, what will the result of this challenge be? Depending on your answer, you can work backwards from there: If Bitcoin is eventually to be destroyed by powerful adversaries, eventually this will become clear, and perhaps it will never even make it that far. If Bitcoin is going to eventually be embraced by world powers, this would also become clear in time, and one can predict what actors will do if they anticipate this.

The most obvious terminal state for Bitcoin is centralization. Those who want a store of value hedging against that risk should hope that the economic predominance of counterparties in the network is also indifferent to centralization.

18.6.1 Asymmetric risk premium

In the first 13 years of Bitcoin, there has been a compelling argument for speculating, summarized as follows: "If decentralized currency becomes a Schelling point for value, it is most likely to be Bitcoin. Even if the probability of such an event is considered low, the multiples of return make such a risk prudent from an expected value computation." Essentially, if you give event "\$750,000 bitcoins by 2028" a 1 in 10 chance, Bitcoin is a fantastic bargain at \$18,000.

This has played out well for speculators who bought in at \$1, \$3, \$60, \$150, \$900, \$2500, etc. As the price has grown, the expectations have shifted upwards; values that were once considered ridiculous were obtained, and new targets were set. This leads to compounding effects, as the asymmetric risk premium drives the price up, and the expectations are adjusted upwards, driving the price further up.

There is a related set of arguments surrounding store of value. One may suggest that today it would not be unreasonable that investors put 0.1 % of their net worth into Bitcoin, as this represents a negligible risk. As a wide set of investors do this, this endeavor is validated by a rise in price, perhaps investors then feel emboldened: 1 % of all investments then go towards Bitcoin, and so forth. This bootstrapping also relies on a set of beliefs about beliefs, in particular the belief that the risk appetite will increase as the price appreciates.

However, there are only so many orders of magnitude that can be achieved before this same process works in reverse. Reality provides limits on how high the price can climb before snapping back. As we approach a limit, the asymmetric risk premium decreases, and expectations on beliefs can just as easily compound downwards. This leads to an interesting backward induction thought process that could become quite relevant if the price surges past \$100,000. While \$10,000 bitcoin enjoys a positive probability of a 50x surge to \$500,000, can the same be said of \$500,000 bitcoin? If it does, it is certainly not due to the same reasoning: 50x gains will not be expected to compound on \$2.5 million. We can conclude that the asymmetric risk premium has decreased significantly. Modeling would necessarily involve some hand-waiving, but heuristically, if one has in mind a "maximum speculative value," one would think that most Bitcoin holders would begin to take profits as this value is approached. This would, of course, lead other speculators to take profits slightly before the value was beginning to be approached, and so forth.

If Bitcoin is largely a speculative game, the dynamics are unpredictable but resemble clock games. Speculation games are as difficult to model as they are to win, but they have been studied by game theorists. Clock games provide a reasonable model for timing the speculative bubble if there are large numbers of players involved [8]. The larger number of smaller players in the game may mollify the herding behaviors inherent in these games; but if larger speculators are involved, some degree of herding is to be expected. Indeed, each of the last three price peaks has been followed by a significant pullback, suggesting that speculators are hoping to time the top and then reduce exposure once it becomes clear the moment has passed.

Certainly, a nation or large corporation does not want to be the last to invest; they prefer to be the first of many. The first can conjecture that there will be more, but the latest must also have some hope that there will be more to follow, each motivated by the more to follow those who follow. With retail investors, this game seems reasonable enough; it is not foolish to have been the 5 millionth person to invest thousands of dollars in Bitcoin, certainly among the billions of people left on Earth; there will be more to follow. With large players, the number is much more finite; there are only so many billionaires. For a large player who moves in early, there is a limited expectation that other large entities will continue to ape into the game.

18.6.2 Failure as a functional store of value

Bitcoin performs much better as a speculative asset than as a store-of-value asset. Price bulls will often discuss historical average returns; these heavily weigh the gains of early whales who achieved 1,000x or 10,000x gains, which are clearly not available to anybody today. If one were to assume that dollars are invested in Bitcoin at a rate proportional to the market cap, one would find that historically, well over half of the dollars invested in Bitcoin had a negative return 6 months from the date of investment.

A decentralized finite supply currency lacks stabilizing mechanisms; thus, the only way Bitcoin can function to preserve value is if the price goes up forever, which clearly it cannot do and does not do. This does not preclude intervention by larger players, perhaps central banks, to stabilize the price, although it is unclear the motivation for doing so.

The truth remains that most parties who hold currency around the world are generally risk-averse. Bitcoin's buying power is a risk; this limits its attractiveness to those not seeking risk. The best way to remove the risk is to remove the speculators, but these seem to be along for the ride. Speculators leaving would lower the price, which would increase the asymmetric risk, which would attract more speculators, and so forth.

If this dynamic continues, hyperbitcoinization has a bit of a chicken-and-egg problem. In order to become a widely accepted unit of account, speculation would have to become less attractive. But as long as Bitcoin remains volatile, speculation will remain attractive.

While store of value remains a hopeful use case designed for residents of nations undergoing hyperinflation to protect their value, in practice this will always face the same speculative cycles. If people start to place their money into Bitcoin, financial institutions who are paying attention will front-run the movement and join the speculative frenzy, only to sell when the frenzy runs its course. For the time being, this appears unavoidable.

18.7 Beasts in the jungle: centralization, capture, and collapse

As Bitcoin moves from a nascent and emerging technology, convergent expectations will begin to form and govern its path forward. For the narrative of a global store of value to reify, it must become clear to everybody that it behaves like a store of value. For the narrative of a global, efficient currency to take hold, people must use it widely as a currency. How people use it depends on their expectations on how it will behave in the future, and so forth. Narratives provide psychological guardrails along the path.

However, almost the only thing certain about Bitcoin's future is uncertainty. A continued "up only" trajectory will eventually lead to either speculative attack or attempts to capture or centralize the protocol. At this point the Schelling point provided by Nakamoto consensus could be challenged. How one feels Bitcoin will navigate this determines how one will treat Bitcoin leading up to this, and how people in aggregate feel Bitcoin will navigate this will determine the behavior of Bitcoin.

The dynamics of expectations suggest that the set of future paths will begin to narrow as it gravitates towards more static roles. At the end of the day, Bitcoin is a social consensus and a negotiation over how value is preserved and distributed, which is enforced by aggregate behavior, not by any particular technology. No Schelling point is sacred or eternal.

18.7.1 Centralization

Bitcoin could become overtly centralized. There are many paths that could lead to this. One of the simplest is a response to a major attack. If a persistent 51 % attack on the network occurs, an economic predominance may find that the only acceptable solution is to defer to a leader or group of leaders. Choosing a leader is an absolute defense against a 51 % attack. Once this occurs, the network may discover most things continue to work well, and may begin to build a governance model around the leader. Centralization would then be in effect.

Once a clear shift towards to hyperbitcoinization occurs, there would be a widely held understanding that the technology is worthy and unlikely to go away. This would lead corporations and nation-states to ask questions about security. The answer could be unclear at first, as security concerns are inherent to a decentralized currency. Recalling that many parties may be indifferent to centralization, the threat of attack will not scare these parties out of Bitcoin. On the other hand, those who are sensitive to centralization may be filtered out by the threat of attack. If the threat of attack is real, over time the preponderance of economic activity could be undertaken by those who are indifferent to centralization.

The suggestion of centralization would arise, while this could be resisted at first; the larger economic players will find their security and longevity concerns assuaged by the idea – this could snowball into an understanding held among the larger players that to continue to operate securely and with the widest participation, Bitcoin should be governed by some central authority. Once centralization becomes inevitable, more parties will be able to use it without fear of certain risks, and growth will continue. Provided individuals or larger parties do not wish to go afoul of the wishes of the centralized governing authority, it will remain available and useful to them.

Centralization is by far the most bullish case and goes hand-in-hand with an "up only" general philosophy about Bitcoin. Provided an economic majority of nation-states and corporations believe that a hard currency is for the common good and should be protected without wasting resources, a centralized fork would afford a large amount of participation.

Centralization could begin as a direct takeover and then switch to a permissioned mining arrangement. If there is wide enough will to move directly to permissioned mining, this would follow. An economic majority of miners merely needs to make a threat of a direct takeover via hashrate in order to begin negotiations. The governing board could be run as a business, a non-profit, or an international organization like the United Nations. Depending on how this organization is run, this could lead to a variety of economic realities which run the gamut from "much like today" to full-blown corporatocratic dystopia.

From a national perspective centralization may be modeled like a coalitional game. Nice, well-behaved nations with strong economies will ally themselves with nice well-behaved nations with strong economies, to the exclusion of underperforming or discountenanced nations, forming a coalition that brings good value to its members. They can make their own rules and make sure the rules are enforced.

There could be many signs along the way indicating the centralization is becoming more likely. Mining operations could be consolidated, as there will be a competitive advantage to scale, energy prices, and vertical integration of the mining industry. As strategic mining becomes a larger issue, there will be non-linear returns from centralization, inducing miners to form teams. Nation-states would forge cozy arrangements with mining firms as executives follow revolving-door patterns that are well trodden in other areas of finance. Such relationships between the Bitcoin industry and regulators increase the aperture for coercion.

Large mining corporations could also have a footprint in both finance and energy sectors. Already in 2022, miner corporations in Texas are on the beneficial end of contracts not afforded to the average home Bitcoin miner. When negotiating with energy grids and using energy or other derivative markets, centralization is an effective way to cut costs, increase negotiation power, and even make profits in energy futures. This could lead to a situation in which only the largest mining corporations are able to reap benefits only indirectly related to the block reward, leading to decreased profits and opportunities for smaller miners, who eventually make the decision to drop out and leave mining to the larger players.

18.7.2 Capture

Capture is somewhat of a weak form of centralization. Under capture, Bitcoin is not explicitly centralized, but is dominated by a number of corporations, who in turn are beholden to the authorities in their jurisdiction. The authorities implicitly dangle the threat of regulation over the corporations, ensuring that Bitcoin never gets too out-of-line or takes a path contrary to the interests of the nation-state. This may look not too much unlike the relationship between financial institutions and regulators enjoyed today.

The path toward capture could begin with regulators and lawmakers who take a friendly approach to Bitcoin, working in tandem with corporations. One easy way for this process to start would be for regulators to enact regulations which are damaging to Bitcoin competitors and favorable to Bitcoin. For example, suppose a decision is made to regulate other coin projects as securities, while excepting Bitcoin. While this may be cheered by some Bitcoin advocates, this outside-the-market power play diminishes Bitcoin's claims as a true market-based solution and increases the likelihood of becoming another piece of government machinery. Regulators who support Bitcoin retain power to pull the plug: This is not to say they want to, as they may be drawn to Bitcoin for speculative reasons and not wish to get in the way of any exuberance.

Capture would entail heavy KYC/AML. Corporations would be allowed to use Bitcoin at scale provided they obtain information on their customers. This information would be provided to counterterrorism, law enforcement, and intelligence agencies when asked (perhaps via a court order, perhaps not). Provided corporations and national intelligence are able to keep tabs on much of Bitcoin use, they may be very happy with Bitcoin. For reference, in 2021 Facebook received over 400,000 requests for information from law enforcement and produced data in over 70 % of these requests [16].

Tainting, unlike on-chain transaction censorship, can be bootstrapped in and does not require a 51 % takeover. Authorities can start by encouraging financial institutions to enforce taintlists. Transactions on the taintlist lose access to much of the network, so they are no longer a store of value that rises with investment by mainstream investors. Rational agents will not accept a tainted transaction at face value as these bitcoins will not be salable on exchanges. Taintlists could begin with transactions associated to known malefactors and proceed to any transactions that are non-KYC. Tainted coins become particularly useless if the tainting extends to coinbase transactions that recycle fees from tainted UTXOs: In this case tainted coins would be priced out of the market for blockspace.

Rogue nations, criminals, and other unsavory characters would be aware of this; they would seek out other ways to transact. Similarly, privacy advocates (with nothing to hide, of course) could refuse to bless such a panopticon by their participation and seek other payment networks.

Under capture, speculative frenzy could be controlled if necessary. It would be understood that if the price becomes large (for example, if a speculative attack takes place) the regulators would enact taxes (perhaps even on unrealized gains) and have the cooperation of corporations to enforce this. The governing authorities could also demand that large mining corporations forming economic and hashrate majority fork to an increased supply.

The scenario here has a limited role for "off-grid" Bitcoin. Regulators cannot touch Bitcoin that is stored in a 12-word phrase that resides only in one's memory. However, the laws and practices could make it difficult for the off-grid Bitcoin to be used on par with the KYC Bitcoin. This could even lead to a bifurcation, with tainted Bitcoin commanding a different market value on a different network. The tainted coins would have their value only as black market currency. This could mean tainted Bitcoin could return to the function and value of Bitcoin in the Silk Road era of 2011–2013.

18.7.3 Collapse

Bitcoin could collapse. There are a number of paths to this. They all rely on a common realization that the Bitcoin experiment was a failure that cannot be salvaged. This could happen in a precipitous event: A major attack could shake the faith and destroy narratives that prop it up. Once the asymmetric risk proposition is no longer in effect, investors could race to the door, leaving infrastructure projects abandoned. But this need not return to a 2011 state; now that ASICs have been produced in huge quantities, anyone who chooses to preserve these machines can take control of the blockchain if they are willing to spend the energy to do so.

This need not require an attack; if it becomes widely believed that an attack is imminent or inevitable, the system could be abandoned by its users to the same effect. Failure to generate enough transaction fees to supply a functioning security budget may be sufficient, so would a secular bear market in which speculative appetite has taken a clear turn downward, cementing Bitcoin's role as the tulip bulb of the 21st century. Security budget FUD could cause larger players to reduce their exposure, and this in turn could lead to further price decays and less security.

Downside speculators, given the opportunity, would go in for the kill; if the hash deployment rate drops low enough so that a hedge fund could launch a mining attack, the collapse could be hastened and seen as inevitable.

Bitcoin was not originally sold as a store-of-value asset; it was sold as a peer-to-peer payment network. Today, it aspires to be a global Schelling point for decentralized store of value. Since 2009, however, many other decentralized payment technologies have emerged. If Bitcoin loses value, it is no longer the only game in town for peer-to-peer payments. So the collapse of the decentralized store-of-value narrative for Bitcoin would likely be a death knell for Bitcoin. A selling point of proof of work is that it was virtually impossible to brute-force attack. If this is proved false, alternatives such as proof of stake will win the day.

Due to the build-up of ASICs, a collapse in interest or price leads to a particularly vicious cycle. A significant loss of price or transaction volume diminishes the security budget, meaning the rewards paid to miners dry up, leaving millions of otherwise useless machines with no purpose but to attack the network. This would discourage any serious use of the network, leading to further abandonment.

This becomes particularly final if the dysphoria caused by a market crash provided governments with enough will to enforce a ban or other heavy crack-down measures on Bitcoin. Prices would drop as the above-board corporations would distance themselves. The governments could then purchase ASICs at distressed prices and keep these around; as long as Bitcoin can be shut down by the governments at will it would require significant risk and effort to build up again.

Once ASICs fall into the hands of hostile parties, there is little that can be done to revive Bitcoin, without a change of consensus algorithm.

These are all somewhat violent deaths, but these need not be the case; the end could be with a whimper. Time will tell. Perhaps the phenomenon was a historical anomaly. The last 15 years have been full of surprises: a financial crisis, a wave of populism, and a global pandemic met with unprecedented government response, all which seem conducive to the flourishing of Bitcoin. The next 15 years will be full of new surprises. The distractions and exciting social movements presented by technology and the Internet in the coming decades are hard to predict. Distracted by the latest thing, there may be a day in the future when many of us wake up, eat our breakfasts, go about our respective business, and sooner or later realize that we have not thought about Bitcoin in a while.

Bibliography

[1] Lyn Alden. A look at the lightning network, 2022. https://www.lynalden.com/lightning-network/.
[2] Andreas Antonopoulos, Olaoluwa Osuntokun, and René Pickhardt. Mastering the lightning network: A second layer blockchain protocol for instant bitcoin payments. O'Reilly, 2021.
[3] Raphael Auer. Beyond the Doomsday Economics of "Proof-of-Work" in Cryptocurrencies. Globalization Institute Working Papers 355, Federal Reserve Bank of Dallas, 2019.
[4] Andrew M. Bailey, Troy Cross, Josh Hendrickson, William Luther, Bradley Rettler, and Craig Warmke. Bitcoin works in practice, but does it work in theory? 2022.
[5] J. Bier. The Blocksize War. Independently Published, 2021.
[6] Ken Binmore, Ariel Rubinstein, and Asher Wolinsky. The Nash bargaining solution in economic modelling. The Rand Journal of Economics, 17(2):176–188, 1986.
[7] Joseph Bonneau. Hostile blockchain takeovers. In The 2018 IFCA Workshop on Bitcoin and Blockchain Research, 2018.
[8] Markus K. Brunnermeier and John Morgan. Clock games: Theory and experiments. Games and Economic Behavior, 68(2):532–550, 2010.
[9] Eric Budish. The economic limits of bitcoin and the blockchain. Working Paper 24717, National Bureau of Economic Research, 2018.
[10] Miles Carlsten, Harry Kalodner, S. Matthew Weinberg, and Arvind Narayanan. On the instability of bitcoin without the block reward. In Proceedings of the 2016 ACM SIGSAC Conference on Computer and Communications Security, CCS'16, pages 154–167. Association for Computing Machinery, New York, NY, USA, 2016.
[11] Philip Daian, Steven Goldfeder, Tyler Kell, Yunqi Li, Xueyuan Zhao, Iddo Bentov, Lorenz Breidenbach, and Ari Juels. Flash boys 2.0: Frontrunning in decentralized exchanges, miner extractable value, and consensus instability. In IEEE Symposium on Security and Privacy, pages 910–927. IEEE, 2020.
[12] Morton Davis and Michael Maschler. The kernel of a cooperative game. Naval Research Logistics Quarterly, 12:223–259, 1965.
[13] Robert L. Devaney. A first course in chaotic dynamical systems. Addison-Wesley Studies in Nonlinearity. Addison-Wesley Publishing Company, Advanced Book Program, Reading, MA, 1992. Theory and experiment, With a separately available computer disk.
[14] Mark Erhardt and Clara Shikhelman. Improvement on the current block building algorithm, 2020.
[15] Ittay Eyal and Emin Gün Sirer. Majority is not enough: Bitcoin mining is vulnerable. Communications of the ACM, 61(7):95–102, 2018.
[16] Facebook. Government requests for user data | transparency center. https://transparency.fb.com/data/government-data-requests/.
[17] Alex Gladstein. Check Your Financial Privilege. BTC Media LLC, 2022.
[18] Tiantian Gong, Mohsen Minaei, Wenhai Sun, and Aniket Kate. Undercutting bitcoin is not profitable, CoRR, 2020. abs/2007.11480.
[19] Tiantian Gong, Mohsen Minaei, Wenhai Sun, and Aniket Kate. Undercutting bitcoin is not profitable, CoRR, 2020. abs/2007.11480.
[20] Cyril Grunspan and Ricardo Pérez-Marco. Double spend races. International Journal of Theoretical and Applied Finance, 21(8):1850053, 2018.
[21] Cyril Grunspan and Ricardo Pérez-Marco. On profitability of selfish mining, CoRR, 2018. abs/1805.08281.
[22] Cyril Grunspan and Ricardo Pérez-Marco. The mathematics of Bitcoin. Newsletter - European Mathematical Society, 115:31–37, 2020.
[23] Cyril Grunspan and Ricardo Pérez-Marco. On Profitability of Nakamoto Double Spend. Probability in the Engineering and Informational Sciences, 36(3):732–746, 2022.
[24] Hasu. Purge attacks (spin on sabotage attacks), 2020. https://lists.linuxfoundation.org/pipermail/bitcoin-dev/2020-January/017606.html.

https://doi.org/10.1515/9783110772845-019

[25] Hasu, James Prestwich, and Brandon Curtis. A model for bitcoin's security and the declining block subsidy, 2019.
[26] Aljosha Judmayer, Nicholas Stifter, Alexei Zamyatin, Itay Tsabary, Ittay Eyal, Peter Gazi, Sarah Meiklejohn, and Edgar Weippl. Pay to win: Cheap, crowdfundable, cross-chain algorithmic incentive manipulation attacks on pow cryptocurrencies, 2019. Cryptology ePrint Archive, Paper 2019/775. https://eprint.iacr.org/2019/775.
[27] Aljosha Judmayer, Nicholas Stifter, Alexei Zamyatin, Itay Tsabary, Ittay Eyal, Peter Gaži, Sarah Meiklejohn, and Edgar Weippl. Sok: Algorithmic incentive manipulation attacks on permissionless pow cryptocurrencies, 2020. Cryptology ePrint Archive, Paper 2020/1614. https://eprint.iacr.org/2020/1614.
[28] James P. Kahan and Amnon Rapoport. Theories of Coalition Formation. Routledge, 1984.
[29] Joe Kelly. Wide and shallow double-spend attacks on bitcoin (+ why hardware cost is not a reliable deterrent), 2019. https://joekelly100.medium.com/double-spend-attacks-are-much-worse-than-we-thought-dfd016ff8f3f.
[30] John Maynard Keynes. The general theory of employment, interest, and money. Palgrave Macmillan, Cham, 2018. Reprinted from the 1936 original with a new introduction by Paul Krugman and an afterword by Robert Skidelsky.
[31] Joshu Kroll, Ian Davey, and Edward Felten. The economics of bitcoin mining or bitcoin in the presence of adversaries, 2013.
[32] Jiri Lebl. Notes on diffy qs: Differential equations for engineers, 2016.
[33] William F. Lucas. The proof that a game may not have a solution. Transactions of the American Mathematical Society, 137:219–229, 1969.
[34] Michael Maschler, Eilon Solan, and Shmuel Zamir. Game theory. Cambridge University Press, Cambridge, 2013. Translated from Hebrew by Ziv Hellman and edited by Mike Borns.
[35] Hass McCook. The cost & sustainability of bitcoin, 2018.
[36] Daniel J. Moroz, Daniel J. Aronoff, Neha Narula, and David C. Parkes. Double-spend counterattacks: Threat of retaliation in proof-of-work systems, 2020.
[37] Abhinay Muthoo. Bargaining Theory with Applications. Cambridge University Press, New York, NY, USA, 1999.
[38] Satoshi Nakamoto. Bitcoin: A peer-to-peer electronic cash system, 2009.
[39] John Nash. Two-person cooperative games. Econometrica, 21(1):128–140, 1953.
[40] J. von Neumann. Zur theorie der gesellschaftsspiele. Mathematische Annalen, 100:295–320, 1928.
[41] Martin J. Osborne and Ariel Rubinstein. A course in game theory. MIT Press, Cambridge, MA, 1994.
[42] Matthew Pines. Bitcoin and U. S. National Security, 2022.
[43] J. B. Rosen. Existence and uniqueness of equilibrium points for concave n-person games. Econometrica, 33:520–534, 1965.
[44] Meni Rosenfeld. Analysis of hashrate-based double spending, CoRR, 2014. abs/1402.2009.
[45] Alvin Roth, Vesna Prasnikar, Masahiro Okuno-Fujiwara, and Shmuel Zamir. Bargaining and market behavior in Jerusalem, Ljubljana, Pittsburgh, and Tokyo: An experimental study. The American Economic Review, 81(5):1068–1095, 1991.
[46] Ariel Rubinstein. Perfect equilibrium in a bargaining model. Econometrica, 50(1):97–109, 1982.
[47] Ariel Rubinstein, Zvi Safra, and William Thomson. On the interpretation of the nash bargaining solution and its extension to non-expected utility preferences. Econometrica, 60(5):1171–1186, 1992.
[48] Thomas C. Schelling. Strategy of conflict. Harvard University Press, London, England, 1960.
[49] Okke Schrijvers, Joseph Bonneau, Dan Boneh, and Tim Roughgarden. Incentive compatibility of Bitcoin mining pool reward functions. In Financial cryptography and data security, volume 9603 of Lecture Notes in Comput. Sci., pages 477–498. Springer, Berlin, 2017.
[50] Martin Shubik. The dollar auction game: a paradox in noncooperative behavior and escalation. Journal of Conflict Resolution, 15(1):109–111, 1971.

[51] Elias M. Stein and Rami Shakarchi. Functional analysis, volume 4 of Princeton Lectures in Analysis. Princeton University Press, Princeton, NJ, 2011. Introduction to further topics in analysis.

[52] Nicholas Stifter, Aljosha Judmayer, Philipp Schindler, Alexei Zamyatin, and Edgar Weippl. Agreement with satoshi—on the formalization of nakamoto consensus, 2018. Cryptology ePrint Archive, Paper 2018/400. https://eprint.iacr.org/2018/400.

[53] Micah Warren. Notes on mining and asics markets, 2022.

[54] Sam M. Werner, Dragos Ioan Ilie, Iain Stewart, and William J. Knottenbelt. Unstable throughput: When the difficulty algorithm breaks, CoRR, 2020. abs/2006.03044.

[55] Aviv Yaish, Gilad Stern, and Avid Zohar. Uncle maker: (time)stamping out the competition in ethereum, 2022.

Index

https://doi.org/10.1515/9783110772845-020